CHANNEL CATFISH FARMING HANDBOOK

CRAIG S. TUCKER
AND
EDWIN H. ROBINSON
Mississippi State University

CHAPMAN & HALL

 International Thomson Publishing
Thomson Science
New York • Albany • Bonn • Boston • Cincinnati • Detroit • London • Madrid • Melbourne
Mexico City • Pacific Grove • Paris • San Francisco • Singapore • Tokyo • Toronto • Washington

Mention of a trademark or proprietary product does not constitute a warranty of the product by the Mississippi Agricultural and Forestry Experiment Station and does not imply its approval to the exclusion of other products that may be suitable.

Copyright © 1990 by Van Nostrand Reinhold

This edition published by Chapman & Hall, New York, NY

Printed in the United States of America

For more information contact:

Chapman & Hall
115 Fifth Avenue
New York, NY 10003

Chapman & Hall
2-6 Boundary Row
London SE1 8HN
England

Thomas Nelson Australia
102 Dodds Street
South Melbourne, 3205
Victoria, Australia

Chapman & Hall GmbH
Postfach 100 263
D-69442 Weinheim
Germany

International Thomson Editores
Campos Eliseos 385, Piso 7
Col. Polanco
11560 Mexico D.F.
Mexico

International Thomson Publishing - Japan
Hirakawacho Kyowa Building, 3F
1-2-1 Hirakawacho-cho
Chiyoda-ku, 102 Tokyo
Japan

International Thomson Publishing Asia
221 Henderson Road #05-10
Henderson Building
Singapore 0315

All rights reserved. No part of this book covered by the copyright hereon may be reproduced or used in any form or by any means--graphic, electronic, or mechanical, including photocopying, recording, taping, or information storage and retrieval systems--without the written permission of the publisher.

3 4 5 6 7 8 9 XXX 01 00 99 98 97

Library of Congress Cataloging-in-Publication Data

Tucker, C. S. (Craig S.), 1951 --
    Channel catfish farming handbook / by Craig S. Tucker and Edwin H. Robinson.
        p.   cm.
    Includes bibliographies and index.
    ISBN 0-412-12331-2
    1. Channel catfish -- Handbooks, manuals, etc. 2. Fish-culture -- Handbooks, manuals, etc. I. Robinson, Edwin H. (Edwin Hollis), 1942 -- . II. Title.
    SH167.C44T83   1991   90-37855
    639.3'752--dc20   CIP

Visit Chapman & Hall on the Internet http://www.chaphall.com/chaphall.html

To order this or any other Chapman & Hall book, please contact **International Thomson Publishing, 7625 Empire Drive, Florence, KY 41042.** Phone (606) 525-6600 or 1-800-842-3636. Fax: (606) 525-7778. E-mail: order@chaphall.com.

For a complete listing of Chapman & Hall titles, send your request to **Chapman & Hall, Dept. BC, 115 Fifth Avenue, New York, NY 10003.**

To Claude Boyd and Tom Lovell

Contents

Preface xi

PART 1
BIOLOGY OF THE CHANNEL CATFISH 7

1.
General Biology 9

Morphology 9 Respiration and circulation 13 Osmoregulation 15 Sensory function 16 Immune function 16 References 18

2. Life History and Reproductive Biology 19

General life history 19 Male reproductive biology 20 Female reproductive biology 22 Spawning and fertilization 23 Early life history 25 References 25

3. Genetics 27

Basic genetics 28 Qualitative traits 32 Quantitative traits 33 References 38

4. Environmental Requirements 39

Salinity 40 Temperature 41 Dissolved oxygen 44 Alkalinity and hardness 48 Carbon dioxide 49 pH 50 Ammonia 51 Nitrite 53 Hydrogen sulfide 55 Suspended solids and turbidity 55 Total gas pressure 56 Copper and zinc 58 References 59

5. Nutrition 61

Energy 62 Nutrients 66 Digestion 98 References 106

PART 2
CULTURAL PRACTICES 109

Contents **vii**

6.
Breeding 111

The need for planned breeding programs 111 Guidelines for a minimal breeding program 113 Breeding for qualitative traits 114 Breeding for quantitative traits 115 Nontraditional genetic improvement programs 131 References 133

7.
Egg and Fry Production 135

Brood fish management 136 Methods of propagation 142 Hatchery design 151 Hatchery practices 160 Fry inventory methods 162 References 163

8.
Fingerling and Food-Fish Production in Ponds 165

Pond culture systems 166 Fingerling production 174 Food-fish production 180 Record keeping 195 Investments and costs 195 Water use in ponds 199 Fish-eating birds 206 Polyculture systems 207 References 215

9.
Water Quality Management in Ponds 217

Dissolved oxygen and aeration 218 Total alkalinity and hardness 244 Carbon dioxide and pH 245 Ammonia and nitrite 250 Off-flavor 260 Aquatic weed control 268 Turbidity 282 Hydrogen sulfide 285 Toxic algae 285 Pesticides 287 References 288

10. Feeds and Feeding Practices 291

Feedstuffs 292 Feed formulation 297 Feed processing 305 Feeding practices 308 References 314

11. Infectious Diseases 317

Role of environmental conditions 319 Clinical signs of fish diseases 321 Disease diagnosis 324 Viral diseases 328 Bacterial diseases 333 Fungal diseases 345 Protozoan parasites 347 Metazoan parasites 356 Diseases of uncertain origin 356 Diseases of eggs 361 Disease treatments 362 Treatment rate calculations 372 References 378

12. Harvesting and Transporting 381

Considerations before harvest 382 Harvesting fish from ponds 383 Transporting fish 395 References 402

13. Alternative Culture Systems 405

Cage culture 407 Raceway culture 412 Closed, water-recirculating systems 417 References 426

Glossary 429

Appendices 437

Index 447

Preface

Although catfish have been farmed for about 30 years and catfish farming is the most successful aquacultural enterprise in the United States, there are those who contend that catfish farming is still as much of an "art" as it is a science. This position is difficult to refute completely, particularly considering that some practices used in catfish farming appear to have little scientific basis. Skill coupled with a small dose of mysticism certainly play a role in the culture of catfish, and the catfish producer is faced with the unenviable task of rearing an animal in an environment that requires considerable management. Certain aspects may still be an "art" because research and technical information needed to support the industry have lagged behind industry growth; however, the basic principles underlying catfish farming are based on sound scientific evidence whose foundation was laid in the 1950s by work conducted at state and federal fish hatcheries in the southeastern and midwestern United States. Since that time, several university and government laboratories have expanded the scientific base for catfish farming. As a result, considerable information is available, but it is generally fragmented and exists in a multitude of diverse scientific and trade journals. The material is often too technical or abstract to be comprehensible to fish culturists and personnel in allied industries.

This book fits the definition of the term *handbook* in the sense that it is intended as a book of instruction or guidance as well as a reference. We have attempted to present an orderly, integrated summary of the commercial culture of channel catfish. In addition, we feel that a summary of the biology of the catfish is mandatory to provide new producers and students with an understanding of the animal being cultured. We hope that we have linked technical and practical information in a concise and readable manner that will serve a diverse audience. To that end, we have tried to keep the material at a level that will be understandable to the catfish producer, yet still be of interest and value to students and researchers. By necessity, certain sections are presented in more detail and some are more technical than others. We have chosen to use English units of measure unless convention dictated otherwise. Scientific names have been omitted from the text but are presented in the Appendices. Excessive reference citations, which often detract from readability, have been avoided.

This book could not have been prepared without the help of many friends and colleagues. We are grateful for the advice and discussion of subject matter outside our primary interests provided by Dr. Cheryl Goudie, Dr. Ron Thune, Jim Steeby, and Dr. Rob Busch. Thanks are also due to Drs. Mike Johnson, Gary Carmichael, David Crosby, Bob Durborow, and Pete Taylor for reviewing various chapters. Dr. Tom Schwedler of Clemson University read the entire draft of the book and offered many useful suggestions. We appreciate the help of Elizabeth Cook, who assisted with the typing, and of Melanie Tucker, who drew some of the figures. Finally, a very special thanks to Laura Clark, to whom we are forever indebted for her patience in editing and typing this book.

Introduction

Per capita consumption of fishery products in the United States increased 25 percent in the 1980s. The increased consumption of fish and shellfish is largely due to recognition by consumers of the health benefits of low-fat protein sources. Increases in both per capita consumption and population have caused demand for fishery products to increase faster than production.

Fishery products come from three sources: commercial harvest of wild stock, recreational catch, and aquaculture. Domestic commercial harvest has been relatively stable over the last 15 years (Fig. I.1). The increased demand for fish and shellfish has been supplied by imports and aquaculture production. Aquaculture is the fastest growing source of fishery products, accounting for less than 1 percent of total supply in 1970 to over 7 percent in 1988.

In 1988, aquaculture production in the United States exceeded 750 million pounds with a farm gate value of over $600 million. Much of the growth in domestic aquaculture is attributable to increased production of farm-raised catfish in the southeastern United States. Catfish accounts for about half of the total aquaculture production and farm gate value (Fig. I.2). Much of the growth in other segments of domestic aquaculture is indirectly attributable to catfish farming as the success of that industry stimulated interest in commercial production of other fish.

2 Introduction

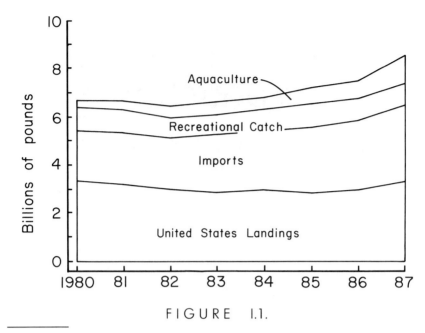

FIGURE I.1.

Supply of edible fishery products in the United States. United States landings are total landings minus exports (Dicks and Harvey 1988).

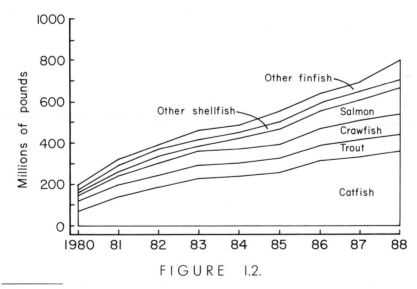

FIGURE I.2.

Aquaculture production of various edible products in the United States (Dicks and Harvey 1988).

COMMERCIALLY IMPORTANT CATFISH

The channel catfish accounts for almost all the commercial production of catfish in the United States. Channel catfish possess many qualities desired in a fish for commercial production. The fish usually does not reproduce in culture ponds, so the farmer has control over pond populations. The animal is, however, easy to spawn and large numbers of fry are easily obtained. Fry will accept relatively simple, prepared feeds at first feeding. Channel catfish are hardy, live over a wide range of temperatures, and adapt well to all commonly used culture systems.

Blue catfish are not widely cultured but are potentially valuable as a commercial species. Blue catfish grow at about the same rate as channel catfish to the size desired by commercial processors (0.75 to 1.50 pounds), after which the blue catfish grow faster. Blue catfish are resistant to infection by the bacterium *Edwardsiella ictaluri*, which is the cause of enteric septicemia of catfish, the most serious disease problem in the industry. Blue catfish also have a higher dress-out percentage and are easier to seine than channel catfish. The interspecific hybrid between female channel catfish and male blue catfish also exhibits a number of superior characteristics and has considerable commercial potential.

The white catfish, the bullhead catfishes, and the flathead catfish have also received some attention as candidates for commercial culture, but each species has one or more characteristics that limit its appeal. White catfish and bullhead catfishes grow slower and have a poorer dress-out percentage than channel catfish. Flathead catfish are highly cannibalistic and thus are unsuitable for commercial culture.

CHANNEL CATFISH FARMING

The foundations for catfish farming were laid prior to 1960 by work conducted at state and federal fish hatcheries in the southeastern and midwestern United States. Channel and blue catfish are popular sport fishes, and work at government hatcheries resulted in the technology necessary for efficient production of fingerlings for stocking reservoirs and sport fishing ponds. Many of the techniques developed at those facilities are still used to produce fry and fingerlings for large-scale commercial culture.

Interest in catfish culture as a commercial venture was stimulated by pioneering work conducted at Auburn University in Alabama. In the 1950s, Homer Swingle and his students explored the potential for small-scale aquaculture of channel catfish in farm ponds. Subsequent research at Auburn and other universities in the Southeast led to improvements in feeding practices and disease control that were necessary for establishment of culture on a larger scale.

From 1955 to about 1970, most of the development in commercial catfish culture occurred in Arkansas. Buffalo (a type of sucker) were widely cultured in Arkansas in the 1950s and sold to markets throughout the Midwest. In the early 1960s culture of channel catfish expanded, aided in large part by the long-standing interest in aquaculture among Arkansas farmers and the aquaculture infrastructure (although limited) already existing for buffalo and bait minnow culture. By 1963, channel catfish had supplanted buffalo as the major food fish produced in Arkansas. In 1966, about 10,000 acres of ponds were devoted to channel catfish production in Arkansas. Annual production was likely about 15 million pounds. Oddly, commercial production of catfish in Arkansas dramatically declined from that point, about the time interest in channel catfish farming in neighboring Mississippi began to increase. It appears that channel catfish farming in Arkansas failed to achieve the "critical mass" necessary to expand from a small enterprise with decentralized production and local marketing to a larger, integrated industry with regional or national markets. Production of channel catfish in Arkansas has recently rebounded and more than 25 million pounds are now produced annually.

Growth of the channel catfish industry in Mississippi has been rapid, particularly since 1978 (Fig. I.3). Mississippi now accounts for about 75 percent of the total channel catfish production (Fig. I.4). The development of Mississippi as the leading producer of channel catfish is somewhat paradoxical. Most of the early research and development necessary for large-scale commercial channel catfish culture was conducted in other states, notably Alabama, Louisiana, Arkansas, and Oklahoma. Furthermore, sizable commercial catfish farms were in existence in other states before significant activity occurred in Mississippi. Certainly the Mississippi River flood plain in northwest Mississippi—locally called the Delta—is endowed with the physical resources required for pond aquaculture (flat land, clay soils, and abundant water), but so are many other locations throughout the Southeast. Socioeconomic factors, rather than physical resources, probably played a major role in the development of the channel catfish industry in Mississippi.

Large-scale channel catfish farming requires a relatively large initial capital investment. Such capital is not routinely available to owners of small farms. Furthermore, the decision to invest large amounts of capital in an untried and risky enterprise such as aquaculture re-

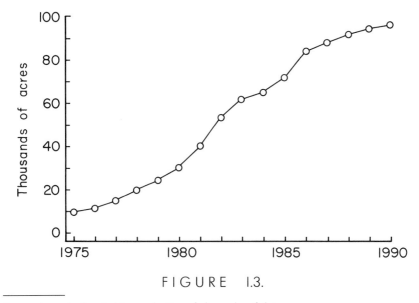

FIGURE 1.3.

Acres of ponds devoted to production of channel catfish in Mississippi.

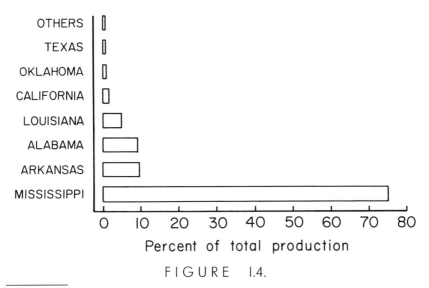

FIGURE 1.4.

Relative production of channel catfish in different states (Dicks and Harvey 1989).

quires a unique attitude. Delta farmers, many of whom own large farms, were willing to take that risk, particularly during the late 1970s and 1980s when profits from traditional row crops were declining. As catfish production expanded in the 1970s, cooperative efforts among farmers resulted in a relatively unified industry. Development of large local feed mills and processing plants, coupled with the large size of individual fish farms, allowed Delta farmers to capture economies of size and dominate regional markets.

The impetus for the phenomenal growth of channel catfish farming in Mississippi and other states in the 1980s was the active and extraordinarily effective marketing efforts by the industry. Channel catfish farming in the Southeast was initially possible because the fish was highly esteemed as a food item in that region. Thus, local markets afforded a reliable outlet for the product. In other areas of the country, particularly in the Northeast and far West, channel catfish were regarded as a scavenger and of little value as a food item. National advertising campaigns by industry-supported marketing groups have dramatically improved the image of channel catfish. The delicate flavor of the farm-raised fish, its nutritional value, and its year-around availability appeal to retailers and consumers. Channel catfish are now the fourth most frequently consumed fish in the United States behind tuna, pollock, and salmon.

REFERENCES

Dicks, M., and D. Harvey. 1988. *Aquaculture Situation and Outlook Report*.
    Washington D.C.: Commodity Economics Division, United States Department of Agriculture.
Dicks, M., and D. Harvey. 1989. *Aquaculture Situation and Outlook Report*.
    Washington D.C.: Commodity Economics Division, United States Department of Agriculture.

PART I
BIOLOGY OF THE CHANNEL CATFISH

The next five chapters provide an overview of the biology of the channel catfish. Many catfish farmers have entered fish culture from a background of traditional agriculture and have little knowledge of how a fish lives. We feel it is necessary that catfish farmers understand the basics of fish biology so that rational management decisions can be made. Three of the chapters in Part 1 have direct counterparts in Part 2, Cultural Practices. Chapter 3 (Genetics) provides a theoretical foundation for the breeding programs discussed in Chapter 6. The environmental requirements of the channel catfish are discussed in Chapter 4; management of the pond environment is then discussed in Chapter 9. Likewise, the discussion of fish nutrition in Chapter 5 complements the summary of feeding practices in Chapter 10. These pairs of chapters should be read as units to derive their full benefit.

CHAPTER 1
General Biology

MORPHOLOGY

Like other animals, channel catfish are composed of several diverse biological systems that function to integrate and regulate life processes: the circulatory, respiratory, digestive, integumentary, endocrine, muscular, nervous, reproductive, excretory, and skeletal systems. To understand how these systems work, it is essential to know how the catfish is assembled. A detailed description of the morphology of the catfish is beyond the scope of this book; however, a general description of the external and internal anatomy of the catfish is presented. For a more comprehensive treatment of the morphology of channel catfish, the reader is referred to the book *Anatomy and Histology of the Channel Catfish* (Grizzle and Rogers 1976).

External Anatomy

The appearance of channel catfish differs, depending on age, size, sex, and environment. Generally, their body shape is elongate with a slightly depressed head (Fig. 1.1). The skin, which covers all exterior surfaces and is scaleless, varies in color from grayish blue on the top and sides of the catfish with irregular dark spots on young fish to al-

10 General Biology

FIGURE 1.1.

Channel catfish weighing about 3 pounds.

most completely black in older fish. Their undersides are generally white to silvery but may be darkened in older fish.

Channel catfish have soft-rayed fins; however, the pectoral and dorsal fins contain "spines" that are sharp and hard. The anal fin, which is used to distinguish channel catfish from other closely related species, is rounded and contains 24 to 30 rays. Another distinguishing feature is the deeply forked caudal fin. The adipose fin is characteristic of Ictalurid catfishes. Barbels, which are arranged in a definite pattern around the subterminal mouth, are typical of catfish (Fig. 1.2).

Channel catfish contain a lateral line that runs longitudinally from the caudal fin to the head; it is distinguished by its lighter color and by pores at intervals along the line.

Internal Anatomy

Location of the major internal organs of the channel catfish is illustrated in Figures 1.3 and 1.4. The visceral organs of channel catfish are generally typical of vertebrates, but with some notable exceptions. The digestive tract is rather short and the intestine is not well defined into small and large intestines. The pancreas consists of white nodules scattered in the mesenteries near the bile duct, along the hepatic portal vein between the liver and spleen, and in the liver. The kidney is completely separated into the head and trunk kidneys. The head kidney is

Morphology **11**

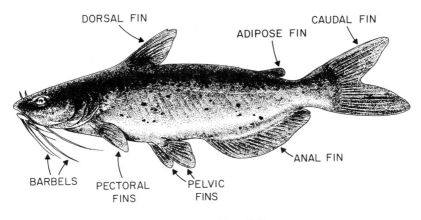

FIGURE 1.2.

External anatomy of channel catfish.

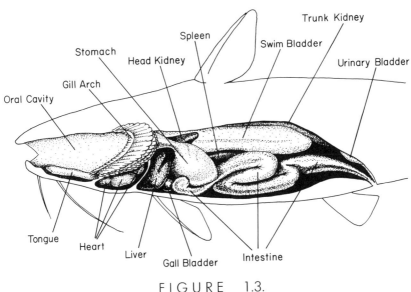

FIGURE 1.3.

Internal anatomy of channel catfish.

12 General Biology

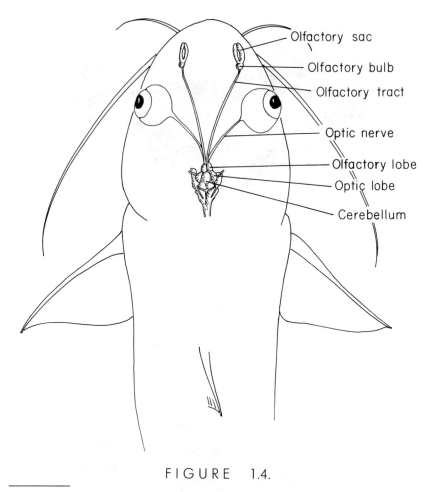

FIGURE 1.4.

Dorsal view of a channel catfish indicating location of eyes, brain, and olfactory organs.

involved in endocrine and hemopoietic functions and the trunk kidney functions primarily as part of the excretory system. The thyroid gland is not a distinct organ; rather it consists of follicles found around arteries ventral to the pharynx. Parathyroid glands are not present, but there are paired glands, which are not grossly visible, found between the liver and heart that appear to function similarly to parathyroid glands. Additional discussion of the internal anatomy of channel catfish will be confined to sections where further definition is essential to the clarity of the material presented.

RESPIRATION AND CIRCULATION

Although the respiratory and circulatory systems may be separated into two distinct body components, functionally they are interrelated in providing oxygen to and eliminating carbon dioxide from the body. Oxygen is essential for life processes, and efficient transfer of oxygen from the water to the fish is necessary because the dissolved oxygen concentration in the water is low. Water contains approximately one-thirtieth by volume the amount of oxygen in air.

Respiration in catfish takes place at the gills, which are specialized organs that enable the fish to extract oxygen from the water efficiently. Four gills, which are covered by a protective flap called the *operculum*, are located on each side of the head. The gills consist of bony or cartilaginous arches to which the gill filaments are attached. Numerous small lamellae, which are the actual respiratory site, protrude from the gill filaments (Fig. 1.5).

The circulatory system functions in respiration by transporting gases between body tissues and the gills. This is accomplished by pumping unoxygenated blood from the heart through the ventral aorta to the gills, where it is distributed to the gill filaments and lamellae via the afferent branchial arteries. Blood flow through the lamellae is in a direction opposite to that of the water flowing over the lamellae. Flow of blood and water in opposite directions provides for countercurrent exchange, which increases the efficiency of gas exchange. As blood passes through the thin lamellae, oxygen diffuses from the water into the blood, where it is bound to the respiratory pigment hemoglobin. Hemoglobin is found in red blood cells (erythrocytes) and has a strong affinity for oxygen. The oxygenated blood enters the efferent branchial arteries that form the carotid artery and dorsal aorta, which carry the blood to various tissues. The blood is returned to the heart from the capillary systems through various veins (Fig. 1.6).

Diffusion of oxygen from the water into the gill lamellae is dependent on a continuous supply of oxygenated water passing over the gills. This is accomplished by a ventilation process during which water is taken into the mouth and pushed over the gills by concurrent contraction of the buccal cavity and expansion of the operculi. After the water passes from the opercular openings, the process is repeated. Change in ventilation volume is one method that fish can use to influence gas exchange at the gills. This mechanism allows for some adjustment to periods of low dissolved oxygen.

Carbon dioxide is eliminated from the catfish's body by counter-

14 General Biology

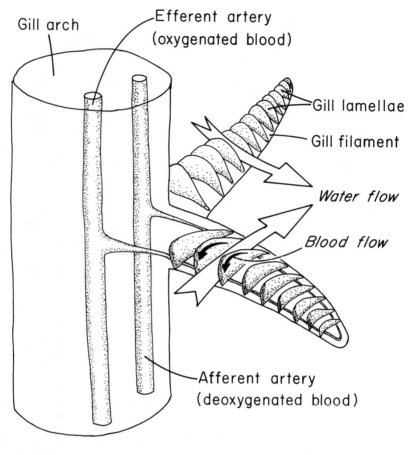

FIGURE 1.5.

Schematic illustration of blood and water flow through a section of gill.

current exchange in a manner similar to that in which oxygen enters the body. Blood high in carbon dioxide enters the gills and carbon dioxide diffuses into the water from the gill lamellae.

In addition to its respiratory function, the circulatory system is involved in several other functions, but generally each function relates to the transport of substances to and from various tissues and organs. A major function involves the distribution of blood throughout the body. This is accomplished by the cardiovascular system, which consists of the heart as a pump in line with branchial (gill) and systemic capillaries connected by arteries and veins. The blood consists primar-

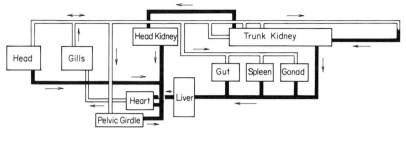

FIGURE 1.6.

Schematic illustration of blood flow in the channel catfish. White vessels are arteries; black vessels are veins. Arrows indicate direction of blood flow.

ily of cells suspended in plasma. The most abundant cells found in blood are the red blood cells, which function to carry oxygen to the tissues. Various types of white blood cells (leukocytes) are also present. They function in blood clotting and as a nonspecific defense against invading microorganisms.

OSMOREGULATION

Maintenance of an optimal internal salt (ionic) composition is important to the well-being of an animal. Organisms that live in an environment that contains a salt concentration different from that of their body must expend energy to maintain the proper internal ionic balance. The process of maintaining an internal salt balance different from that which occurs in the environment is termed *osmoregulation*.

Catfish have an internal ionic environment that is more concentrated than that of the surrounding water; thus they continually gain water and lose salts (ions) by diffusion, primarily at the gills. Also, certain salts may be lost in the urine. One of the functions of the excretory system, the kidney and gills, is to eliminate excess water and to regain lost salts. The kidney functions to remove excess water as well as to conserve certain salts. Some salts are reabsorbed by the kidney or gained from the food, but most are replaced by transport across the gills. Specialized cells (chloride cells) located at the base of the gill filaments function in the transport of ions. Sodium and chloride ions are regained at the gills and are exchanged for the ammonium ion and the bicarbonate ion, respectively.

SENSORY FUNCTION

Generally, sensory perception in fish is similar to that of other vertebrates. That is, fish possess organs for sight, hearing, smell, and taste much like other animals, although, in an aquatic environment distinction between "taste" and "smell" is nebulous compared to their perception in terrestrial animals. Catfish also have sensory means of detecting water currents and electrical stimuli; they include eyes, ears, olfactory organs, taste buds, and the lateral line. Information collected by the sensory organs is transmitted to the brain for interpretation.

The eye is used to gather light (sight) and is like the eyes of other animals except that they are adapted for seeing clearly in water. The acousticolateralis system, which consists of the inner ear and lateral line, contains numerous sensory cells that detect sound and vibrations in the environment. The inner ear is made up of membranous passages (labyrinths) that are connected to the brain. The inner ear is involved in sound reception as well as orientation and balance. Sensitivity to sound is increased by a mechanism in which the inner ear is connected to the swim bladder by a series of small bones called the *Weberian ossicles*.

The lateral line contains sense receptors (neuromasts) that are located between adjacent lateral line pores that are open to the environment. These receptors respond to water currents and vibrations and act as a sort of "distance touch" mechanism.

Detection of odors by the olfactory organs and taste by the gustatory organs in catfish is stimulated by contact of receptors with chemicals dissolved in the water. The olfactory organs appear to be the more sensitive means of chemoreception and have the ability to detect more dilute odors. The olfactory organs in the catfish are located in olfactory sacs located on the head. Water flows through the sacs and odors are perceived as the dissolved chemical contacts the receptors embedded in the olfactory sac.

Taste buds (gustatory organs) are found over the entire surface of the catfish as well as inside the mouth, pharynx, and part of the esophagus. Taste buds are particularly concentrated on the barbels and gill arches. These organs are useful in sensing food and harmful substances dissolved in the water, which provide the fish with a "close range" sense.

IMMUNE FUNCTION

Since catfish are continually exposed to pathogens in pond waters, it is important to minimize stressors that may predispose the fish to infectious diseases by impairing the function of natural defense mechanisms. Stress-induced susceptibility to disease may be the result of several factors, including poor environment, inadequate nutrition, crowding, and age and size of fish. Stress may be manifested in several ways, ranging from shock and sudden death to more subtle behavioral changes with underlying physiological and biochemical alterations in blood serum and other tissues. Catfish have two lines of defense against invading organisms that may be categorized into nonspecific and specific defense (immune response).

Nonspecific Defense

Nonspecific defense includes intrinsic permanent barriers that serve as the first line of defense against nonliving and living invasive substances. The outer layer of the skin (epidermis) and the mucus secreted by specialized cells in the epidermis are physical barriers to foreign materials. Mucus is continually replenished, and when it sloughs off, it carries debris and microorganisms with it. Mucus also appears to contain bactericidal and antiviral factors, as well as substances that inhibit the growth of parasites on external surfaces. Certain irritants such as parasites cause an increase in mucus production. Maintenance of the structural integrity of the barriers is important because an injury that results in a break in the skin provides a port of entry for pathogens.

Inflammation is another form of nonspecific defense against invasive organisms and other types of injury. The inflammatory response occurs after injury and attempts to localize the damage, thereby preventing further intrusion into the body. Serum clotting factors initiate the formation of a clot that helps to prevent the escape of fluids from the body and reduces the spread of bacteria or other foreign substances. In addition, white blood cells migrate to the site of injury and destroy foreign particles. Signs of inflammation are varied, but generally an accumulation of fluid (edema) as well as hemorrhage occur at the site of injury.

Specific Defense

The immune system of fish has not been as well defined as that of mammalian species. Lymph nodes and bone marrow comparable to those found in mammals are not present in fish. In fish, immunity is related to hematopoietic (blood-forming) organs including the anterior and posterior kidneys, the spleen, and possibly the thymus. Although differences exist, the immune response of fish appears to be basically the same as that of other animals. The immune response is a delayed defensive reaction initiated by exposure to an invading agent (antigen), such as bacteria, viruses, toxins, or other foreign substances. An immune response may occur because of previous contact with the pathogen itself or artificial exposure through vaccination. If an animal survives the initial exposure, an acquired immunity is maintained. Exposure to an antigen elicits two basic types of immune reaction, humoral and cellular.

Humoral immunity is an acquired immunity in which the role of the circulating antibody (immunoprotein formed by lymphocytes in response to an antigen) is predominant. Antibodies formed in response to a particular antigen react specifically only with that one type of antigen. The antibody attaches to the antigen and either destroys or inactivates the pathogen or its toxins. Antibodies can also activate a group of nonantibody serum proteins called *complement*. Complement then acts with the antibody to destroy the antigen. Complement is nonspecific; that is, it works with various antibodies.

Cellular immunity is an acquired immunity in which lymphocytes become sensitized against the foreign agent. Sensitized lymphocytes act as whole-cell "antibodies" and attach to a specific antigen, destroying it. Cellular immunity imparts a longer-lasting immunity than does the humoral response. Also, cellular immunity may respond to lower levels of antigen, thus initiating an immune response when the humoral mechanism cannot.

REFERENCE

Grizzle, John, and Wilmer Rogers. 1976. *Anatomy and Histology of the Channel Catfish.* Auburn: Alabama Agricultural Experiment Station.

CHAPTER 2
Life History and Reproductive Biology

GENERAL LIFE HISTORY

Channel catfish were originally native to Mexico and states bordering the Gulf of Mexico as well as states within the Mississippi Valley. The species has been introduced throughout the United States and in many other countries during the past century; thus the channel catfish is widely distributed.

Since channel catfish have been produced by state and federal hatcheries for stocking, they can be found in most freshwater environments and some brackish waters in the United States. Their natural habitat is in moderate- to swift-flowing streams; however, they are at times found in slow-moving streams. They are found in turbid water but generally prefer clear water. Channel catfish are bottom dwellers that prefer a substrate of sand, gravel, and rubble.

Channel catfish are usually found in deep protected holes during the day. Activity, which is primarily associated with feeding, increases immediately after sunset and just before sunrise. Young fish feed in shallow areas while adult fish feed in deeper waters. Adults do not often move from area to area as do young fish. Although most feeding appears to occur at night, feeding does occur during the day. Channel catfish are typically bottom feeders but will take food at the surface. Young channel catfish feed mostly on aquatic insects, whereas adults feed on insects, snails, crawfish, algae, aquatic plants, and small fish. Small fish contribute a large portion of the diet of channel catfish larger than 18 inches in total length.

The optimal temperature for channel catfish growth is about 85°F. Growth declines as the temperature deviates from the optimum, and below about 50°F feeding activity essentially stops, as does growth. In nature, growth of channel catfish is relatively slow. A channel catfish may take from 2 to 4 years to reach 1 pound. Growth is dependent on several factors; two of the more important are environmental temperature and food availability. Channel catfish have been reported to live for up to 40 years and reach 58 pounds.

Channel catfish generally reach sexual maturity at 3 years of age, but the period to maturity may vary from 2 to 5 years. Once fish sexually mature, spawning (discussed in the section Spawning and Fertilization) can occur. Channel catfish are cavity spawners in that they spawn in secluded, semidark areas in nature, such as holes, hollow logs, or undercut banks. Spawning occurs when water temperatures are between 75 and 85°F. The male prepares the spawning site, fertilizes the eggs laid by the female, and cares for the eggs after fertilization. After the eggs hatch, the newly hatched fish (fry) use nutrients stored in their yolk sac for a few days until they are fully developed and are capable of feeding. A newly hatched catfish fry is a replicate of the adult; that is, there are no free-swimming larval stages of development as in many other species of fish.

MALE REPRODUCTIVE BIOLOGY

The channel catfish male can be distinguished from the female internally by examination of the gonads. The testes of the males are paired structures located in the rear of the body cavity and are suspended by the mesenteries below the trunk kidney and swim bladder. There is an increase in testicular size as spawning nears. For example, the gonosomatic index (gonadal weight expressed as a percentage of body weight) increases from about 0.1 to about 0.3 in July in a 2- to 3-year-old channel catfish (Guest, Avault, and Roussel 1976). The testes are lobed and consist of numerous wormlike projections. Each testis is separated into anterior and posterior sections that differ in appearance and function. The anterior region comprises about three-fourths of the testis and is lighter-colored and thicker than the posterior region. Seminiferous tubules, which are the site of sperm formation (spermatogenesis), are found in the anterior region of the testes. The size of these tubules as

TABLE 2.1.
Hormones Involved with Fish Reproduction

Hormone	Source	Type
Androgens	Gonads	Steroid
Estrogens	Gonads	Steroid
Progestins	Gonads	Steroid
Maturational	Pituitary	Glycoprotein
Vitellogenic	Pituitary	Protein
GnRH[a]	Brain	Oligopeptide
Urotensin	Urophysis	Oligopeptide
Isotocin	Neurophypophysis	Oligopeptide
GRIF[b]	Brain	Amino acid derivative
Melatonin	Pineal, eye	Amino acid derivative

[a]Gonadotropin-releasing hormone.
[b]Gonadotropin release-inhibitory factor.
Source: Adapted from Grizzle (1985).

well as the number of spermatozoa increase as spawning nears. Spermatogenesis in the channel catfish is apparently similar to that of other vertebrates (Grizzle 1985). The function of the posterior section of the testes is not clearly defined, but it appears to produce a fluid that is involved in the transfer of sperm or perhaps in other aspects of reproduction. A sperm duct that ties the testes to the urogential pore (located on the end of the urogential papilla) is used to expel sperm and urine. The urogential papilla, a small fleshy nipplelike structure located behind the anus, can be used to distinguish males from females externally (see the section Brood Fish Management). The male generally has a broad, muscular head that is wider than the body. The sexually mature male also has dark mottled gray pigmentation on the undersides of the jaw and the abdomen.

Several hormones are directly involved in fish reproduction. These include the steroids that are produced by the gonads and head kidney and peptide hormones produced by the pituitary, brain, and urophysis (Table 2.1). The androgens produced in the testes are the most important hormones involved in regulating male reproduction and development of secondary sex characteristics, which include widening of the head and darkening of skin color in sexually active males. Specialized cells (Leydig and Sertoli cells) have been identified as sources of androgens produced in the testes of certain fish (Grier 1981; Nagahama 1983). Although these cells have yet to be found in channel catfish testicular tissue, it is likely that they are present. A possible

22 Life History and Reproductive Biology

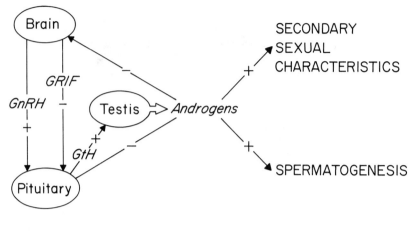

FIGURE 2.1.

Endocrine regulation of secondary sexual characteristics and spermatogenesis in channel catfish. A (+) sign indicates stimulation; a (−) sign indicates inhibition. GnRH = gonadotropin-releasing hormone; GRIF = gonadotropin release-inhibiting factor; GtH = gonadotropin.

mechanism of endocrine regulation of spermatogenesis and secondary sexual characteristics in channel catfish is depicted in Figure 2.1. Other steroids produced by the testis include progestins and estrogens; their role in male reproduction is not well defined.

FEMALE REPRODUCTIVE BIOLOGY

The ovaries of the channel catfish are paired yellowish glandular structures of approximately the same size that are attached by the mesenteries to the trunk kidney and swim bladder. The ovaries can make up a considerable portion of body weight. The gonosomatic index has been shown to range from 0.2 in a 2-year-old female in midwinter to 15.6 for prespawning females (Brauhn and McCraren 1975). As a result of the increase in ovarian size, the abdomen of the female increases in size. Otherwise, sexually developed females change little in appearance. The ovary connects to the oviduct, which terminates at the genital pore, which is used to expel eggs. Urine is released from the urinary pore just posterior to the genital pore. The presence of the two pores can be used to determine the sex externally. The urinary pore is located in a groove and may not be easily seen; however, the groove and surrounding folds of skin form a slit that is easily recognizable. The ovaries are a source of estrogen and progestin hormones as well as the hormonelike substances the prostaglandins, which function in regulating female reproduction.

Oogenesis (egg development) generally occurs annually in channel catfish. The process of oogenesis is continuous, but it has been divided into arbitrary stages, which have been described in detail by Grizzle (1985). Endocrine and neural regulation control reproduction. In the female the process results in maturation of the egg, ovulation, and spawning. A summary of the regulation of female reproduction is presented in Table 2.2.

SPAWNING AND FERTILIZATION

Spawning, which is the release of eggs and sperm, is under neural and hormonal control but can be affected by a number of environmental factors as well as nutritional status of the fish. Water temperature is critical. Spawning may occur at from 70 to 86°F, but 78 to 80°F appears to be optimal. At water temperatures above 86°F, egg development and fry survival are adversely affected. Before spawning, the male chooses a suitable spawning site. Since catfish are cavity spawners, the site is usually a hole, burrow, or submerged log in natural waters. In commercial catfish culture, spawning containers are provided (see Chapter 7). The male begins to clean the spawning site and to guard the surrounding territory. The female ready to spawn also defends the territory against invaders. The female is apparently attracted to the spawning site by olfactory or social signals. Males are attracted to the female by pheromones (Timms and Kleerekoper 1972). A few hours before spawning, the female completes cleaning of the spawning site by moving over the location where the eggs are to be deposited with pelvic and pectoral fins beating against the bottom of the site. This action may also serve to stimulate the male. There is some aggression by the male against the female, but if the female remains near the bottom of the spawning site the aggression is stopped (Clemens and Sneed 1957).

When the fish are ready to spawn and positioned head-to-tail, the male tries to mate with the female. If they spawn, the male's body quivers and the pectoral fins move for a few seconds and the sperm is released. At about the same time the female's body quivers and eggs are released. Then the male and female remain still for a few seconds until

TABLE 2.2.
Summary of the Regulation of Female Teleost Reproduction

Event	Definition	Mediator
Oogonial proliferation	Increase in cells from which eggs develop	Hormone may not be required
Previtellogenic growth of oocytes	Formation and early growth of eggs	Gonadotropin not required or is inhibitory
Vacuoles form in oocytes	Formation of cavities in eggs containing glycoprotein	Estrogens and gonadotropin
Vitellogenesis	Production of yolk protein deposits in eggs	Vitellogenic hormone, maturational hormone, estrogen
Maturation of oocytes	Ripening of eggs	Corticosteroids, progestins
Ovulation	Expulsion of mature egg from follicular envelope	Corticosteroids, progestins, prostaglandins
Spawning	Release of eggs from body	Nervous system, urotensin II, neurohypophyseal hormones, prostaglandins

Source: Adapted from Grizzle (1985).

the female positions herself over the eggs, using the beating pelvic fins to move water over the eggs. The process is repeated until spawning is complete. Spawning has been observed to last for up to 6 hours with approximately nine releases of eggs per hour (Clemens and Sneed 1957). Females spawn once a year but males may spawn three to four times each year. The number of eggs deposited by a female channel catfish is dependent on fish size; approximately 3,000 to 4,000 eggs are deposited per pound of body weight. Busch (1985) estimated egg numbers in spawns from channel catfish weighing 4 to 11 pounds and found the average number of eggs per pound of body weight to be about 3,800. Clemens and Sneed (1957), working with 1- to 4-pound channel catfish, found the average number of eggs per pound of body weight to be about 4,000, and about 3,000 for fish larger than 4 pounds.

The eggs and sperm of the channel catfish are released near each other so that they are likely to make contact and fertilization can occur. The sperm are motile for about 4 minutes in water (Guest, Avault, and Roussel 1976). When the sperm penetrates and fertilizes the egg, a number of reactions occur, which result in the formation of a zygote, which divides repeatedly and develops into the embryo. The actual mechanisms involved in the union of the male and female pronuclei to form the zygote nucleus are not known for the catfish.

EARLY LIFE HISTORY

The fertilized egg mass is adhesive and contains several thousand eggs, depending on the condition and size of the female (see Chapter 7). The eggs, which average about 3/16 inch in diameter, are initially light yellow but become brownish yellow with age. The male protects the eggs by driving away other fish, including the female. The male also aerates the eggs by fanning the mass with the pelvic fin and continues to guard the fry for a few days after they hatch.

The eggs hatch within 5 to 10 days after fertilization, depending on water temperature. Several stages (discussed in Chapter 7) are apparent as the embryo develops. At hatching, the embryo emerges from the egg after the outer egg membrane (the chorion) ruptures, presumably by enzymatic and mechanical means.

The newly hatched channel catfish fry uses nutrients contained in the yolk sac for nourishment. Initially some body organs, including the digestive system, swim bladder, and fins, are not fully developed. About 5 to 16 days after hatching the yolk is depleted and the fry become juveniles that have external and internal features (except for the reproductive system) similar to those of adult catfish. At this time, they swim to the surface seeking food.

REFERENCES

Brauhn, J. L., and J. McCraren. 1975. Ovary maturation in channel catfish. *Progressive Fish-Culturist* 37:209–212.

Busch, R. L. 1985. Channel catfish culture in ponds. In *Channel Catfish Culture*, ed. C. S. Tucker, pp. 13–84. Amsterdam: Elsevier.

Clemens, H. P., and K. E. Sneed. 1957. The spawning behavior of the channel catfish *Ictalurus punctatus*. U.S. Fish and Wildlife Service, Special Science Report in Fisheries. No. 219.

Grier, J. H. 1981. Cellular organization of the testis and spermatogenesis in fishes. *American Zoology* 21:345–357.

Grizzle, John M. 1985. Reproductive biology. In *Channel Catfish Culture*, ed. C. S. Tucker, pp. 229–282. Amsterdam: Elsevier.

Guest, W. C., J. W. Avault, Jr., and J. D. Roussel. 1976. A spermatology study of channel catfish, *Ictalurus punctatus*. *Transactions of the American Fisheries Society* 105:463–468.

Nagahama, Y. 1983. The functional morphology of teleost gonads. In *Fish Physiology*, Vol. IX, part A, eds. W. S. Hoar, D. J. Randall, and E. M. Donaldson, pp. 223–264. New York: Academic Press.

Timms, A. M., and H. Kleerekoper. 1972. The locomotor responses of male *Ictalurus punctatus*, the channel catfish, to a pheromone released by the ripe female of the species. *Transactions of the American Fisheries Society* 101:302–310.

CHAPTER 3
Genetics

The result of genetic variation is obvious to every channel catfish farmer. Differences in growth, body shape, and color can be seen even in fish from the same spawn and reared in the same environment. In nature, genetic variation is important because it enables populations to adapt to changing environments and survive. Genetic variation is important in channel catfish farming because the heritable component of variation can be exploited through breeding programs to improve commercially important traits.

The success of genetic research and breeding programs is evident in the cattle, poultry, and swine industries. Substantial improvements in yields, feed conversion, and disease resistance are directly attributable to genetic improvement programs. Channel catfish culture is a young industry and lags far behind terrestrial animal husbandry in genetic research background as well as broad application of fundamental principles of breeding.

This chapter outlines some basic genetic principles to clarify the basis for breeding programs discussed in Chapter 6. Individuals interested in more complete accounts of practical fish genetics should consult the excellent works of Tave (1986) and Kapuscinski and Jacobson (1987).

BASIC GENETICS

Genes and Chromosomes

The basic unit of inheritance is the gene. Genes contain the biological code for the production of a trait (phenotype). A gene is a sequence of subunits on a large molecule called deoxyribonucleic acid (DNA). Most of the DNA is packaged in paired structures called *chromosomes* that are located in the nucleus. Each parent contributes one set of chromosomes to the progeny (N represents the number of chromosomes an individual parent contributes). Individuals with two full sets of chromosomes are called *diploid* and their number of chromosomes is denoted by $2N$. Channel catfish have 29 pairs of chromosomes; therefore $2N = 58$ in this species. Unlike in most animals, in channel catfish sex chromosomes are not distinguishable from the other chromosomes (autosomes), but the sex-determining genes are probably located on a single pair of autosomes.

Because chromosomes occur in pairs, there are two representatives of each gene in diploids. Each gene representative is called an *allele* and is located in the same position (*locus*) on each of the pairs of homologous chromosomes. Different alleles produce variations of a phenotype. If the pair of alleles is identical, the individual is said to be *homozygous* for that gene. An individual is heterozygous for a gene if the alleles at a given locus are different. The *genotype* is the set of alleles an individual carries at one or more loci.

Although most channel catfish are diploid, variation in ploidy (number of chromosome sets) can occasionally occur naturally or can be induced. In fact, production of polyploids (more than $2N$ chromosomes) may have commercial value. Triploid organisms have $3N$ chromosomes and tetraploids have $4N$ chromosomes. Triploid individuals are usually sterile because the genetic material cannot be split into equal parts. Tetraploid individuals may be able to reproduce because the genetic material can be split into equal parts (each $2N$).

Chromosome Replication and Cell Division

During growth of an organism, somatic cells (all cells except egg and sperm) must divide to increase in number or replace old cells. When cells divide, the whole chromosome set is duplicated to produce daughter cells identical to each other. Division of the cell nucleus (and chromosomes) is called *mitosis*. Mitosis of a 2N cell produces two 2N daughter cells from a single cell division.

In gametogenesis (production of sperm or eggs) the number of chromosomes must be reduced from the diploid state, 2N, to the haploid state, N, so that when sperm and egg unite, the resulting zygote is again 2N. Meiosis also results in a reshuffling of genetic material among gametes that increases phenotypic variation of progeny.

Meiosis is illustrated in Figure 3.1 for a hypothetical organism with two pairs of homologous chromosomes (N = 2; 2N = 4). Three significant events with important consequences for genetic variability occur during meiosis. After replicated pairs of homologous chromosomes form tetrads, pieces of chromosome are exchanged between homologous chromosomes in the process called *crossing over*. Genes of maternal and paternal origin can thus come to reside on the same chromosome. This creates new combinations of genes on a chromosome and increases genetic and phenotypic variation in a population. Segregation and independent assortment (known as *Mendel's principles*) are the other two consequences of meiosis. Segregation occurs during the first meiotic division. Each of the secondary gametocytes contains a single replicated chromosome from each homologous pair: alleles are separated into different cells. Independent assortment means that nonhomologous chromosomes separate randomly so that each gamete receives a random mixture of chromosomes of maternal and paternal origin.

Crossing over, segregation, and independent assortment greatly increase variation in the genetic makeup of sperm and eggs. The number of possible different gametes produced by a channel catfish is astronomical.

Each primary male gametocyte (spermatocyte) produces four haploid sperm, but each primary female gametocyte (oocyte) normally produces only one haploid egg and three nonfunctional haploid cells. When the oocyte divides, cytoplasm (yolk) is unequally segregated; this results in one functional cell and one with nuclear material and little cytoplasm. This nonfunctional cell is called a *polar body*. The first polar body is produced during reduction division and subsequently divides during equational division to form two cells. Secondary oocytes do not undergo equational division until activated by contact with the sperm. Activation stimulates equational division, and one haploid egg and a second polar body are produced. Equational division of activated eggs can be stopped by temperature or pressure shock applied at the appropriate time after fertilization. Consequently, an extra haploid set of maternal chromosomes is retained and triploids result when the egg nucleus and sperm nucleus fuse.

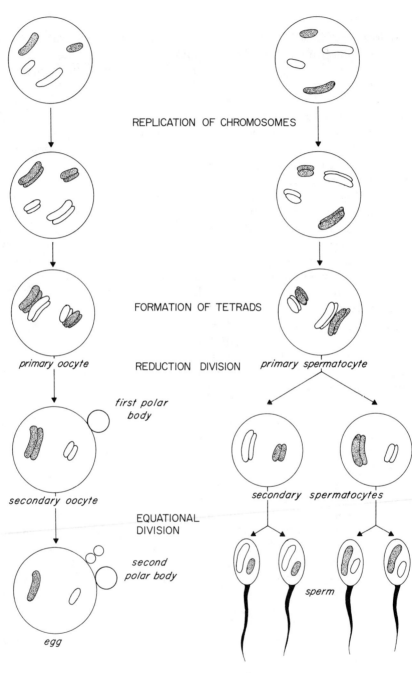

FIGURE 3.1.

Schematic diagram of meiosis. Shaded chromosomes are of maternal origin; white chromosomes are of paternal origin.

Basic Genetics **31**

Development

Fusion of nuclei of normal, haploid sperm and egg produces a diploid zygote. First mitosis (cell division) occurs after about 90 minutes at 80°F. Successive divisions occur at 30- to 40-minute intervals up through the 64-cell stage. Subsequent development of the embryo is described by Saksena, Yamamoto, and Gibbs (1961). Suppression of first mitosis with high pressure or temperature shock causes a doubling of the ploidy in cells of subsequent divisions, and the normal diploid zygote becomes a tetraploid (4N) embryo.

Sex Determination

Sex genes are the principal determinants of gender. The chromosomal sex-determining system in channel catfish is the XY system, so called because, in humans, one of the pair of sex chromosomes looks like an X and the other, the one that determines maleness, looks like a Y. Females are homogametic XX (*homogametic* means homozygous for sex chromosomes), whereas males are heterogametic XY. Eggs can possess only an X chromosome, but sperm can have either X or Y chromosomes.

Normal females are produced when sperm carrying the X chromosome fertilizes an egg. Females can also be produced when the eggs are fertilized with sperm whose DNA has been inactivated with radiation. The sperm contributes no DNA to the embryo, but contact of the sperm and egg membrane results in activation of the egg. If pressure or cold shock is applied to the egg just before the second meiotic division, the egg retains both maternal haploid sets of chromosomes. The resulting diploid zygote is not only homogametic XX but also nearly homozygous for all genes since both chromosomes of a pair came from a single chromosome that was replicated before formation of tetrads. The zygote is not entirely homozygous because some genetic material is reshuffled among chromosomes during tetrad formation before the first meiotic division of the oocyte (crossing over). This procedure for producing all-female progeny, called *gynogenesis*, is an important experimental tool in genetic research. Gynogenesis may also have commercial value as a method to produce monosex breeding populations.

The functional sex of an individual can also be influenced by genes on autosomes or can be manipulated by exposure of early life stage individuals to certain hormones. For instance, normal XY male embryos or juveniles exposed to estrogen become functional females even though they still retain XY sex chromosomes. When these XY functional females are mated with normal XY males, the progeny comprise XX normal females, XY normal males, and YY "supermales." It is not known yet whether the YY condition is an advantage or disadvantage to the individual fish. However, YY fish can be used to produce all-male offspring that may have some commercial value (see the section Nontraditional Genetic Improvement Programs).

QUALITATIVE TRAITS

Qualitative traits are those that exhibit "all-or-nothing" expression of a phenotype: individuals fall into one of two or more discrete categories. Coloration (albinism versus normal pigmentation) is a qualitative trait. The genetics of qualitative traits often is simple because phenotypic expression is controlled by only a few genes. Since individuals fall into categories, the incidence of qualitative traits within a population can be described by ratios. For example, the ratio of pigmented fish to albino fish might be 3 to 1. This means that, on average, 3/4 of a population is normal, 1/4 is albino.

A variety of deformities and pigmentation patterns have been found in channel catfish (Dunham and Smitherman 1985). Most of the deformities appear to have no genetic basis and may be related to poor environmental conditions during egg or fry development. The various odd color patterns of channel catfish, such as piebald or solid black, are little more than curiosities. Albinism, however, is a potentially important commercial trait. Certain processed fish products, such as whole dressed fish, are more attractive to consumers when produced from albino channel catfish rather than normally colored fish.

Albinism (Fig. 3.2) is an example of complete dominant gene action. *Dominance* describes the interaction of alleles at the same locus. A dominant allele is one whose phenotype is always expressed regardless of the genotype. The other allele is recessive. Although three combinations of alleles (genotypes) are possible (dominant-dominant, dominant-recessive, recessive-recessive), only two phenotypes are produced. The allele for normal pigmentation (designated "+") is dominant; the allele for albinism (designated "*a*") is recessive. The three possible genotypes are + +, +*a*, and *aa*. Normal pigmentation develops in individuals with either + + or +*a* genotypes. Albinism develops only in individuals homozygous for the recessive allele (*aa*).

Pigmented parents homozygous for the dominant allele (+ +) will produce only + gametes. Albino parents must be homozygous *aa* and thus will produce only *a* gametes. Heterozygous pigmented parents (+*a*) will produce 50 percent + gametes and 50 percent *a* ga-

Quantitative Traits **33**

FIGURE 3.2.

An albino channel catfish. Compare this fish to the normally pigmented fish in Figure 1.1.

metes. Therefore, there are a variety of outcomes for possible matings among pigmented and albino channel catfish. For example, assume one parent is + +, the other *aa*. The four possible combinations of gametes all result in +*a* zygotes, so all progeny are heterozygous and normally pigmented. Now assume mating between two +*a* parents. The four possible combinations of gametes will produce + +, +*a*, *a*+, and *aa* zygotes. On average, 3/4 of the progeny will be normally pigmented (1/4 homozygous and 1/2 heterozygous) and 1/4 will be albino. Outcomes of all possible matings are shown in Table 3.1. The important point is this: only matings between albino parents can be used to produce 100 percent albino progeny.

QUANTITATIVE TRAITS

Quantitative traits or phenotypes exhibit a continuous range of expression. These traits are measured rather than described, and individuals

TABLE 3.1.

Outcomes of All Possible Mating Combinations among Normally Pigmented and Albino Catfish

Mating	Progeny Ratios	
	Genotype	Phenotype
Normal (++) × normal (++)	All ++	All normal
Normal (++) × normal (+a)	1 ++ : 1 +a	All normal
Normal (++) × albino (aa)	All +a	All normal
Normal (+a) × normal (+a)	1 ++ : 2 +a : 1 aa	3 Normal : 1 albino
Normal (+a) × albino (aa)	1 +a : 1 aa	1 Normal : 1 albino
Albino (aa) × albino (aa)	All aa	All albino

Source: Tave (1986).

do not fall into distinct categories such as albino versus pigmented. Most of the important commercial traits are quantitative: growth rate, feed conversion efficiency, percentage body fat, dress-out percentage, fecundity, disease resistance, tolerance to environmental stressors, and seinability. Because quantitative traits are not discrete, their distribution within a population cannot be described by simple ratios as can that of qualitative traits. The distribution of quantitative phenotypes is described by measures of central tendency (mean or mode) and variation (range, variance, and so on).

Although qualitative traits can be controlled by a single gene, a large number of genes typically control most quantitative traits. Quantitative traits also differ from qualitative traits in that the environment can have a substantial impact on expression of a phenotype.

Components of Phenotypic Expression

The phenotype of an individual for a quantitative trait is determined by genes, the environment, and the interaction between genes and environment:

$$P = G + E + GE \tag{3.1}$$

where P is the phenotypic value (the expressed phenotype), G is the effect due to genes alone (the genotypic value), E is the environmental effect, and GE is the genotype-environment interaction. Genotype and environment interact because different environments affect the same genotype differently and different genotypes are affected differently by the same environment.

The genotypic value (G) can be broken down into additive effects, A, and nonadditive effects. Nonadditive effects are due to dominance, D, and epistasis, I:

$$G = A + D + I \qquad (3.2)$$

Dominance effects are due to interactions of alleles at the same locus (this use of the term *dominance* does not refer to dominant gene action as described for the qualitative trait albinism). Remember that pairs of alleles are separated during meiosis into different gametes and new pairs of alleles are formed after fertilization. Thus, any particular dominance effect expressed by the parents is not transferred to the offspring except by chance: dominance effects are created anew each generation.

Epistatic effects are caused by the complex interaction of alleles at different loci. Like dominance effects, most specific epistatic effects are disrupted during meiosis as alleles are separated into different gametes. Therefore, epistatic effects are also created anew each generation.

Additive effects are due to the cumulative contribution of individual alleles at all loci that affect a given trait. Additive effects do not depend on specific interactions or combinations of alleles. Individual alleles are transmitted to offspring, so additive effects are passed on to progeny in a predictable manner.

Breeding programs exploit the genetic variation within a population or stock of fish. The phenotypic variance within a population is due to the same factors that determine phenotypes of individuals. Thus, the phenotypic variance (V_P) observed for a given trait can be partitioned into genetic variance (V_G), environmental variance (V_E), and the variance due to the interaction between genetic and environmental differences ($V_{G \times E}$):

$$V_P = V_G + V_E + V_{G \times E} \qquad (3.3)$$

Genetic variance can likewise be partitioned into variance components due to additive genetic variance (V_A), dominance genetic variance (V_D), and epistatic genetic variance (V_I):

$$V_G = V_A + V_D + V_I \qquad (3.4)$$

Genetic Variance Components and Breeding Programs

Epistatic genetic variance is extremely difficult to exploit because of the staggering number of possible interactions between alleles at different loci. For this reason, it is often assumed when trying to measure genetic inheritance that epistatic genetic variance does not contribute to quantitative phenotypes. While this is untrue—epistasis is an important component of genetic variance—this assumption simplifies matters considerably.

Additive genetic variance and dominance genetic variance are the components of phenotypic expression most often exploited in breeding programs. The relative contribution of these components to expression of a phenotype determines the type of breeding program most likely to produce the best results. Recall that additive effects are passed from parent to offspring and dominance effects are created anew each generation. If additive genetic variance for a trait is large within a population, superior fish can be selected and bred and the alleles contributing to the phenotype will be passed on to the progeny. Superior progeny will be the expected result. If additive genetic variance for a particular trait is small, selection within the population may not result in improved phenotype. Thus, improvement will depend on a new combination of alleles. The combination of alleles that will result in improvement is impossible to predict and must be found through hybridization trials. Hybrids can be produced by breeding genetically isolated strains of the same species (intraspecific hybrids) or different species (interspecific hybrids). In either case, new combinations of alleles, some of which may improve a particular phenotype, are formed.

Hybrids generally should not be used as brood stock because their superiority is due to interaction among alleles and that interaction is not passed on to progeny. Selection should be used to improve brood stock because selection exploits additive genetic variance that can be passed on to progeny.

Heritability

Heritability (h^2) is an experimentally derived value describing the proportionate contribution of additive genetic variance to a particular phenotype. In essence, *heritability* describes the percentage of a phenotypic expression that is inherited in a predictable manner. Values for heritability range from 0 to 1. If $h^2 = 0$, additive genetic variance does not contribute to expression of phenotype; if $h^2 = 1.0$, expression of phenotype is entirely due to additive genetic variance. In practice, h^2 values usually range from 0.1 to 0.9 for commercially important traits.

When phenotypes or traits have heritabilities greater than about 0.2, additive genetic variance is usually sufficient to allow effective selection programs. If heritability for a trait is less than about 0.15,

TABLE 3.2.

Some Estimates of Heritability (h^2) for Commercially Important Traits of Channel Catfish

Trait	h^2	Reference
Weight at 40 weeks	0.58	Bondari (1983)
Weight at 18 months	0.41	Reagan (1979)
Length at 18 months	0.40	Reagan (1979)
Feed conversion	0–0.38	Burch (1986)
Percent fat	0.61	Reagan (1979)

dominance genetic variance is more important and that trait will be difficult to change by selection.

Heritability estimates for several commercially important traits are listed in Table 3.2. Heritabilities apply only to the population or strain that was tested and are not necessarily representative of channel catfish in general. Heritability estimates also vary from generation to generation as fish with different genotypes mate and may also differ when fish are grown in different environments. For estimates of h^2 to be of use in a selection program, they must be derived in the same environment in which selection will occur.

Inbreeding

Inbreeding is the mating of related individuals. Related individuals share common alleles through common ancestors. When related individuals mate, the common alleles can be paired, producing progeny that are homozygous at one or more loci. Like alleles can also be paired when unrelated individuals mate, but the probability of homozygosity at any one locus will be higher and the number of homozygous loci will, on average, be greater when related individuals mate because they have more alleles in common. Inbreeding creates more homozygosity than would occur if unrelated individuals were mated.

Inbreeding can have dire consequences for practical breeding programs. All populations carry hidden genetic variation in the form of rare deleterious recessive alleles. The alleles are "hidden" because in the heterozygous state they are not expressed. Inbreeding increases the chances that detrimental recessive alleles will be expressed because relatives are more likely to have the same recessive alleles and, because inbreeding increases homozygosity, these alleles are more likely to be paired. The pairing of detrimental recessive alleles by inbreeding can result in increased probability of abnormal or subviable offspring. Uncontrolled inbreeding in populations of fish has been found to reduce growth, survival, and reproductive performance and to increase the number of physical abnormalities. The detrimental effects of inbreeding are called *inbreeding depression*, and the more intense the inbreeding, the more harmful are the effects. The implications of unintentional inbreeding and methods for minimizing inbreeding are thoroughly discussed by Tave (1984) and Tave (1986). Brood stock management practices to minimize inbreeding are also discussed in Chapter 6.

REFERENCES

Bondari, K. 1983. Response to bidirectional selection for body weight in channel catfish. *Aquaculture* 35:73–81.

Burch, E. P. 1986. Heritabilities for body weight, feed consumption, and feed conversion and the correlations among these traits in channel catfish, *Ictalurus punctatus*. Master's Thesis, Auburn University, Alabama.

Dunham, R. A., and R. O. Smitherman. 1985. Genetics and breeding. In *Channel Catfish Culture*, ed. C. S. Tucker, pp. 283–321. Amsterdam: Elsevier.

Kapuscinski, A. R., and L. D. Jacobson. 1987. *Genetic Guidelines for Fisheries Management*. Minneapolis: University of Minnesota/Minnesota Sea Grant.

Reagan, R. E., Jr. 1974. Heritabilities and genetic correlations of desirable commercial traits in channel catfish. Mississippi Agricultural and Forestry Experiment Station Research Report Vol. 5, No. 4, Mississippi State, Mississippi.

Saksena, V. P., K. Yamamoto, and C. D. Riggs. 1961. Early development of the channel catfish. *Progressive Fish-Culturist* 23:156–161.

Tave, D. 1984. Effective breeding efficiency: an index to quantify the effects that different breeding programs and sex ratios have on inbreeding and genetic drift. *Progressive Fish-Culturist* 46:262–268.

Tave, D. 1986. *Genetics for Fish Hatchery Managers*. Westport, Conn.: AVI Publishing.

CHAPTER 4
Environmental Requirements

Channel catfish grow fastest and are healthiest when reared under an optimal set of environmental conditions. When any factor deviates from the optimum, the general well-being of the fish is affected. Fish respond to suboptimal conditions (stressors) by modifying metabolism or behavior in an attempt to adapt to the new conditions. Initial reactions (stress responses) include release of hormones that prepare the animal for emergency action by mobilizing energy reserves and increasing respiratory efficiency. If stress is severe or maintained over a long period, the capacity of the fish to adapt is exceeded and some of the initial responses may actually have deleterious consequences. The ultimate results of severe or prolonged stress include reduced growth, impaired immunity to infectious disease organisms, poor reproductive success, or death (Pickering 1981).

Optimal and tolerated ranges for some water quality variables are listed in Table 4.1. This list is of limited value because tolerance to environmental stressors varies with life stage, nutritional status, general health, and other modifying factors. Values in Table 4.1 are for fingerling or adult fish. Sac fry and early swim-up fry are the life stages most sensitive to most stressors. Environmental requirements for hatchery waters are thus more stringent than for waters used to rear larger fish. Hatchery water quality requirements are further detailed in Chapter 7. Sick or malnourished fish also are less tolerant of poor water quality. Tolerance to stressors is further complicated by interactions

Environmental Requirements

TABLE 4.1.

Optimal and Tolerated Levels of Some Water Quality Variables for Growth of Channel Catfish

Variable	Optimal Level	Tolerated Level
Salinity	0.5–3 ppt	<0.1–8 ppt
Temperature	80–85°F	32–104°F
Dissolved oxygen	5–15 ppm	2 ppm–300 percent saturation
Total alkalinity	20–400 ppm as $CaCO_3$	<1 to >400 ppm as $CaCO_3$
Total hardness	20–400 ppm as $CaCO_3$	<1 to >400 ppm as $CaCO_3$
Carbon dioxide	0 ppm	Depends on dissolved oxygen concentration
pH	6–9	5–10
Un-ionized ammonia	0	< 0.2 ppm as N
Nitrite	0	Depends on chloride concentration
Hydrogen sulfide	0	<0.01 ppm as S

Note: Prolonged exposure to nonoptimal conditions may be tolerated but can result in reduced growth, impaired reproductive performance, or increased susceptibility to disease.

among stressors. A combination of several stressors, even though individual values are in the tolerable range, can kill fish.

SALINITY

Salinity is the dissolved salt content of water and is often expressed as parts of salt by weight per thousand parts of water by weight (ppt). Fresh water is arbitrarily defined as having a salinity of less than 0.5 ppt; the salinity of seawater is about 35 ppt. Waters of intermediate salinity are termed *brackish*.

The concentration and composition of substances dissolved in channel catfish body fluids must be maintained within fairly narrow limits to buffer cells against changes that would disrupt function or integrity. This process is called *osmoregulation* (see the section Osmoregulation) and requires expenditure of metabolic energy. Channel catfish are adapted to live in environments within a certain range of salinity. When they are forced to inhabit environments outside this range, excess metabolic energy is spent on osmoregulation at the expense of other physiological functions.

Channel catfish can survive in waters up to about 11 to 14 ppt salinity, but fish grow slowly at salinities greater than about 6 to 8 ppt (Perry and Avault 1970). This is approximately the salinity of channel catfish blood serum. The optimal salinity for channel catfish growth, reproduction, and immune function is not known but lies in the range of 0.5 to 3 ppt. Although increased energy is required for osmoregulation at lower salinities, fish achieve good growth at salinities lower than 0.5 ppt as some or all of certain ion requirements are met by minerals in feeds and natural foods.

Salinities less than 3 ppt are also optimal for egg and fry development. Eggs can tolerate up to 16 ppt, but tolerance declines to 8 ppt at hatching. Tolerance increases to 9 to 10 ppt at swim-up and then to 11 to 14 ppt by 5 to 6 months of age (Allen and Avault 1971).

TEMPERATURE

Temperature controls the rate of all chemical reactions including those constituting metabolism. Channel catfish are poikilothermic ("many temperatured"), meaning that body temperature is very nearly that of the water. Water temperature thus controls all aspects of the metabolism of channel catfish. Water temperature is also a primary factor affecting the economic feasibility of commercial channel catfish culture. Control of temperature is impossible in ponds, and large-scale pond culture is limited to those areas affording a sufficiently long growing season to produce a market-sized fish from egg in less than 18 months. This is generally in areas where average daily water temperatures are above 70°F for at least 180 days a year, with water temperatures above 80°F for at least 120 of those days.

Water temperature affects all aspects of channel catfish culture. Discussed here are the effects of water temperature on survival, growth and immune function, and reproduction. Interactions of water temperature with other environmental variables are mentioned throughout this chapter.

Critical Temperatures

Most strains of channel catfish can survive for long periods at temperatures ranging from just above freezing to about 104°F. Survival at or near critical temperatures depends on the rate of temperature change and acclimation temperature. Other environmental factors, such as dissolved oxygen concentration, fish health, and nutritional status, also affect survival at extremes of temperature.

When temperatures change slowly (a few degrees per day), physiological changes occur to allow the fish to adapt and survive at the final temperature. This process, called *acclimatization*, involves many biochemical changes in enzyme systems that allow metabolic reactions to operate at the new temperature. Changes in cell membrane lipids also occur to compensate for the effect of temperature on membrane integrity (Reynolds and Casterlin 1980). Changes in temperature of more than a few degrees per day do not afford time for physiological adaptation and fish are stressed. Fish may eventually adapt, but initially they are more susceptible to infectious diseases and may feed poorly. Very rapid temperature changes (more than 1°F/minute) cause thermal shock and possibly death. Channel catfish should not be moved directly from one water to another if temperatures differ by more than 5°F. If the waters differ by more than this, temper the fish by changing the temperature of the water holding the fish by no more than 1°F/minute to match the temperature of the receiving water.

Upper and lower lethal temperatures are also affected by the temperature at which fish are acclimated. This relationship is depicted in Figure 4.1. The higher the acclimation temperature, the higher are both the upper and lower lethal limits. For example, fish acclimated to water at 33°F will eventually die if the water temperature is raised 2°F/hour to a final temperature of 85°F, and fish acclimated at 85°F will survive to about 104°F. The area within the lethal limits in Figure 4.1 is called the *zone of tolerance*. Of course, acclimation temperature becomes less important as the rate of temperature change decreases. When temperatures change very slowly, channel catfish can live over the entire range (freezing to about 104°F) regardless of the initial acclimation temperature.

Growth and Immune Function

Most strains of channel catfish grow fastest and convert feed most efficiently at temperatures between 80 and 85°F. Potential growth decreases rapidly at temperatures above 90°F or below 70°F (Fig. 4.2). Channel catfish feed and grow poorly, if at all, at temperatures below about 50°F.

The immune system of channel catfish functions most effectively at temperatures from about 70 to 90°F, with the optimum being around 85°F. At temperatures below about 60°F, immune function is severely

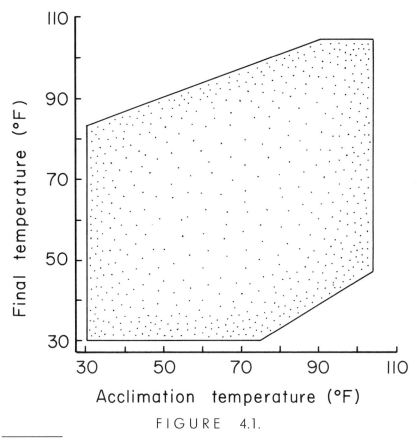

FIGURE 4.1.

Thermal tolerance diagram for channel catfish. To use this diagram, find the temperature at which fish are acclimated on the x axis. Then move vertically to the new temperature. If the new temperature falls within the stippled area (called the "zone of tolerance"), fish will survive. If the new temperature falls outside the stippled area, fish will eventually die. This diagram is applicable for temperature changes of about 2°F/hour.

impaired and nonspecific defense factors are the primary means of preventing infectious diseases. Rapid changes in temperature may also impair immune function even if the changes occur within the range considered optimal.

Reproduction

Seasonal changes in temperature exert primary control over the reproductive cycle in channel catfish (Davis et al. 1986). Cold water temper-

44 Environmental Requirements

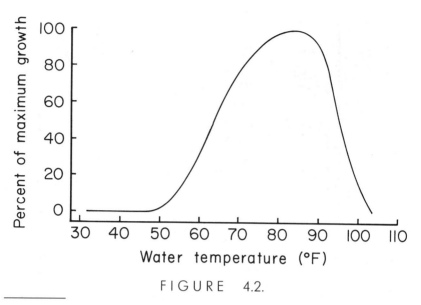

FIGURE 4.2.

Relationship between temperature and potential growth of channel catfish.

atures are apparently required to stimulate gametogenesis (development of mature sperm and eggs) in channel catfish. Conditions required for stimulation are not known with certainty, but a month or more of temperatures lower than 50 to 55°F may be required. A subsequent rise in water temperatures stimulates spawning. A week or more of average daily water temperatures of 70 to 75°F usually initiates spawning in the spring. Water temperatures of 78 to 80°F are considered optimal spawning temperatures (Busch 1985).

DISSOLVED OXYGEN

The availability of dissolved oxygen frequently limits the activities and growth of fish because water contains only small amounts of oxygen available for respiration. Fresh water saturated with oxygen contains twenty to forty times less oxygen by volume than air. Also, the energetic costs of breathing water are greater than for air because water is much denser and more viscous than air. Compounding the problems caused by the limited amount of dissolved oxygen in water is the fact that the oxygen content varies much more in aquatic environments than in air. The oxygen content of water can vary greatly because the availability of oxygen is limited to begin with and small differences in

TABLE 4.2.
Oxygen Consumption Rates for Channel Catfish of Different Weights at Five Temperatures

	Average Fish Weight (Pounds)					
°F	0.05	0.1	0.25	0.5	1.0	1.5
35	50	48	42	35	25	22
50	92	88	77	63	47	41
65	167	160	140	114	86	75
80	326	311	254	224	168	147
95	589	365	493	405	306	267

Note: Oxygen consumption rates expressed as milligrams of oxygen/pound of fish/hour.
Source: Adapted from Boyd (1979).

the metabolism of the aquatic community can rapidly change dissolved oxygen concentrations. This effect is greater in warm water because water holds less oxygen as temperatures rise and the rates of metabolic activities that use or produce oxygen also increase with temperature. Dissolved oxygen is particularly critical in pond culture because the combined respiration of fish, plankton, and mud-dwelling organisms exerts a tremendous demand for oxygen.

Oxygen Consumption Rates

The low availability of dissolved oxygen in water limits the metabolic rate of fish. Maximum sustained oxygen uptake rates of fish are 10 to 100 times lower than those of mammals of comparable size. Rates of oxygen consumption by channel catfish vary with water temperature, fish size, activity, feeding, ambient dissolved oxygen concentration, and other factors. The effects of temperature and size of fish on oxygen consumption rates by channel catfish are illustrated in Table 4.2. Small fish consume much more oxygen for a given total weight than large fish and oxygen uptake increases as water temperature increases. Oxygen consumption rates increase substantially with activity and can double when fish are forced to swim against modest currents of 1 mile/hour or so. Metabolic energy demands for digestion can cause oxygen consumption rates to double 1 to 6 hours after a fish eats. Increased swimming activity during feeding also results in a temporary rise in oxygen consumption rates.

When dissolved oxygen concentrations are constantly near saturation, fish meet increased demands for oxygen during periods of activity or after feeding by increasing the volume of water pumped past the gills. This is accomplished by increasing both the ventilation rate and the volume of water respired with each pumping motion. When dissolved oxygen concentrations decrease, fish attempt to compensate through a series of behavioral and physiological changes. Ventilation volume increases because less oxygen is available in a given amount of water. Channel catfish also minimize extraneous activity to reduce metabolic oxygen demand. Circulatory responses to low dissolved oxygen concentrations also result in an increased capability for gas exchange at the gills (Holeton 1980).

As dissolved oxygen concentrations continue to drop, a point where these compensatory responses are no longer sufficient is reached and the oxygen demand of tissues exceeds the amount that can be supplied. At about this point, channel catfish swim to the surface in an attempt to exploit oxygen in the surface film. The dissolved oxygen concentration is near the lethal level when channel catfish are seen at the surface. Fish can live for a brief time under these conditions because metabolic energy demands are supplied in part by glycolysis or anaerobic metabolism. However, the acidic end-products of glycolysis lower blood pH, thus decreasing the affinity of hemoglobin for oxygen. This further reduces the amount of oxygen reaching tissues. Eventually the energy requirements for metabolism in the brain are not met and fish die.

Dissolved Oxygen Requirements

Critical dissolved concentrations are difficult to assign for fish because responses are not all-or-nothing; rather there is a continuum of effects as dissolved oxygen concentrations change. Further complications arise because effects depend on exposure time, fish size and health, water temperature, and other environmental conditions.

Fish perform best and are healthiest when dissolved oxygen concentrations are near saturation. Healthy channel catfish can survive for days when dissolved oxygen concentrations are above 2 parts per million (ppm), but they feed poorly, grow more slowly, and are more susceptible to infectious diseases if concentrations are below about 5 ppm. Adult channel catfish can live for several hours at dissolved oxygen concentrations as low as 0.5 ppm, and fingerlings may survive short exposure to concentrations even lower. Small fish consume more oxygen per unit weight than large fish, but this is somewhat offset by a higher ratio of gill area to body weight. Small fish are also more effective at using oxygen in surface films than larger fish.

Low dissolved oxygen concentrations can prevent gametogenesis or spawning, but critical levels are unknown. In commercial brood ponds, spawning success decreases dramatically if dissolved oxygen concentrations are routinely below 4 ppm (Steeby 1988).

Channel catfish eggs are particularly vulnerable to low dissolved oxygen concentrations because they are nonmotile and cannot seek waters of higher oxygen concentrations. In nature localized oxygen deficiencies are prevented by the male parent as he fans and circulates water over the eggs. The oxygen requirements of eggs increase as they mature and embryos become more active. Eggs will begin to die after several hours at dissolved oxygen concentrations of 2 to 3 ppm. Sac fry and young swim-up fry are also relatively intolerant of low dissolved oxygen concentrations because they have limited motility and their respiratory and circulatory systems are not fully developed.

Concentrations of dissolved oxygen above saturation provide little or no benefit to fish because the hemoglobin in arterial blood leaving the gills is very nearly 100 percent saturated with oxygen when environmental dissolved oxygen concentrations are at saturation. Higher dissolved oxygen levels result in little additional oxygen carried in the blood. Extremely high dissolved oxygen concentrations can be harmful. When dissolved oxygen concentrations exceed about 300 percent of saturation (25 to 40 ppm, depending on water temperature), fish may develop gas bubble trauma. Gas bubble trauma occurs in fish inhabiting waters supersaturated with dissolved gases. Gases can come out of solution in the blood and form bubbles (emboli) that block blood flow and possibly result in death (see the section Total Gas Pressure).

Factors Affecting Lethal Levels

Asphyxiation of fish is not always due simply to low dissolved oxygen concentrations. It can be caused by any set of conditions that increase oxygen demand of the fish, decrease the rate of diffusion of oxygen from water to blood, or decrease the amount of oxygen carried by blood. Asphyxiation can occur even when dissolved oxygen concentrations are near saturation. If channel catfish show signs of respiratory distress, such as lethargy or swimming at the surface, yet dissolved oxygen concentrations are above 2 ppm, a secondary complicating factor should be suspected.

Factors that increase metabolic oxygen demand have already been mentioned. Of considerable practical importance, recent feed intake and forced activity will raise lethal levels of dissolved oxygen. Fish should not be fed when dissolved oxygen concentrations are below 3 to 4 ppm; preferably concentrations should be above 5 ppm. Fish forced to swim for long periods in currents near large aerators may receive inadequate dissolved oxygen to meet the higher demand for oxygen brought about by increased activity. Fish may ultimately tire and die even though dissolved oxygen concentrations near the aerator appear adequate.

Certain infectious diseases and environmental conditions affect the efficiency of oxygen diffusion across the gill surfaces. Heavy infestations of parasites on gill surfaces can physically impede the passage of oxygen from water to blood. Infestations of some gill parasites result in proliferation of cells (hyperplasia) and fusion of gill lamellae, which increases the distance between water and blood and decreases gill surface area. Other organisms, such as the bacterium *Flexibacter columnaris*, actually destroy gill tissue. Many waterborne toxicants cause gill lesions that reduce respiratory efficiency. Lesions are caused by certain pesticides, metals (copper and zinc, for example), un-ionized ammonia, or pH extremes. Common lesions include excessive mucus secretion, hyperplasia, hypertrophy (enlargement of cells), lamellar fusion, and lifting of the epithelium from the underlying pillar cells. These alterations probably serve a defense function by slowing the rate of diffusion of toxicant from water to bloodstream (Mallatt 1988), with the undesirable consequence of slowing oxygen uptake as well.

Hemoglobin carries oxygen to the tissues, and substances or conditions that affect the amount or function of hemoglobin can worsen the effect of low dissolved oxygen concentrations. Anemia (decreased hemoglobin or red blood cells) caused by malnutrition, infectious disease organisms, or toxicants can lead to insufficient oxygen reaching tissues. Nitrite changes hemoglobin to a form incapable of carrying oxygen (methemoglobin), causing a functional anemia. Carbon dioxide decreases the affinity of hemoglobin for oxygen, thereby reducing oxygen uptake at the gills. A similar change in hemoglobin function is caused by decreased blood pH (acidosis). Blood acidosis can result from exposure to un-ionized ammonia, low environmental pH, high carbon dioxide concentration, or excessive activity.

ALKALINITY AND HARDNESS

Total alkalinity is a measure of the buffering capacity of water. In most fresh waters, alkalinity is attributable to the presence of bicarbonate (HCO_3^-), carbonate (CO_3^{2-}), and, to a much smaller degree, hydroxide (OH^-). *Total hardness* refers to the combined concentrations of calcium (Ca^{2+}) and magnesium (Mg^{2+}). Although alkalinity and hardness refer to different properties of water, both are expressed as ppm as $CaCO_3$.

Channel catfish thrive in waters with total alkalinities and hardnesses ranging from less than 5 ppm as $CaCO_3$ to over 400 ppm as $CaCO_3$. At alkalinity or hardness values greater than 400 ppm, effects on growth or performance may be due to elevated salinity rather than to specific ions associated with either property. Poorly buffered waters of low alkalinity (<20 ppm as $CaCO_3$) are undesirable in pond culture (Chapter 9) and water-reuse systems (Chapter 13) because pH is difficult to manage. However, low alkalinity has little direct effect on fish.

Calcium and magnesium are essential minerals for growth, and channel catfish absorb significant quantities from water. However, channel catfish can be reared in calcium- and magnesium-free water if sufficient quantities of the two minerals are present in the diet (see the section Nutrients). Environmental calcium is, however, required for hardening of eggs and fry development. Calcium also decreases the toxicity of low pH, un-ionized ammonia, and dissolved metals such as copper and zinc. In general, water used to rear channel catfish should have a total hardness greater than about 20 ppm as $CaCO_3$ with most of the hardness attributable to calcium.

CARBON DIOXIDE

Carbon dioxide is produced by cells during respiration and carried in the blood mainly as bicarbonate. At the gills, bicarbonate is converted to carbon dioxide by the enzyme carbonic anhydrase, and the carbon dioxide diffuses into the water. An increase in environmental carbon dioxide concentration reduces the concentration gradient necessary for diffusion; blood carbon dioxide levels increase, thus lowering blood pH. These conditions decrease the amount of oxygen that hemoglobin can carry and reduce the affinity of hemoglobin for oxygen. This phenomenon, known as the *Bohr-Root effect*, may cause respiratory distress even when normally sufficient dissolved oxygen is present in the water.

Channel catfish have a relatively small Bohr-Root effect and tolerate environmental carbon dioxide concentrations better than most fish. Carbon dioxide concentrations less than 10 ppm appear to be well tolerated even at low dissolved oxygen concentrations. Channel catfish can survive in waters with carbon dioxide concentrations at least as high as 50 ppm provided dissolved oxygen concentrations are high.

pH

pH expresses the intensity of the acidic or basic character of water. Effectively, pH is the negative logarithm of the hydrogen ion concentration. The pH scale is usually represented as ranging from 0 to 14. Conditions become more acidic as pH values decrease and more basic as they increase. At 77°F, pH 7.0 is the neutral point.

When considered alone, the effects of pH on channel catfish can be summarized as follows: The optimal pH range is in the range from 6 to 9. Osmoregulation, maintenance of blood pH, and respiration become more difficult as pH decreases below 6. This stress results in reduced growth, reduced reproductive performance, and lower disease resistance. At pH values below 4 to 5, fish may die from massive osmoregulatory failure. Osmoregulatory distress also affects performance at pH values greater than 9, with death occurring at pH 11 to 12.

Simply stating that the optimal pH for channel catfish is in the range 6 to 9 is somewhat misleading because the pH of most waters is a function of carbon dioxide content and total alkalinity. Because carbon dioxide can interfere with respiration, optimal pH must be defined not only in terms of the direct effect of pH on fish but also relative to the carbon dioxide content of the water. For instance, at pH 7.0, water with a total alkalinity of 200 ppm as $CaCO_3$ will contain about 40 ppm carbon dioxide at 77°F (see Table 9.2). Although pH 7 is in the range considered optimal from the standpoint of pH alone, the carbon dioxide in that particular water at pH 7 may be a significant stressor. From a practical standpoint, *optimal pH* can be defined as the pH of the particular water in question when that water is in equilibrium (saturated) with atmospheric carbon dioxide. Water in equilibrium with atmospheric carbon dioxide contains from 0.5 to 1 ppm carbon dioxide. The pH of waters saturated with carbon dioxide depends on the temperature, salinity, and alkalinity of the water. For present purposes, the effects of temperature and salinity can be ignored. The approximate pH range of waters of various total alkalinities in equilibrium with atmospheric carbon dioxide is as follows:

Total alkalinity (ppm as $CaCO_3$)	pH
1	6.3–6.4
10	7.3–7.5
50	7.9–8.1
100–500	8.3–8.5

The pH values listed can be taken as the optimal pH for fish in waters of different alkalinities. At pH values below those listed for a particular total alkalinity, significant carbon dioxide may exist.

AMMONIA

Ammonia* in water establishes an equilibrium that can be written as:

$$NH_3 + H_2O = NH_4^+ + OH^- \qquad (4.1)$$

The relative proportions of NH_3 and NH_4^+ depend primarily on pH and temperature. For a given concentration of total ammonia, the concentration of un-ionized ammonia increases as pH and temperature increase (see Table 9.3).

Ammonia is the major product of protein degradation in fish. The pH of channel catfish blood is 7.3 to 7.6, so most of the ammonia is transported in blood as NH_4^+. Ammonia is lost from the gill as NH_3, which freely diffuses across gill membranes (Fig. 4.3). As NH_3 is lost, reestablishment of equilibrium in the blood continually brings more ammonia into the diffusible un-ionized form. This system provides a steady loss of metabolic ammonia to the environment provided external un-ionized ammonia concentrations are low. After crossing the gill membrane, un-ionized ammonia reestablishes an equilibrium with ionized ammonia based on the pH and temperature of the water. Some metabolic ammonia is also excreted from fish as NH_4^+ in exchange for sodium (Na^+).

Ammonia is toxic to channel catfish, and un-ionized ammonia is much more toxic than the ionized form. As the concentration of environmental un-ionized ammonia increases, the concentration gradient between blood and environment is reduced and the rate of ammonia excretion decreases. This results in increased blood and tissue ammonia levels, which can have serious physiological consequences (Colt and Armstrong 1981).

The physiological mechanism of severe, acute ammonia toxicosis appears to be suppression of metabolic energy production in the central nervous system. These effects can occur within minutes to hours of exposure to more than about 1 ppm un-ionized ammonia-nitrogen. Symptoms of acute ammonia toxicosis include hyperactivity, convulsions, lethargy, loss of equilibrium, and coma.

Long-term exposure to lower concentrations of un-ionized ammonia causes osmoregulatory disturbances and reduced respiratory efficiency. This results in poor growth and impaired immune function. A "no-effect" level for un-ionized ammonia apparently does not exist (Colt and Tchobanoglous 1978), meaning that any measurable un-

---

*The following convention will be used: NH_3 will be referred to as un-ionized ammonia; NH_4^+ will be referred to as ionized ammonia; the sum of $NH_3 + NH_4^+$ will be referred to as total ammonia or simply ammonia. All concentrations will be expressed in terms of nitrogen.

52   Environmental Requirements

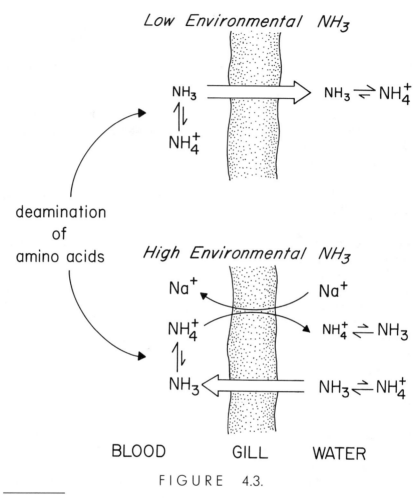

FIGURE 4.3.

Schematic diagram of ammonia movement across fish gill membranes. See text for details.

ionized ammonia will affect growth or performance of channel catfish. For practical purposes, however, a maximum level of 0.05 ppm unionized ammonia-nitrogen is acceptable for long-term exposure. This criterion is applicable to channel catfish in raceways and water reuse systems where concentrations are fairly constant with time. Un-ionized ammonia concentrations can vary dramatically over a 24-hour period in ponds because pH cycles diurnally. During periods of high afternoon pH, a large fraction of total ammonia is present in un-ionized ammonia. When pH declines at night, un-ionized ammonia concentrations decline. Because fish are exposed to constantly changing concentrations, it is difficult to assign a critical level for un-ionized ammonia in ponds. Observations of pond-raised fish indicate that un-ionized ammonia up to 0.2 ppm is tolerated for periods of several hours without severe consequences as long as fish are otherwise healthy.

NITRITE

Environmental nitrite (NO_2^-) can be toxic to channel catfish at relatively low concentrations. Waterborne nitrite enters the fish circulatory system through the gills. Nitrite in the blood oxidizes hemoglobin to a product called *methemoglobin*, which is incapable of reversibly combining with oxygen. Thus, exposure to high levels of nitrite may cause considerable respiratory distress because of the impairment of oxygen transport by the blood.

Methemoglobin has a characteristic brown color, and the color of the blood (and gills) changes as the amount of hemoglobin in the methemoglobin form increases. Methemoglobinemia ("brown blood disease") becomes noticeable as a slight brownish cast to the blood and gills when about 20 to 30 percent of total hemoglobin is present as methemoglobin. At methemoglobin levels of 30 to 70 percent, the blood is a definite brown-red. When methemoglobin exceeds about 70 percent of total hemoglobin, the blood is chocolate-brown.

Obviously, the severity of the condition worsens as the percentage of methemoglobin increases. Channel catfish blood normally contains less than 10 percent methemoglobin, and methemoglobin levels less than 20 to 30 percent seem to be well tolerated. At higher levels, a significant proportion of the blood hemoglobin cannot carry oxygen, and this compounds the effects of exposure to low dissolved oxygen. If methemoglobin levels are greater than 30 to 40 percent for a week or more, fish also may become anemic because red blood cells are destroyed (Tucker, Francis-Floyd, and Deleau 1989). This further reduces the hemoglobin available to carry oxygen. When methemoglobin levels are greater than 70 to 80 percent, fish become inactive in an attempt to reduce metabolic oxygen demand. Dissolved oxygen concentrations must remain near saturation if fish are to survive for more than a few hours. Brown blood disease is easy to diagnose from the characteristic color of the blood and gills. However, the first indication of the condition is usually fish behavior characteristic of oxygen deficiencies even when dissolved oxygen concentrations are above 2 to 3 ppm.

Nitrite enters the bloodstream primarily as nitrous acid (HNO_2)

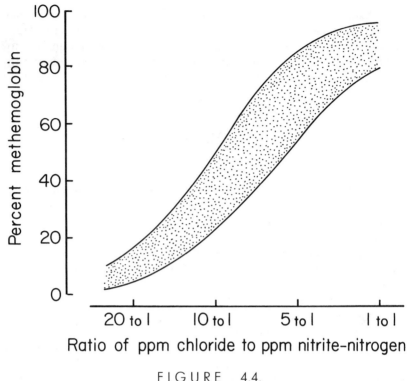

FIGURE 4.4.

Relationship between the ratio of environmental chloride to nitrite-nitrogen and the amount of methemoglobin (as a percentage of total hemoglobin) formed in channel catfish blood. Methemoglobin formation varies with water temperature and among strains of fish; the range of values usually encountered is indicated by the stippled area (adapted from Schwedler and Tucker 1983).

when pH values are below about 5. At higher pH values, nitrite is actively concentrated from water into blood by the same gill anion uptake mechanism responsible for transporting chloride. The transport mechanism cannot discriminate between the two ions because both carry one negative charge and are of similar ionic size. The two ions compete for uptake and the amount of nitrite entering the bloodstream (and hence the amount of methemoglobin that is formed) is related to the ratio of chloride to nitrite in the water (Fig. 4.4) rather than the nitrite concentration alone. If sufficient chloride is present, little nitrite will enter the blood even if the environmental nitrite concentration is high. Addition of chloride, as common salt (NaCl), thus provides a convenient, inexpensive treatment for high nitrite levels.

Past recommendations called for maintenance of at least a 10:1 ratio of environmental chloride to nitrite-nitrogen. At that ratio, methemoglobin levels are usually 20 to 40 percent and fish are afforded fair protection from asphyxiation at low dissolved oxygen levels. To provide more protection and prevent other long-term side effects such as anemia, maintenance of at least a 20:1 ratio of chloride to nitrite-nitrogen is presently recommended. This results in methemoglobin levels of less than 20 percent and largely eliminates the adverse effects of nitrite on channel catfish. Calculation of salt treatment rates for ponds is described in the section Ammonia and Nitrite.

HYDROGEN SULFIDE

Sulfides are produced by certain bacteria under anaerobic conditions. Sulfides exist as H_2S, HS^-, or S^{2-}, depending on pH and temperature. The proportion of total sulfide present as hydrogen sulfide increases as pH and temperature decrease.

Hydrogen sulfide is extremely toxic to channel catfish, but HS^- and S^{2-} are considered nontoxic. Hydrogen sulfide blocks one of the final steps in the biochemical pathway of energy production during oxidative respiration. The net effect is similar to acute oxygen deficiency.

Hydrogen sulfide concentrations less than 0.05 ppm kill fish after brief exposure. Fry are the most sensitive life stage to hydrogen sulfide toxicosis, and concentrations less than 0.01 ppm can inhibit reproduction (Smith et al. 1976). Because of the extremely toxic nature of hydrogen sulfide, any detectable concentration should be considered a health hazard.

SUSPENDED SOLIDS AND TURBIDITY

Suspended solids are inorganic and organic particles found in the water column and include clay, silt, phytoplankton, and organic detritus. *Turbidity* refers to the decreased penetration of light through water. Turbidity is caused by the presence of suspended solids as well as soluble colored compounds. The suspended solids concentration in most natural waters is seldom greater than 20,000 ppm. Even waters that visually appear "muddy" usually have less than 2,000 ppm suspended solids (Boyd 1979).

Channel catfish tolerate high levels of suspended solids. Fingerlings and adults can inhabit waters with at least 100,000 ppm suspended solids for several weeks or more, although behavioral changes may be noticed in water with 20,000 ppm suspended solids (Wallen 1951).

Excessive turbidity is undesirable in hatcheries because solids may settle in tanks or troughs and smother eggs or clog the gills of fry. In hatcheries and raceways, excessive turbidity hinders observation of fish behavior and generally leads to poor hygiene. Water sources for these systems should contain less than about 100 ppm suspended solids (Piper et al. 1982).

TOTAL GAS PRESSURE

Total gas pressure is the sum of the partial pressures of all gases dissolved in water. The difference between total gas pressure and barometric pressure is called ΔP. Equilibrium conditions occur when total gas pressure is equal to local barometric pressure ($\Delta P = 0$). When total gas pressure is greater than barometric pressure ($\Delta P > 0$), the water is supersaturated and gases tend to come out of solution. When total gas pressure is less than barometric pressure ($\Delta P < 0$), the water is undersaturated and gases tend to enter the water from the atmosphere. Gas saturation is sometimes reported as a percentage of local barometric pressure (BP):

$$\% \text{ Total gas pressure} = [(BP + \Delta P)/BP] \times 100 \qquad (4.2)$$

For example, assume the local barometric pressure is 760 mmHg (millimeters of mercury)* and ΔP is 38 mmHg. The total gas pressure is 105 percent of saturation.

Gas bubble trauma (also called gas bubble disease) occurs in fish living in supersaturated waters. Supersaturation is an unstable condition, and as gases come out of solution they form bubbles. If the gases in solution diffuse across the gill and into the blood before coming out of solution, bubbles will be formed in the vascular system and other

---

*Gas pressures are reported here in metric units because instruments used to measure pressure in water are usually calibrated in mmHg. To convert to English units, 760 mmHg = 29.92 inches of mercury.

tissues. Acute gas bubble trauma in fingerlings and adult channel catfish occurs at high levels of supersaturation, usually at ΔP values greater than 76 mmHg (about 110 percent saturation at sea level). Acute gas bubble trauma is caused by bubbles in the bloodstream that restrict blood flow and cause oxygen deficiencies in tissues. Mortality rate may be high over short (a few days) exposure. Chronic gas bubble trauma may occur after long-term exposure to positive ΔP values less than 76 mmHg. Chronic gas bubble trauma is associated with hyperinflation of the swim bladder and bubbles in the gut and buccal cavity. Mortality is usually less than 5 percent over extended periods and may be related to secondary, stress-related infections. Sac fry and early swim-up fry are particularly sensitive to gas supersaturation. Water in hatcheries should have a ΔP of less than 38 mmHg (about 105 percent saturation) to prevent gas bubble trauma. Sac fry exposed to excessively supersaturated water develop bubbles in the yolk sac that prevent normal swimming and trap the fry at the surface. If they survive to swim-up, fry may not feed and will eventually die. Eggs are the most resistant life stage to gas supersaturation and may survive ΔP greater than 150 mmHg (about 120 percent of saturation). Of course, fry would be quickly killed at this level of gas supersaturation soon after hatching.

One of the most important factors affecting the development of gas bubble trauma is the depth at which the fish is positioned in the water column. The deeper the fish the less actual ΔP the fish experiences because of the hydrostatic pressure of the overlying water. In fresh water at 70°F, ΔP is reduced by 67 mmHg for every 3 feet beneath the surface. For this reason, channel catfish confined to shallow water, such as fish in hatcheries or net cages, are particularly susceptible to gas bubble trauma. These fish cannot increase their depth so that hydrostatic pressure can reduce the effect of a high ΔP.

Gas supersaturation can be caused by a variety of processes. Many groundwaters are supersaturated because the water in the aquifer is under considerable hydrostatic pressure and atmospheric gases in the recharge zone are driven into solution. The ΔP of groundwaters ranges from negative values to more than 500 mmHg. Surface waters become supersaturated when water passes over a dam or through penstocks and is carried to depth in a plunge basin. At depth, the increased hydrostatic pressure increases the solubility of atmospheric gases and the air goes into solution. Supersaturated conditions also develop when a water source is rapidly heated in a closed system such as occurs in cooling waters for steam electrical generating plants. Gas solubility decreases with increasing temperature, but the volume of gas dissolved in the water remains the same since these systems are closed to the atmosphere. Supersaturation can also occur if there is an air leak on the low-pressure side of a pump in a water supply line. The air drawn in through the leak is mixed with the water, which is then pressurized after moving through the pump, and some of the air is passed into solution. Colt (1986) has reviewed these and numerous other causes and occurrences of gas supersaturation in water.

Gas bubble trauma is most likely in hatcheries because holding tanks are shallow and water quickly flows through the system with little time afforded for degassing. Supersaturated water added to ponds rapidly reaches equilibrium with the atmosphere, and gas bubble trauma related to a supersaturated water supply never occurs. However, rapid rates of photosynthesis by phytoplankton or submersed weeds can cause dissolved oxygen concentrations in ponds that far exceed saturation. Gas supersaturation due to oxygen alone is tolerated better by fish than that due to the sum of all atmospheric gases. The ΔP of oxygen is somewhat moderated within the fish by metabolic consumption during cellular respiration. Dissolved oxygen concentrations must exceed 300 percent of saturation (24 to 45 ppm, depending on temperature) before gas bubble trauma develops. Although such dissolved oxygen concentrations are not uncommon in channel catfish culture ponds, gas bubble trauma is rare. Photosynthetic oxygen production by phytoplankton is most rapid at the surface, where light is most intense and only surface waters may be supersaturated. Fish inhabiting bottom waters are also protected by hydrostatic pressure. Young fry are most at risk because they may spend long periods swimming near the surface.

COPPER AND ZINC

Copper and zinc can be highly toxic to channel catfish. These elements are relatively common in the earth's crust, but within the pH range of most waters compounds of copper and zinc are of low solubility. Consequently, concentrations of copper and zinc are extremely low in water supplies otherwise suited for use in channel catfish culture. Acid-mine drainage and waters receiving wastes from smelting or electroplating plants may contain high concentrations of copper or zinc (and other toxic metals, such as cadmium, nickel, and chromium) and should be avoided as water supplies for channel catfish culture. The toxicity of copper and zinc is of interest, however, because both metals are present in certain materials used to fabricate pipes or tanks used in hatcheries or intensive culture systems. Copper is also the active component in copper sulfate, a widely used algicide and fish disease treatment.

The toxicity of copper and zinc decreases as pH, alkalinity, and hardness increase. At high pH and alkalinity, copper and zinc are precipitated or complexed as largely nontoxic substances. In waters of high hardness, calcium and magnesium compete with the metals for absorption sites on the gill membrane. Alkalinity and pH exert the greatest influence on toxicity, but in practice separating the effects of the three variables is difficult because they vary together in most waters. Suggested maximum levels for long-term exposure in waters of low alkalinity (<100 ppm as $CaCO_3$) are 0.01 ppm for total dissolved copper and 0.03 ppm for total dissolved zinc (Piper et al. 1982). In waters of higher alkalinity at least twice those amounts are tolerated.

The treatment rate commonly used for copper sulfate as an algicide or disease treatment is calculated by dividing the total alkalinity by 100. Copper sulfate pentahydrate ($CuSO_4 \cdot 5H_2O$) is about 25 percent copper, so a pond with an alkalinity of 100 ppm as $CaCO_3$ would be treated with 0.25 ppm of copper, well above the "safe" limit recommended. The reason this rate can be used safely is that copper from a single addition of copper sulfate is rapidly precipitated and total dissolved copper concentrations will decrease to virtually undetectable levels within a day or two after treatments. Copper persists much longer in waters of low alkalinity and pH, and waters with a total alkalinity less than 50 ppm as $CaCO_3$ should not be treated with copper sulfate.

Copper and zinc can be leached from pipes, fittings, or tanks containing the metals. If the water is acidic and low in alkalinity, considerable metal may be present, particularly if the water has long been in contact with the material. Copper and zinc can accumulate in water-reuse culture systems or in water standing in pipes overnight. Leaching of copper or zinc should not be a problem in waters with a pH greater than 8 and an alkalinity greater than 50 ppm as $CaCO_3$. In these waters copper and zinc form highly insoluble oxides and hydroxides that actually coat the metal and prevent leaching. To be safe, however, copper and brass pipe and fittings and zinc galvanized steel should be entirely avoided in fish culture facilities.

REFERENCES

Allen, K. O., and J. W. Avault, Jr. 1971. Effects of salinity on growth and survival of channel catfish, *Ictalurus punctatus*. *Proceedings of the Southeastern Association of Game and Fish Commissioners* 23:135–139.

Boyd, C. E. 1979. *Water Quality in Warmwater Fish Ponds*. Auburn: Auburn University/Alabama Agricultural Experiment Station.

Busch, R. L. 1985. Channel catfish culture in ponds. In *Channel Catfish Culture*, ed. C. S. Tucker, pp. 13–84. Amsterdam: Elsevier.
Colt, J. 1986. Gas supersaturation—impact on the design and operation of aquatic systems. *Aquacultural Engineering* 5:49–85.
Colt, J., and D. Armstrong. 1981. Nitrogen toxicity to crustaceans, fish and molluscs. *Bio-Engineering Symposium for Fish Culture* (FCS Publ. 1):34–47.
Colt, J., and G. Tchobanoglous. 1978. Chronic exposure of channel catfish, *Ictalurus punctatus*, to ammonia: effects on growth and survival. *Aquaculture* 15:353–372.
Davis, K. B., C. A. Goudie, B. A. Simco, R. MacGregor, and N. C. Parker. 1986. Environmental regulation and influence of the eyes and pineal gland on the gonadal cycle and spawning in channel catfish *(Ictalurus punctatus)*. *Physiological Zoology* 59:717–724.
Holeton, G. F. 1980. Oxygen as an environmental factor of fishes. In *Environmental Physiology of Fishes*, ed. M. A. Ali, pp. 7–32. New York: Plenum Press.
Mallatt, J. 1985. Fish gill structural changes induced by toxicants and other irritants: a statistical review. *Canadian Journal of Fisheries and Aquatic Sciences* 42:630–648.
Perry, W. G., Jr., and J. W. Avault, Jr. 1970. Culture of blue, channel, and white catfish in brackish water ponds. *Proceedings of the Southeastern Association of Game and Fish Commissioners* 23:592–597.
Pickering, A. D., ed. 1981. *Stress and Fish*. New York: Academic Press.
Piper, R. G., I. B. McElwain, L. E. Orme, J. O. McCraren, L. G. Fowler, and J. R. Leonard. 1982. *Fish Hatchery Management*. Washington, D.C.: United States Fish and Wildlife Service.
Reynolds, W. W., and M. E. Casterlin. 1980. Role of temperature in the environmental physiology of fishes. In *Environmental Physiology of Fishes*, ed. M. A. Ali, pp. 497–518. New York: Plenum Press.
Schwedler, T. E., and C. S. Tucker. 1983. Empirical relationship between percent methemoglobin in channel catfish blood and dissolved nitrite and chloride in ponds. *Transactions of the American Fisheries Society* 112:117–119.
Smith, L. L., P. M. Oseid, L. L. Kimball, and S. M. El-Kaudelgy. 1976. Toxicity of hydrogen sulfide to various life history stages of bluegill *(Lepomis macrochirus)*. *Transactions of the American Fisheries Society* 105:442–449.
Steeby, J. A. 1987. Effect of spawning container type and placement depth on channel catfish spawning success in ponds. *Progressive Fish-Culturist* 49:308–310.
Tucker, C. S., R. Francis-Floyd, and M. H. Beleau. 1989. Nitrite-induced anemia in channel catfish, *Ictalurus punctatus*. *Bulletin of Environmental Contamination and Toxicology* 43:295–301.
Wallen, I. E. 1951. The direct effect of turbidity on fishes. *Oklahoma A&M College Bulletin* 48:1–27.

CHAPTER 5
Nutrition

Nutrition may be defined simply as the process by which an organism takes in and assimilates food. As such, it involves the ingestion, digestion, absorption, and transport of various nutrients to body cells where the elements of foods are transformed into body tissues and activities. Nutrition also encompasses the removal of excess nutrients and unused metabolites. Thus it is a complex biological science and, like other sciences that deal with a biological species, it is an inexact science because of the natural variability among individuals of a given species. Also, nutritional requirements are affected by sex, feed intake, energy density of the diet, interactions of nutrients in the diet, availability of nutrients to the fish, presence of toxins or mold in the diet, expected level of performance, desired carcass composition, and environmental factors. In addition, the aquatic animal nutritionist is further constrained because fish live in an environment where dissolved substances, including excretory products, may affect nutrient requirements.

As a result of research efforts over the past 30 to 40 years, forty nutrients have been identified as being necessary for the normal metabolic function of catfish. Also, the quantitative requirements for many of the nutrients are known. Tremendous progress has been made in catfish nutrition during recent years; however, this is not to imply that unknowns do not exist. Little is known concerning the energy requirements of catfish, the interrelations between disease and nutrition, changes in nutrient needs with size or age, interactions of various nutrients, nutrient availability in feed ingredients, or impact of feed processing on nutrient availability.

The following sections will present the "state of the art" of catfish nutrition. Energy, nutrients, and digestion will be discussed. For a more detailed treatment of some of the subjects presented herein, the reader is referred to the National Research Council (1977, 1983) and to articles by Robinson and Wilson (1985), Halver (1989), Lovell (1989), and Robinson (1989). In addition, almost any modern text on animal nutrition will be a valuable resource on basic nutrition principles.

ENERGY

Energy, which is defined as the capacity to do work, is essential to life processes during all stages of an animal's life. Quantitatively energy is the most important component of the diet, because animals generally eat to satisfy an energy requirement. Feeding standards for many animals are based on energy needs. Considerable information exists on energy needs of ruminant animals; less is known about the energy requirements of nonruminant species, particularly fish. Thus it is difficult to base feeding standards for catfish on energy needs.

Solar energy is the ultimate source of energy. It is stored by plants and is available to animals to the extent they are able to digest plant materials. Animals that consume plants store part of the ingested energy and thus become sources of energy for carnivorous species that consume them. Animals derive energy from oxidation of organic compounds ingested in food or from stored lipid, protein, and to a lesser extent stored carbohydrate. These compounds are catabolized to yield energy, which is usually expressed in calories. A *calorie* is the amount of heat required to raise the temperature of 1 gram (g) of water by 1°C (e.g., from 14.5 to 15.5°C). One kilocalorie = 1,000 calories. The average total caloric values for protein, lipid, and carbohydrate are 5.65, 9.45, and 4.15 kilocalories/gram, respectively (Maynard et al. 1979). However, all of the calories inherent in these compounds are not available to the body. The amount available is dependent on the digestibility of each component. For example, when the caloric values given are corrected for human digestibility and for the calories present as nitrogen in protein they become 4.0, 9.0, and 4.0 kilocalories/gram for protein, lipid, and carbohydrate, respectively (Maynard et al. 1979). These values are termed *physiological fuel values* (PFVs) and have been used

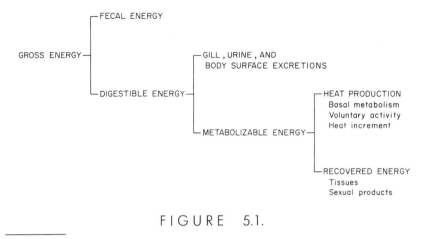

FIGURE 5.1.

Partitioning of energy in food consumed by channel catfish.

as caloric values for various food components in studies with fish. Though PFVs are not exact measures of the usable energy in food for various fish species, they are useful estimates of available energy when no data on the energy requirements of a particular species exist. Total energy should be corrected to usable energy by applying the digestion coefficients determined for the animal of interest.

Utilization and Expression

The manner in which energy is partitioned into various fractions by fish, presented schematically in Figure 5.1, is similar to energy flow in other animals. Perhaps the most notable difference in the nutrition of fish and farm animals is in energy utilization; more specifically less energy is required for protein synthesis in fish. The protein gain per megacalorie (Mcal) (1 Mcal = 1,000,000 calories) of metabolizable energy (ME) consumed is 47, 23, 9, and 6 for catfish (ME estimated), broiler chickens, swine, and beef cattle, respectively (Lovell 1989). Maintenance energy needs are lower for fish than for warm-blooded animals because fish do not have to maintain a constant body temperature and they expend less energy to maintain their position in space. Losses of energy in urine and gill excretions are lower in fish because most nitrogenous waste is excreted as ammonia instead of urea or uric acid, which are excreted by mammals and birds, respectively. Also, the increase in energy cost associated with the assimilation of ingested food, heat increment, is less in fish.

TABLE 5.1.

Gross Energy Values for Sources of Carbohydrates, Fats, and Proteins Determined by Bomb Calorimeter

Substrate	Kilocalories/Gram
Glucose	3.77
Cornstarch	4.21
Triglyceride	
Beef fat	9.44
Soybean oil	9.28
Casein	5.84

Dietary energy should be expressed in a manner that reflects available (utilizable) energy. Gross energy (IE), which is a measure of the heat liberated on complete oxidation of a food (Table 5.1), is not a practical indicator of usable energy because certain foods are less digestible than others. As an example, IEs for starch and cellulose are similar but the digestible energy (DE), defined as IE minus fecal energy losses, from starch for catfish is about 2.5 to 3.0 kilocalories/g and essentially zero for cellulose. Since IE is of little practical value in expressing usable energy values for catfish, DE is often used to express the dietary energy of catfish feeds. Metabolizable energy (ME), DE minus energy losses in gills and urine, is used to express energy content of feeds for livestock. Theoretically using ME to express dietary energy may be more desirable than using DE, since ME is a more precise measure of available energy for metabolism. Also, ME has been adopted by the National Research Council for use in formulating animal feeds. However, in a practical sense, there is little advantage in using ME values rather than DE values in formulating fish feeds because losses in digestion account for most of the variation in losses of IE. Also, energy losses through the gills and urine by fish are smaller than nonfecal losses in other animals, and these losses do not vary among feedstuffs as much as fecal losses. Additionally, there is less stress on the fish in experimentally determining DE than in determining ME.

Although DE is useful for expressing dietary energy for catfish, there are a limited number of DE values available for use. As a result, ME values determined with other animals or mammalian PFVs (discussed previously) have been used to express the caloric density of experimental catfish feeds. Metabolizable energy values based on studies with terrestrial animals are not necessarily accurate for catfish but are more reflective of usable energy than IE. Also, the accuracy with which PFVs reflect usable energy for catfish is questionable. Digestible energy values are available for most commonly used feedstuffs for catfish, and these values should be used in formulating practical catfish feeds. If DE values are not available for a particular feedstuff, a DE value can be estimated by using proximate composition of the feedstuff and digestion coefficients derived from a similar type of feedstuff.

Requirements

Energy requirements of catfish were largely neglected in the early stages of catfish feed development, primarily because an imbalance in dietary energy does not appreciably affect the health of the fish. Also, feed prepared from feedstuffs typically used in catfish feeds, such as soybean, corn, and fish meal, is unlikely to be extreme in respect to energy balance. As it turns out, these assumptions were more or less true. However, correct balance of dietary energy is an important consideration when formulating catfish feeds, because too much energy can result in a reduction in food intake and thus reduce nutrient intake. Also, excess dietary energy may result in an increased deposition of body fat. If the dietary energy level is too low, protein can be catabolized for energy instead of being used for tissue synthesis.

Energy requirements for catfish have been based on weight gain or on composition of gain and have been reported as a ratio of calories to dietary protein. Energy requirements for catfish are tentative because in the various studies dietary energy was generally estimated from ME values of other species or calculated by PFVs. In some studies the fish were not fed to satiation and diet composition varied from study to study. Also, in pond studies the contribution of natural food was not accounted for.

On the basis of current information, it appears that a DE level of 8 to 9 kilocalories/g of protein is adequate for use in catfish feeds. Increasing dietary energy may result in increased weight gain, but body fat is also likely to increase. Also, catfish, like other fish, require less energy for maintenance than terrestrial animals. Energy requirements change with fish size and environmental temperature. Increasing the ratio of digestible energy to protein when environmental temperature deviates from optimal appears to be beneficial. Additional research to define the energy requirements of catfish more precisely is needed, particularly considering the assumptions made in previous energy requirement studies with catfish.

TABLE 5.2.

Classification of Selected Carbohydrates

Monosaccharides		Oligosaccharides	Polysaccharides
Pentoses ($C_5H_{10}O_5$)	**Hexoses** ($C_6H_{12}O_6$)	**Disaccharides** ($C_{12}H_{22}O_{11}$)	**Hexosans** $(C_6H_{10}O_5)_n$
Ribose	Glucose	Sucrose	Starch
	Fructose	Maltose	Dextrin
	Galactose	Lactose	Glycoyen
			Cellulose

NUTRIENTS

Carbohydrates

Carbohydrates are compounds of carbon, hydrogen, and oxygen that include sugars, starch, cellulose, gums, and other closely related compounds. Carbohydrates are the major constituent of plants, comprising 50 to 80 percent of the dry weight of various plants. They form the structural framework of plants and are the primary form of energy stored in seeds, roots, and tubers. Plants synthesize carbohydrates from solar energy, carbon dioxide, and water through the process of photosynthesis. Although the photosynthetic process is complex, it can be represented by the following expression:

$$6CO_2 + 6H_2O + 673 \text{ kcal} = C_6H_{12}O_6 + 6O_2 \quad (5.1)$$

The reaction is essential to all animal life, since it provides energy for life processes and free oxygen needed for utilization of energy and nutrients.

Animal tissues contain small amounts of stored carbohydrate. Glucose in the blood of animals is relatively constant at about 0.05 to 0.1 percent. Circulating glucose is used for energy and is replenished from stores of glycogen in the liver. Generally glycogen stores in the liver are small, representing only about 3 to 7 percent of liver weight in most animals. Excess ingested carbohydrate is converted to and stored primarily as lipid.

Classification and structure. Nutritionally important carbohydrates are classified in Table 5.2. Classification is based on the number of carbon atoms in a molecule of carbohydrate and on the number of molecules of sugar in a particular carbohydrate. For example, a monosaccharide contains only one molecule of sugar, and a disaccharide consists of two sugar molecules. Carbohydrates containing from two to ten sugar molecules are termed *oligosaccharides* and those containing more than ten sugar units are called *polysaccharides*. Plants contain a mixture of several different carbohydrates; however, from a practical feeding standpoint polysaccharides are the most important carbohydrates ingested by fish.

Chemically, carbohydrates are polyhydroxy aldehydes and ketones or substances that yield those compounds on hydrolysis. Thus they contain an aldehyde ($-\underset{|}{C}= O$) or ketone ($C-\underset{\|}{C}-C$) group in their structure.

Functions and utilization. Carbohydrates are an inexpensive energy source for animals, including fish. Fish vary in their ability to utilize dietary carbohydrate as an energy source, but catfish appear to use certain carbohydrates relatively well. Efficiency of carbohydrate utilization by catfish is related to the complexity of the carbohydrate. In general, catfish use polysaccharides, such as starch or dextrin (partially hydrolyzed starch), more effectively than disaccharides or simple sugars. Catfish do not effectively utilize dietary mono- and disaccharides, such as glucose, fructose, sucrose, and maltose, as energy sources. The polysaccharides dextrin and starch are well utilized. Additionally, catfish, like certain other fish, resemble diabetic animals in that they have insufficient insulin for maximum carbohydrate use. Glucose is highly digestible (90 percent) by catfish but is not effectively utilized; thus it appears that much of the absorbed glucose is excreted instead of being utilized by various tissues because of the lack of adequate insulin.

Catfish appear to use carbohydrates better than cold-water fish species. This may be because catfish fed high-carbohydrate diets show an increase in the activity of lipogenic enzymes (which function in the synthesis of lipid) in liver and adipose tissue, whereas cold-water fish species do not appear to have this ability. Thus catfish apparently convert excess carbohydrate to lipid and thereby adapt to high dietary carbohydrate.

Practical catfish feeds contain considerable amounts of grains or grain milling by-products, and as a result they contain a mixture of carbohydrates. The two most important carbohydrates in feed are starch and cellulose (fiber). Both starch and cellulose are polymers of glucose, but starch is about 50 to 80 percent digestible to catfish and cellulose is considered to be indigestible. The difference is in the type of glucose present. Starch contains an alpha form of glucose and cellulose contains a beta form. Cellulase enzymes are required to break the beta bond before utilization. Catfish tissues do not contain cellulases, although these enzymes are found in the animal's intestine. Crude fiber (cellulose) is not beneficial in practical catfish feeds and becomes a pollutant in the culture system; thus fiber levels are usually relatively low (< 6 to 8 percent) in catfish feeds. Fiber cannot be completely eliminated in practical diets because it is inherent in plant feed ingredients. Fiber may serve a role in experimental diets as a diluent to regulate nutrient density.

Requirements and sources. Carbohydrates are not essential nutrients, although they are a source of energy, serve as precursors for certain metabolic intermediates necessary for growth, aid in feed manufacture, and spare protein for growth. Fish growth and health are not diminished by the lack of carbohydrate in the diet. However, it is beneficial to include some carbohydrate in catfish feeds because it is an inexpensive source of energy and aids in feed manufacture by helping to bind the feed ingredients together. A typical commercial catfish feed contains about 25 percent soluble (digestible) carbohydrate and 3 to 8 percent indigestible fiber. Primary sources of carbohydrate for catfish feeds include grains, grain milling by-products, and oilseed meals.

Lipids

Lipids (fats and oils) are organic compounds that are insoluble in water but soluble in organic solvents, such as ether, chloroform, and benzene. They serve important physiological and biochemical functions in animals and plants. Lipids should be included in fish diets because they are a source of concentrated energy (providing 2.25 times the amount of energy of an equivalent amount of carbohydrate), supply essential fatty acids, enhance the absorption of fat-soluble vitamins, improve feed palatability, and serve as a precursor for steroid hormones, and prostaglandins and tissue lipids impact flavor and neutral buoyancy. The amount of lipid that can be included in the diet of catfish is not dictated only by nutritional considerations but by constraints of feed manufacture and economics as well.

Classification and structure. Lipids that are important to fish and other animals may be classed as shown in Table 5.3. Nutritionally the most important components of lipids are fatty acids, glycerol, glycerides, and phospholipids. Other lipids may serve roles in metabolism, but they are nutritionally quantitatively unimportant or poorly utilized by catfish.

Fatty acids are composed of chains of carbon atoms that range in length from two to twenty-four carbons and contain a carboxyl group

TABLE 5.3.

Classification of Lipids

Type of Lipid	Chemistry	Example	Comments
Simple	Esters of fatty acids with various alcohols	Fats Oils Waxes	Most abundant lipids in nature Fats are solid room temperature Oils are liquids room temperature
Compound	Esters of fatty acids containing groups in addition to an alcohol and fatty acid	Phospholipids Glycolipids Lipoproteins	Contain phosphoric acid Contain carbohydrate Lipids bound to protein
Derived	Substances derived from other groups by hydrolysis	Fatty acids Glycerol Other alcohols	Usually contain an even number of carbons
Miscellaneous	Multiples of isoprene units Contain phenanthrene-type ring structure	Terpenes Sterols	Vitamin A is an important terpene Cholesterol, bile acids, sex hormones, vitamin D, and cortisol are important sterols

Source: Adapted from Robinson and Wilson (1985).

(COOH) on one end. The basic chemical structure may be represented by RCOOH, where R represents carbon chains of various lengths. Fatty acids are either *saturated* (all carbons in the chain contain hydrogens) or *unsaturated* (one or more of the carbon pairs in the chain are connected by double bonds, and hydrogen has been removed). Fatty acids that contain two or more double bonds are often referred to as *polyunsaturated fatty acids* (PUFAs). More recently, the term *highly unsaturated fatty acids* (HUFAs) has been used to refer to fatty acids that contain several unsaturated carbon pairs. The position of the unsaturated carbon pairs in the carbon chain of fatty acids is of nutritional importance, which will be discussed in the section Essential Fatty Acids. Representative fatty acids are presented in Table 5.4.

Glycerides are formed by the reaction of the alcohol glycerol with fatty acids. A mono-, di-, or triglyceride contains one glycerol group and one, two, or three fatty acids, respectively. Glycerol is a component of all triglycerides found in plants and animals.

The fatty acid composition of a triglyceride may vary or all fatty acids may be the same. Chain length and degree of unsaturation (number of double bonds) of the fatty acids composing a triglyceride determine its chemical and physical characteristics. Triglycerides containing fatty acids of fewer than ten carbons are usually liquids at room temperature, whereas those containing saturated fatty acids of ten or more carbons are solids. Triglycerides containing only long-chain saturated fatty acids are usually solids at room temperature, whereas those containing primarily long-chain unsaturated fatty acids are liquids.

Two measurements commonly used to characterize the chemical properties of lipids, chain length and degree of unsaturation, that are useful in nutrition are saponification number and iodine number. The *saponification number* is defined as the number of milligrams of potassium hydroxide (KOH) required for the hydrolysis of 1 g of lipid. This number represents the average chain length of the fatty acids in a lipid. The saponification number of a lipid composed of short-chain fatty acids is large, and as the chain length increases the saponification number decreases. The *iodine number* is the number of grams of iodine that can be taken up by unsaturated bonds in 100 g of lipid. Thus the iodine number is indicative of the degree of hydrogenation or saturation of the fatty acids contained in the lipid. A completely saturated lipid would have an iodine number of zero, whereas an unsaturated lipid such as sunflower oil would have an iodine number of 129 to 136. Saponification and iodine numbers for selected lipids are given in Table 5.5.

Phospholipids are composed of phosphoric acid, an alcohol (usually glycerol), and a nitrogenous base. Phospholipids contain more

TABLE 5.4.

Representative Fatty Acids

Common Name	Scientific Name[a]	Structure	Shorthand Designation[b]
Saturated			
Lauric acid	Dodecanoic	$CH_3(CH_2)_{10}COOH$	12:0
Myristic acid	Tetradecanoic	$CH_3(CH_2)_{12}COOH$	14:0
Palmitic acid	Hexadecanoic	$CH_3(CH_2)_{14}COOH$	16:0
Stearic acid	Octadecanoic	$CH_3(CH_2)_{16}COOH$	18:0
Unsaturated			
Palmitoleic acid	Hexadecenoic	$CH_3(CH_2)_5CH=CH(CH_2)_7COOH$	16:1 n7
Oleic acid	Octadecenoic	$CH_3(CH_2)_7CH=CH(CH_2)_7COOH$	18:1 n9
Linoleic acid	Octadecadienoic	$CH_3(CH_2)_4CH=CHCH_2CH=CH(CH_2)_7COOH$	18:2 n6
Linolenic acid	Octadecatrienoic	$CH_3CH_2CH=CHCH_2CH=CHCH_2CH=CH(CH_2)_7COOH$	18:3 n3
Arachidonic	Eicosatetraenoic	$CH_3(CH_2)_4CH=CHCH_2CH=CHCH_2CH=CHCH_2CH=CH(CH_2)_3COOH$	20:4 n6

[a]Anoic: no double bond; enoic: one double bond; dienoic: two double bonds; trienoic: three double bonds; tetraenoic: four double bonds.
[b]Number of carbons:number of double bonds and position of first double bond counting from the methyl end.
Source: Robinson and Wilson (1985).

TABLE 5.5.

Saponification and Iodine Numbers for Selected Lipids

Lipid	Saponification Number	Iodine Number
Beef	196–200	35–50
Butter	210–230	26–38
Coconut	253–262	6–10
Corn	187–193	111–128
Cottonseed	194–196	103–111
Lard	195–203	47–67
Peanut	186–194	88–98
Soybean	189–194	122–134
Sunflower	188–193	129–136

Source: Church and Pond (1982).

highly unsaturated fatty acids than triglycerides and are more widely dispersed in body tissues.

Sterols are important lipids of which cholesterol is the most abundant. Bile acids, sex hormones, vitamin D, and cortisol are other sterols important in animals.

Functions. Lipids are important both as a dietary source of energy and essential fatty acids (EFAs) and as a stored form of energy or precursors for various metabolically important compounds. Ingested lipids are a highly digestible form of concentrated energy for catfish, yielding 8 to 9 kilocalories/g. Depot lipids are also a reservoir of energy that can be utilized during times of deprivation. Lipid appears to be more readily mobilized from tissue storage than is carbohydrate in certain fish. Dietary lipids also improve the absorption of fat-soluble vitamins and influence flavor and texture of feeds. Phospholipids and sterol esters are structural components of various membranes aiding in maintenance of the fluidity and flexibility of membranes. Tissue lipids are also important as precursors for synthesis of steroid hormones and hormonelike prostaglandins, provide protective insulation, and affect flavor and texture of edible tissues.

Requirements and sources. Some animals can tolerate relatively high levels of dietary lipid. Although lipid is generally highly digestible by aquatic animals, feed efficiency and weight gain are depressed in certain aquatic species when fed diets containing in excess of 12 to 15 percent lipid. Channel catfish have been fed diets containing up to

16 percent lipid without detrimental effects on feed efficiency and weight gain. However, generally as dietary lipid increases so does lipid deposition, which can result in a reduction in dressed yield, if the lipid is deposited in the visceral cavity. Also, excessive lipid deposition in edible tissues reduces frozen storage life of processed fish and may adversely affect flavor. An additional consideration that influences the optimal level of lipid to include in a catfish feed is that of feed manufacture. Feeds high in supplemental lipid are often difficult to pellet.

Except for negative effects of increased fattiness in catfish fed high-lipid diets, there is no nutritional evidence that can be used to set an optimal dietary lipid level for catfish. Additionally, catfish appear to be able to synthesize most of their fatty acids, and only a small amount of lipid is adequate to supply EFAs. Considering protein sparing, product quality, and constraints of feed manufacture a practical recommendation concerning how much lipid should be included in catfish feeds can be made.

Total lipid levels in commercial catfish feeds used for growing stocking-size fingerlings to a harvestable size typically do not exceed 6 percent. Approximately 3 to 4 percent lipid is inherent in dietary ingredients used to manufacture catfish feeds. The remaining 1 to 2 percent is sprayed on the finished pellet. Spraying ("top dressing") feed pellets with lipid aids in reducing feed dust (fines) and increases feed energy. A mixture of animal (including fish) and plant lipids can be used to "top-dress" catfish feeds.

Essential fatty acids. Essential fatty acids are fatty acids that cannot be synthesized in body tissues; they were identified in rainbow trout in the 1970s. Trout require the omega-3 (n3) linolenic acid (18:3n3), in which the unsaturated bonds begin at the third carbon from the methyl end of the molecule (Table 5.4), rather than omega-6 (n6) fatty acids required by mammals. In recent years, various species of aquatic animals have been shown to require either omega-3, omega-6, or a combination of these fatty acids. It is thought that the omega-3 structure allows for a greater degree of unsaturation, thus permitting maintenance of cell membrane flexibility and permeability at low temperatures.

Signs of EFA deficiency in fish have been demonstrated experimentally, but it does not appear to be a problem in commercial catfish culture. Deficiency signs include reduced weight gain, decreased feed efficiency, high mortality rate, elevated muscle water content, abnormal permeability of membranes, fatty degeneration of liver tissue, decreased hemoglobin levels and red blood cell volume, and elevated tissue levels of eicosatrienoic acid, 20:3n9 (National Research Council

1981,1983). Catfish fed diets considered to be deficient in EFA exhibit reduced weight gain and feed efficiency, and in certain cases tissue eicosatrienoic acid levels increase.

The EFA requirements for catfish have yet to be defined precisely. However, on the basis of data from a number of lipid studies, the EFA requirements of catfish are relatively low, perhaps less than 1.0 percent of the diet. It appears that linoleic acid is not effectively metabolized by channel catfish and thus may not be essential. Linolenic acid is desaturated and elongated and may be required in small amounts. However, when the level of linolenic acid is about 1 percent of the diet, it apparently interferes with fatty acid synthesis.

From a practical standpoint, knowing the EFA requirements of the catfish would not greatly impact current channel catfish feed formulations, since the requirement for EFA is most likely met through practical dietary ingredients. However, if animal protein costs continue to increase and feeds are formulated primarily from plant protein sources, knowledge of EFA requirements becomes more important.

Protein and Amino Acids

Protein accounts for about 65 to 80 percent of the dry weight of the soft tissues that make up catfish organ and muscle tissues. All proteins comprise subunits of amino acids. Thus catfish, like other animals, do not require protein per se but rather have a requirement for amino acids and nonspecific nitrogen. Generally, the most economical source of these chemicals is a mixture of proteins. Since protein metabolism is a dynamic process in that tissue proteins are continually being catabolized and resynthesized, a dietary source of amino acids and nitrogen is required throughout life. Ingested proteins are hydrolyzed to release amino acids that may be used for synthesis of tissue proteins or, if in excess, for energy. Utilization of protein for energy is expensive; thus the nutritionist should balance rations to assure that adequate levels of nonspecific nitrogen, amino acids, and nonprotein energy are supplied in proper proportions necessary for the animal to maximize protein deposition.

Structure and functions. The term *protein* refers to a large group of chemically similar but physiologically distinct molecules. They are similar in that they are composed of polymers of amino acids linked by a peptide bond (Fig. 5.2). Proteins contain carbon, hydrogen, oxygen, and nitrogen, and most contain sulfur. Various combinations and sequential arrangements of amino acids occur in different proteins and thus provide for a wide diversity of function. Twenty-two to twenty-

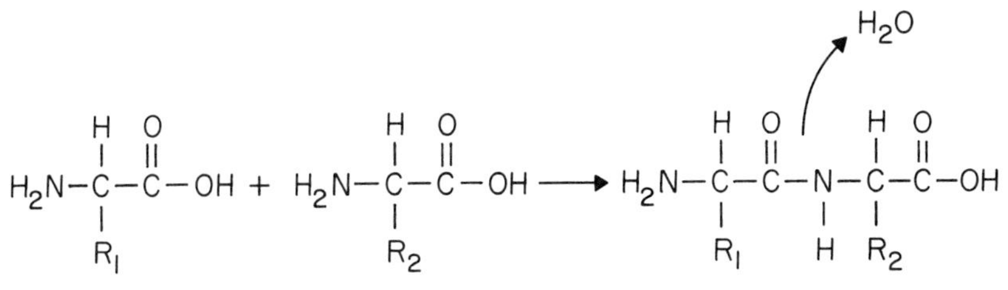

$$\underset{\textit{amino acid 1}}{\mathrm{H_2N-\underset{\underset{R_1}{|}}{\overset{\overset{H}{|}}{C}}-\overset{\overset{O}{\|}}{C}-OH}} + \underset{\textit{amino acid 2}}{\mathrm{H_2N-\underset{\underset{R_2}{|}}{\overset{\overset{H}{|}}{C}}-\overset{\overset{O}{\|}}{C}-OH}} \longrightarrow \underset{\textit{peptide linkage}}{\mathrm{H_2N-\underset{\underset{R_1}{|}}{\overset{\overset{H}{|}}{C}}-\overset{\overset{O}{\|}}{C}-\underset{\underset{H}{|}}{N}-\underset{\underset{R_2}{|}}{\overset{\overset{H}{|}}{C}}-\overset{\overset{O}{\|}}{C}-OH}} \; \nearrow H_2O$$

FIGURE 5.2.

Formation of a dipeptide from two amino acids.

six amino acids are found in proteins of animal or plant origin, although about twenty amino acids are typical of most proteins.

Proteins are important in the structure and function of all living organisms. In animals, they are the primary constituent of many structural and protective tissues, such as bones, ligaments, hair, nails, skin, and the soft tissue that makes up organs and muscles. Proteins also compose enzymes, certain hormones, and various components of the serum fraction of the blood.

Amino acids. Nutritionally amino acids may be classified as indispensable (essential) or dispensable (nonessential). An *indispensable amino acid* is one that the animal cannot synthesize or cannot synthesize in quantities sufficient for body needs; thus indispensable amino acids must be provided preformed in the diet. A *dispensable amino acid* is one that can be synthesized by the animal in quantities sufficient for maximal growth. Most simple-stomached animals, including fish, require the same ten indispensable amino acids: arginine, histidine, isoleucine, leucine, lysine, methionine, phenylalanine, threonine, tryptophan, and valine.

Determining amino acid requirements. Qualitative amino acid requirements for catfish were determined by measuring weight gain and feed efficiency of fish fed crystalline amino acid diets (pH adjusted to 6.8 to 7.2) in which a single amino acid was omitted. Quantitative amino acid requirements for catfish were based on weight gain, feed efficiency, and serum-free amino acid concentrations of fish fed a pH-adjusted diet prepared from a combination of purified proteins and crystalline amino acids whose amino acid profile resembled that of 24 percent whole chicken egg protein. Serum-free amino acid concentrations are not always indicative of the amino acid requirement established from growth data.

There has been some disagreement among fish nutritionists concerning the reliability of quantitative amino acid requirements determined for catfish using highly purified diets. Since fish grow slowly when fed diets containing predominantly crystalline amino acids as compared to those reared on diets with similar amino acid profiles prepared from intact proteins, some nutritionists contend that the requirements are not of use in practical diets. In addition, others contend that the 24 percent protein level used in diets to determine the amino acid requirements of catfish was too low and that amino acid requirements determined at that protein level are not applicable to fish fed under commercial conditions. However, it appears that the amino acid requirement values currently used for formulating commercial catfish feeds are relatively reliable. The low level of protein was used because preliminary studies indicated that protein deposition was maximum in fish fed diets containing 24 percent protein, and in addition low-protein diets accentuated amino acid deficiency. The lysine requirement for catfish has been reaffirmed, using both purified and practical diets containing 30 percent protein. A high correlation exists between the quantitative amino acid requirements of catfish and the indispensable amino acid concentrations of catfish whole body tissue. Further, there are no indications of amino acid deficiencies of catfish fed commercial catfish feeds formulated on an amino acid basis.

Expressing amino acid requirements. Amino acid requirements may be expressed as the amount of amino acid needed per animal per day, as a percentage of the diet, as a percentage of the dietary protein, or on a caloric basis. The best method might be to express amino requirements on amount needed per animal per day, but this is very difficult in growing animals and with catfish information is insufficient to express amino acid requirements accurately in such a manner. Expression of amino acid requirements as a percentage of diet is not precise because of the numerous factors that influence dietary intake (e.g., water temperature and fish size). The lack of sufficient information concerning the effects of caloric density on the amino acid requirements of catfish and the lack of a precisely defined energy requirement for catfish preclude the expression of amino acid requirements on a caloric density basis. Apparently amino acid requirements remain rather constant at a given dietary protein level. For example, catfish fed diets containing 24 or 30 percent protein require 1.53 and 1.78 percent lysine (expressed as percentage of diet), respectively. Those values represent a lysine requirement of about 5 percent of the dietary protein. Pres-

TABLE 5.6.
Indispensable Amino Acid Requirements of Various Fish

Amino Acid	Channel Catfish	Common Carp	Japanese Eel	Chinook Salmon	Tilapia nilotica
Arginine	4.3	4.2	4.2	6.0	4.2
Histidine	1.5	2.1	2.1	1.8	1.7
Isoleucine	2.6	2.3	4.1	2.2	3.1
Leucine	3.5	3.4	5.4	3.9	3.4
Lysine	5.1	5.7	5.3	5.0	5.1
Methionine[a]	2.3	3.1	5.0	4.0	3.2
Phenylalanine[b]	5.0	6.5	8.4	5.1	5.7
Threonine	2.0	3.9	4.1	2.2	3.6
Tryptophan	0.5	0.8	1.0	0.5	1.0
Valine	3.0	3.6	4.1	3.2	2.8

[a]Value is for total sulfur amino acid requirement (methionine + cystine).
[b]Value is for total aromatic amino acid requirement (phenylalanine + tyrosine).
Note: Indispensable amino acid requirements expressed as percentage of dietary protein.
Source: Catfish, carp, eel, and salmon (National Research Council 1983); tilapia (Santiago 1985).

ently, the best way to express amino acids of catfish appears to be as a percentage of dietary protein.

Quantitative amino acid requirements. The quantitative amino acid requirements of catfish and other selected fish are presented in Table 5.6. Some nutritionists argue that little variation should exist in the amino acid requirements among various species of fish, since the amino acid composition of lean fish tissues does not vary much between species. Although the data presented in Table 5.6 show similarities in the requirements of various amino acids among species, differences do exist. It is probable that differences result from variation in the relative proportion of structural proteins among species as well as physiological needs for certain of the amino acids.

Dispensable amino acids. Dispensable amino acids are only dispensable in the sense that they do not have to be provided in the diet, but they are part of the protein and thus are necessary for synthesis of protein. They are metabolically essential. Dispensable amino acids include alanine, asparagine, aspartic acid, cystine, glutamic acid, glutamine, glycine, proline, serine, and tyrosine.

If the dispensable amino acids are supplied in the diet they do not have to be synthesized; thus a savings in energy is realized. Two specific examples of this are the conversion of phenylalanine to tyrosine and of methionine to cystine. The nonessential amino acid can only be synthesized from the essential precursor. Dietary cystine can replace about 60 percent of methionine on a per mole sulfur basis to meet the sulfur amino acid requirement of catfish. Tyrosine can replace 50 percent of the total phenylalanine requirement in catfish.

Amino acid interactions. Amino acid interactions are relatively common among the branched-chain amino acids leucine, isoleucine, and valine for various animals. That is, excesses of leucine or isoleucine fed in diets deficient in one of the branched-chain amino acids cause a reduction in weight gain and feed efficiency. The effects can be reversed by addition of the deficient amino acid. Similar responses have been observed in catfish.

Excesses of dietary tyrosine are detrimental to the growth of certain animals; however, they are not detrimental in catfish. Apparently, catfish are able to deaminate excess tyrosine and utilize the remaining carbon for energy. Catfish also appear to differ from poultry and swine in that the lysine-arginine antagonism that occurs in those species does not occur in catfish.

The effects of excesses of dietary amino acids on catfish nutrition have little practical significance, because the occurrence of large excesses of amino acids in commercial catfish feeds is improbable. Excesses of amino acids may occur in certain experimental diets.

Meeting amino acid requirements. Amino acid requirements may be met by feeding an excess of protein, supplementing deficient proteins with crystalline amino acids, or feeding a mixture of complementary proteins. Feeding an excess of protein is wasteful and may actually reduce growth. The practice of using supplemental amino acids to improve the quality of inferior protein sources in catfish feeds is somewhat questionable. Although there is evidence to indicate that catfish can utilize supplemental lysine and methionine, the efficacy of this practice in commercial catfish culture, where fish are generally fed once daily, is still debatable. Some fish nutritionists feel that fish cannot efficiently use supplemental amino acids when fed once a day because the crystalline amino acids are not absorbed from the intestine on a timely basis. Multiple feedings were used in most of the studies in which supplemental amino acids were evaluated in catfish feeds. A recent study demonstrated that catfish given a single daily feeding could effectively utilize supplemental lysine in a practical feed. Work is continuing to assess the efficacy of using supplemental amino acids in commercial catfish feeds, because it may be possible to improve the quality of inferior proteins on a cost-effective basis.

The best method to provide adequate amino acid nutrition to the catfish is to use a mixture of complementary proteins. To mix proteins effectively to meet the amino acid requirements of catfish, one needs to know the amino acid composition of potential ingredients and the biological availability of amino acids from each feed ingredient. Amino acid composition data are generally available in feed tables for commonly used feedstuffs. Amino acid availability data are not readily obtainable. However, some amino acid availability data from feed ingredients typically used in catfish feeds are available as well as protein digestibility data (see the section Digestion). It is more desirable to use amino acid availabilities because the average protein digestibility and availability of individual amino acids do not always correlate. For example, the average protein digestibility of cottonseed meal to catfish is about 86 percent, whereas lysine from cottonseed meal is only about 66 percent available to catfish. This is a fairly extreme example, but if protein digestibility is used to formulate a feed containing cottonseed meal a deficiency of available lysine could result.

Protein requirements. Although many studies have been conducted to establish the protein requirement of catfish, the requirement is still uncertain and often the subject of debate among fish nutritionists. The level of dietary protein needed for the most economical gain may differ as the cost of ingredients fluctuates. Because of the number of factors that influence the dietary protein requirement, it is difficult to establish a standard that is nutritionally satisfactory and economical under all conditions. The protein requirement is affected by water temperature, feed allowance, fish size and age, dietary protein to energy ratio, protein quality, availability of natural food, and management practices.

Studies to determine the dietary protein requirement of catfish have been conducted with fish reared in ponds or in the laboratory using practical or purified diets. On the basis of these studies, the dietary protein requirement for catfish has been reported to be 25 to 45 percent (National Research Council 1983). The wide variation in the reported protein requirements for catfish is not surprising because of the differing conditions under which the studies were conducted. Fish size, water temperature, feeding regimens, fish stocking densities, and diet composition, all of which impact the dietary requirement, differed in many of the studies.

Most of the early studies conducted to determine the protein requirement of catfish reared in ponds were conducted at relatively low stocking densities, 2,000 to 3,000 pounds/acre. At these stocking densi-

TABLE 5.7.

Protein Requirements of Catfish

Fish Weight (Grams)	Percentage of Diet
0.02–0.25	52
0.25–1.5	48
1.5–5.0	44
5.0–20.0	40
20.0–above	28–36

Source: Fish ≤20 grams (National Research Council 1983); above 20 grams (estimated from various reports).

ties, lower (25 to 30 percent) protein feeds were adequate presumably because natural food contributed significantly to the diet of the fish. In addition, fish fed a restricted ration require more protein than those fed to satiation. Also, small fish generally require higher protein diets. Catfish weighing about 0.1 ounce (3 grams) require approximately ten times the daily intake of protein needed by 8-ounce fish for maximum weight gain. Small catfish grow equally well on low-energy diets containing either 27 or 38 percent protein, but when the energy level of the feed is increased the higher protein feed is needed to maintain maximum weight gain. Apparently, increasing dietary energy results in a decrease in feed consumption. High-protein, high-energy diets appear to be more beneficial to small catfish as the environmental temperature deviates from the optimum. Catfish reared in ponds during the winter do not benefit from high dietary levels of protein. A feed containing 25 percent protein appears to be adequate for winter feeding in temperate zones.

Protein requirements for various sizes of catfish are given in Table 5.7. A 32 percent protein feed is the standard presently used for commercial culture of catfish from advanced fingerling to harvest. However, there are those who argue that the protein level could be reduced without sacrifice of weight gain, dressing percentage, or product quality. Conversely, others assert that the dietary protein level should be increased to improve product quality (e.g., lower body fat). There is evidence that the dietary protein level can be lowered, particularly when the diet is balanced in energy and other nutrients. In contrast, some data indicate that increasing the dietary protein level would be cost-effective. Without additional research on the effects of dietary protein concentration on weight gain and composition of gain in catfish fed to satiation under conditions that reflect those encountered in com-

TABLE 5.8.

Protein Requirements of Selected
Fish Species

Species	Protein (Percentage)
Channel catfish	25–36
Common carp	31–38
Smallmouth bass	45
Largemouth bass	40
Grass carp	41–43
Chinook salmon	40–50
Rainbow trout	35–40
Red drum	35–45
Japanese eel	44

Source: Adapted from Robinson (1989).

mercial catfish culture, it is unlikely that the catfish industry will change from the present 32 percent protein feed.

Protein requirements for selected fishes are given in Table 5.8. The protein requirements of the various species are fairly similar and all require a relatively high level of dietary protein.

Nutritional value of proteins. The nutritional value of proteins, or protein quality, is based on the amino acid composition of the protein source, particularly the indispensable amino acid content and the biological availability of the amino acids. In addition to the determination of protein digestibility and amino acid availability values (discussed in the section Digestion) there are other methods that appear to be applicable for evaluation of protein quality in fish. These include protein efficiency ratio, net protein utilization, and essential amino acid index.

Protein efficiency ratio (PER), which is defined as the grams of wet weight gained per gram protein consumed, is widely used because it is rather simple to calculate and does not require chemical analyses. The major criticism of this assay is that it assumes that all of the protein is used for growth and makes no allowance for maintenance. Diet composition, fish size, and other factors may also impact the results. PER should be determined by using single-protein diets containing a suboptimal level of protein. If a high level of protein is fed, amino acid deficiencies may be masked.

PER = grams wet weight gain/grams crude protein fed   (5.2)

TABLE 5.9.

Essential Amino Acid Content of Selected Proteins

Amino Acid	Requirement Value Catfish	Whole Egg Protein	Casein	Gelatin
Arginine	4.3	6.5	4.0[a]	8.0
Cystine		0.5	0.4	0.0
Histidine	1.5	2.6	3.2	0.8
Isoleucine	2.6	5.5	5.5	1.3
Leucine	3.5	9.1	9.7	2.9
Lysine	5.1	6.9	8.8	3.8
Methionine[b]	2.3	3.4	3.1	0.9
Phenylalanine[c]	5.0	5.8	5.6	2.1
Threonine	2.3	5.2	4.6	1.8
Tryptophan	0.5	1.4	1.2	0.0
Tyrosine[c]		4.6	6.4	0.6
Valine	3.0	6.7	6.9	2.4

[a]Values underlined indicate limiting amino acids.
[b]The methionine or total sulfur amino acid requirement can be met by the sum of the methionine plus cystine content.
[c]The phenylalanine or total aromatic amino acid requirement can be met by the sum of the phenylalanine plus tyrosine content.
Note: Essential amino acid content is expressed as percentage of crude protein.
Source: Adapted from Wilson and Robinson (1982).

Net protein utilization (NPU) is an index of the efficiency of the deposition of dietary protein as body protein:

$$NPU = [B - (B_k - I_k)]/I \qquad (5.3)$$

where B is the total body nitrogen of fish fed a test diet, B_k is the body nitrogen of fish fed a low-protein diet (a protein-free diet should be used but consumption of such a diet by fish is usually poor), I is the nitrogen intake of fish fed the test diet, and I_k is the nitrogen intake of fish fed the low-protein diet.

Essential amino acid index. The essential amino acid index is based on the comparison of the essential amino acid content of protein to the essential amino acid requirement values of the fish. The use of the index in predicting *limiting amino acids*, those amino acids present in the protein at a level below that required by the fish, is illustrated in Table 5.9. Of the three protein sources given in Table 5.9, only whole egg protein contains adequate amounts of each of the essential amino acids to meet the requirements for catfish. Casein contains adequate amounts of all of the indispensable amino acids except arginine. Gelatin is deficient in all indispensable amino acids except arginine.

Vitamins

The term *vitamin* collectively describes a group of compounds that are highly diverse in chemical structure and physiological function. Although vitamins are not chemically related as closely as are other groups of nutrients, they are considered to be organic compounds that are required in small amounts in the diet for normal growth, health, and reproduction by one or more animal species. Some vitamins may be synthesized in the body in quantities sufficient to meet metabolic needs and thus are not required in the diet. For example, vitamin C (ascorbic acid) is synthesized by most animals, but certain species, including humans and catfish, lack the ability to synthesize the vitamin.

Although it was recognized that catfish required vitamins, examination and definition of the vitamin nutrition of catfish began only about 30 years ago. Qualitative and quantitative vitamin requirements have now been relatively well defined. Qualitative vitamin requirements have been largely based on growth assays in which fish were fed purified diets either replete in vitamins or deficient in a specific vitamin. Quantitative vitamin requirements for catfish were determined by feeding fish graded levels of a specific vitamin and measuring weight gain and absence of deficiency signs. Also, vitamin tissue storage was often used as an indicator of the dietary requirement.

Although characteristic signs of vitamin deficiency can be produced in fish in laboratory studies, such deficiencies rarely occur in nature or in commercial catfish culture. Excesses of certain vitamins may be toxic, but this is not a major concern in practice because the occurrence of excesses of the magnitude necessary to cause problems would be rare. However, there is interest in using megadose levels of some vitamins in catfish feeds either to improve product quality or to enhance immune response. Thus it may become important to document the effects of large excesses of certain vitamins on catfish.

Fourteen vitamins are considered to be metabolically essential for catfish. They are classified on the basis of their solubility as either water-soluble or fat-soluble. Characteristic deficiency signs and dietary requirements for each of the water-soluble and fat-soluble vitamins are summarized in Table 5.10.

Water-soluble Vitamins

The water-soluble vitamins consist of the B-complex vitamins and vitamin C. Small amounts of these vitamins are stored in the liver; how-

TABLE 5.10.

Vitamin Deficiency Signs and Minimum Dietary Levels Required to Prevent Signs of Deficiency in Catfish

Vitamin	Deficiency Signs	Units (ppm or IU/kg)	Requireme
Fat-soluble			
A	Exophthalmia, edema, acities	IU	1,000–2,0
D	Low bone ash	IU	250–1,00
E	Skin depigmentation, exudative diathesis, muscle dystrophy, erythrocyte hemolysis, splenic and pancreatic hemosiderosis	IU	50
K	Skin hemorrhage, prolonged clotting time	ppm	R
Water-soluble			
Thiamin	Dark skin color, neurological disorders	ppm	1.0
Riboflavin	Short-body dwarfism	ppm	9.0
Pyridoxine	Greenish blue coloration, tetany, nervous disorders	ppm	3.0
Pantothenic acid	Clubbed gills, anemia, eroded skin, lower jaw, fins, and barbels	ppm	15
Niacin	Anemia, lesions of skin and fins, exophthalmia	ppm	14
Biotin	Anemia, skin depigmentation, reduced liver pyruvate carboxylase activity	ppm	R
Folic acid	None demonstrated	ppm	R
B_{12}	Reduced hematocrit	ppm	R
Choline[a]	Hemorrhagic kidney and intestine, fatty liver	ppm	400
Inositol	None demonstrated	ppm	NR
Ascorbic acid	Reduced hematocrit, scoliosis, lordosis, increased susceptibility to bacterial infections, reduced bone collagen formation, internal and external hemorrhage	ppm	60

[a]Determined by using diets marginal in methionine and based on liver lipid patterns.
Note: Anorexia, reduced weight gain, and mortality are common vitamin deficiency signs; thus, they are n included in the table. R and NR refer to required and not required, respectively.
Source: Adapted from Robinson (1989).

ever, tissue storage is limited and excesses are eliminated by the kidney. Generally, a regular dietary intake of the water-soluble vitamins is needed unless metabolic needs can be met by intestinal microbial synthesis or by synthesis in body tissues. Since certain vitamins (e.g., vitamin B_{12}) may be synthesized by the intestinal microflora, demonstration of a deficiency may be difficult unless the intestinal tract is sterilized or an antimetabolite is fed.

Thiamin. Thiamin or vitamin B_1 was first isolated from rice polishings. It is required for the normal metabolic function of all animal species. Although thiamin is present in many natural feedstuffs of both plant and animal origin, it is added to commercial catfish feeds to ensure adequacy.

Thiamin is absorbed from the intestine and is converted to thiamin pyrophosphate in the liver. The phosphorylated form acts as a coenzyme for the oxidative decarboxylation of alpha-ketoacids that are needed for metabolism of carbohydrates. Thiamin is also involved in the oxidation of glucose in the cytoplasm of cells. The vitamin is essential for good appetite, growth, and normal nervous system function.

Deficiency signs for fingerling catfish fed thiamin-deficient semipurified diets in controlled environments include anorexia, poor growth, dark skin coloration, and increased mortality. The dietary thiamin requirement for maximum growth and prevention of deficiency signs in catfish is 1 ppm of the diet.

Certain fish tissues contain an enzyme, thiaminase, that destroys thiamin. Because thiaminase is inactivated by heat, it is not likely to cause problems in processed feeds.

Riboflavin. Riboflavin, or vitamin B_2, is another of the water-soluble B vitamins. It was first isolated from rice polishing and yeast. Plants and certain microorganisms can synthesize riboflavin. Animals do not synthesize or store large quantities of it; thus a dietary source is required. Commercial catfish feeds are supplemented with riboflavin.

Riboflavin is absorbed from the intestine, phosphorylated in the intestinal wall, and then transported to various tissues. The primary functions of riboflavin involve energy metabolism as part of the coenzymes flavin mononucleotide (FMN) and flavin adenine dinucleotide (FAD). FMN and FAD function in the electron transport system as well as in oxidase and reductase enzyme systems required for the catabolism of fatty acids, amino acids, and pyruvic acid. Riboflavin also is thought to have a function in the respiration of vascular tissues such as the cornea.

Catfish fed riboflavin-deficient diets demonstrate signs of deficiency, including anorexia, reduced weight gain, and short-body dwarfism. The amount of riboflavin needed in the catfish's diet for normal growth and prevention of deficiency is 9 ppm.

Pyridoxine. Pyridoxine, pyridoxal, and pyridoxamine are three forms of vitamin B_6. All have equivalent vitamin activities in animals and are interconvertible. Pyridoxine hydrochloride is added to commercial catfish feeds.

Pyridoxine is absorbed from the intestine, phosphorylated in the liver, and oxidized to the physiologically active pyridoxal phosphate. The vitamin is part of various enzyme systems that are essential for the metabolism of fats, carbohydrates, and particularly proteins.

Signs of pyridoxine deficiency develop as early as 6 weeks in catfish. They include greenish blue coloration, tetany, nervous disorders, anorexia, and mortality. The amount of dietary pyridoxine needed for normal growth and prevention of deficiency in catfish is 3 ppm of the diet.

Pantothenic acid. Pantothenic acid is essential for many animals, including catfish. It is synthesized by plants and microorganisms and is widespread in nature. Although pantothenic acid is found in many feed ingredients, calcium pantothenate is added to commercial catfish feeds to ensure adequacy.

Pantothenic acid is absorbed from the intestine and is found in tissues as a component of coenzyme A (CoA) and acyl carrier protein (ACP), which are important in intermediary metabolism. CoA is required in the synthesis and oxidation of fatty acids, oxidation of pyruvate, conversion of choline to acetylcholine, conversion of oxyloacetic acid to citric acid, and formation of two-carbon units from amino acids and carbohydrates for utilization in the Krebs cycle and for synthesis of steroids. ACP serves a role in fatty acid synthesis.

Deficiency of pantothenic acid is one of the few nutritional diseases known to have occurred in commercial catfish culture. Clubbed gills, anorexia, loss of weight, anemia, mortality, and eroded skin, lower jaw, fins, and barbels are signs of pantothenic acid deficiency in catfish. The dietary requirement for normal growth and prevention of deficiency signs of pantothenic acid in catfish is 15 ppm of the diet.

Niacin. Niacin (nicotinic acid) is an essential nutrient for many animals, including the catfish. Often precise dietary requirements for niacin are difficult to establish because the vitamin can be synthesized from the amino acid tryptophan. The importance of this conversion in catfish is not known. Niacin is added to commercial catfish feeds. Nicotinic acid or nicotinamide may be added to feeds because they have equivalent vitamin activity and both are relatively stable during feed processing.

Niacin is absorbed from the intestine, small quantities are stored in the liver, and excesses are excreted by the kidney. The primary function of niacin is as a component of nicotinamide adenine dinucleotide (NAD) and nicotinamide adenine dinucleotide phosphate (NADP). NAD and NADP are part of enzyme systems that are involved with electron transfer. Thus, they are essential in certain energy-yielding and synthetic pathways, including fatty acid catabolism and synthesis, oxidative deamination, glucose catabolism, glycerol catabolism, and the Krebs cycle.

Catfish fed niacin-deficient diets exhibited slow growth, anemia, skin and fin lesions, hemorrhages, exophthalmia, and mortality. About 14 ppm of dietary niacin is needed for normal growth and prevention of all other deficiency signs. About 6.6 ppm niacin in the diet is adequate to prevent mortality and gross deficiency signs; 11.6 ppm is needed to prevent anemia.

Biotin. Biotin supplements were at first thought not to be necessary for animals, because the vitamin was widely distributed in feedstuffs and could be synthesized by intestinal bacteria. However, more recently biotin supplements have been shown to be necessary for some animals under certain production conditions. The vitamin is essential for catfish, but it is not normally added to feeds because the requirement is apparently very low and can be met from indigenous biotin.

Biotin is an essential component of specific enzyme systems involved in carboxylation and decarboxylation reactions. Thus, it is essential in the biosynthesis of fatty acids, gluconeogenesis, purine synthesis, and protein metabolism.

Biotin deficiency can be induced in catfish by feeding diets containing raw eggs. Egg whites contain avidin, an agent that binds biotin, making the vitamin biologically unavailable. More recently, it has been shown that a biotin deficiency can be induced in catfish by feeding biotin-deficient diets without raw egg. Deficiency signs include reduced growth, lighter skin, hypersensitivity, and reduced liver pyruvate carboxylase activity. A quantitative biotin requirement for catfish has not been determined; however, it appears to be as low as 0.25 ppm of the diet.

Folacin. Folacin is a group of compounds having folic acid activity. It is very abundant in natural feedstuffs and is synthesized by intestinal microflora in some animals. Folic acid is added to catfish feeds, although no dietary requirement has been established.

Folacin is converted into its active form tetrahydrofolic acid in the liver and bone marrow (in animals possessing marrow). Tetrahydrofolic acid is a coenzyme that acts as an intermediate carrier of single-carbon units. It is required for histidine degradation and synthesis of methyl groups, purines, pyrimidines, glycine, serine, and creatine. Its role in prevention of anemia appears to be related to purine synthesis, since purine deficiency results in a deficiency of nucleoproteins necessary for maturation of red blood cells.

Vitamin B_{12}. Vitamin B_{12} is a group of compounds having B_{12} activity. The vitamin is synthesized only by microorganisms and thus is not usually found in plants or feedstuffs derived from plant sources.

Microbial synthesis of vitamin B_{12} occurs in the intestine of many animals, including catfish. Absorption of the vitamin depends on the presence of a carrier termed *intrinsic factor*, which is produced in the gastric mucosa. Vitamin B_{12} is stored in the liver of most animals and can be mobilized as needed. It is involved as a coenzyme in many important functions, including synthesis of labile methyl groups, and in concert with folacin vitamin B_{12} is necessary for the synthesis of nucleic acids. Thus vitamin B_{12} is needed for the maturation and development of erythrocytes, metabolism of lipids, synthesis of methionine, metabolism of propionic acid, and for recycling of tetrahydrofolic acid.

Catfish fed vitamin B_{12}-deficient diets exhibited slow growth and reduced hematocrit. Other vitamin B_{12} deficiency signs are not apparent. Intestinal synthesis of the vitamin, in the presence of cobalt, is adequate for normal growth and red blood cell formation. A quantitative dietary requirement for vitamin B_{12} has not been established. Although it is likely unnecessary to supplement commercial catfish feeds containing generous amounts of animal feedstuffs with vitamin B_{12}, it is often added to catfish feeds to ensure adequacy.

Choline. Choline is considered to be a B-complex vitamin, although its status as a vitamin is uncertain. It does not have a known function as a coenzyme; rather it appears to have a structural function. Choline is added to some commercial catfish feeds, but some feed manufacturers choose not to add supplemental choline because it is widespread in feedstuffs and the catfish may meet its choline requirement from the choline found in natural feed ingredients. However, its bioavailability from feedstuffs is not known.

As a component of phospholipid, choline in involved in membrane structure and in lipid transport. It is also part of acetylcholine and as such functions in transmission of nerve impulses. Choline is involved in the synthesis of methionine.

Catfish fed choline-deficient diets exhibited reduced weight gain and hemorrhagic areas in the kidney and intestine. Deficiency signs may not occur if the diet contains excess methionine. Apparently catfish can utilize methionine to spare part of their need for choline. The amount of choline suggested for catfish diets is 275 ppm.

Vitamin C. Vitamin C (ascorbic acid) can be synthesized by most animals, except humans and other primates, guinea pigs, and certain birds and fish. Catfish require a dietary source of the vitamin. Commercial catfish feeds are generally overfortified with vitamin C to compensate for losses during feed manufacture.

Vitamin C plays an important role in several biochemical reactions. Major functions involve its oxidation and reduction properties. Thus it is required for the formation of collagen, electron transport, and reactions involving the hydroxylation of tyrosine, phenylalanine, tryptophan, and the adrenal steroids. It also appears to be involved in the immune response of various animals.

Catfish reared in confinement and fed vitamin C-deficient diets show characteristic signs, including reduced weight gain, scoliosis and lordosis, increased susceptibility to bacterial infections, internal and external hemorrhage, dark skin color, fin erosion, and reduced formation of bone collagen. About 60 ppm dietary vitamin C is required for normal growth and prevention of deficiency signs. The requirement is dependent on fish size. Small fish (0.35 ounce or 10 grams) require 60 ppm, whereas 1.7-ounce (50-gram) fish require about 30 ppm. The requirement is likely even lower in larger fish. However, 60 ppm is generally considered to be the dietary vitamin C requirement for catfish.

Higher than normal levels of vitamin C may be beneficial in disease resistance in catfish. Mortality of fish fed diets containing 0 to 3,000 ppm vitamin C infected with *Edwardsiella ictaluri* ranged from 0 percent in fish fed the highest level of the vitamin to 100 percent in fish fed vitamin C-deficient diets. A level of 140 ppm vitamin C reduced mortality. Liver vitamin C levels appear to level off at about 500 to 1,000 ppm dietary vitamin C. If liver vitamin C concentration is indicative of the vitamin C status of catfish, maximum resistance to bacterial infections may be in this range.

Although feeds containing high levels (up to 2,000 ppm) are currently used by some catfish producers during early spring when the fish are most susceptible to bacterial infections, the impact of this practice is still uncertain. Existing data derived from studies conducted under practical conditions are contradictory. Also, high dietary vitamin C levels have been reported to depress certain aspects of the immune reaction in catfish. The effect of using vitamin C for disease resistance in catfish may be affected by fish size, severity of infection, and method of infection.

Catfish feeds typically used for growout of fish are fortified with approximately 330 ppm ethylcellulose-coated ascorbic acid. The coated vitamin C is used in attempt to improve stability of the vitamin during feed processing. About 60 percent of the coated vitamin C is lost during extrusion cooking of catfish feeds. A newer, more stable form of vitamin C, ascorbyl polyphosphate (AsPP), is commercially available. Anywhere from 0 to 20 percent of AsPP is lost during extrusion. High-C feeds contain about 2,000 ppm vitamin C.

Inositol. Inositol (myo-inositol) is classified as a water-soluble growth factor. Its status as a vitamin is unclear. It does not appear to be essential for most species, including catfish. It is not added to commercial catfish feeds.

Myo-inositol is a structural component of living tissues as a component of phosphoglycerides. It also appears to have a lipotropic role.

Channel catfish fed inositol-deficient diets do not exhibit deficiency signs. Inositol is apparently synthesized by the intestinal microflora in catfish and synthesis appears to occur in the liver.

Fat-soluble Vitamins

Fat-soluble vitamins A, D, E, and K (Table 5.10) are found in feedstuffs in association with lipids and are absorbed as dietary lipids are. Storage of fat-soluble vitamins can be substantial if dietary intake exceeds metabolic needs, and as a result hypervitaminosis (vitamin toxicity) can result from accumulation of these vitamins in tissues. Most cases of hypervitaminosis in fish have occurred under experimental conditions and are unlikely to occur under normal catfish culture conditions, because commercial catfish feeds do not contain large excesses of fat-soluble vitamins. Qualitative requirements for A, D, E, and K have been established for the catfish, and quantitative dietary requirements for A, D, and E have been determined for catfish.

Vitamin A. Vitamin A is only found in animal tissues, although carotenoids, which are precursors of the vitamin, occur in plants. Vitamin A appears to be essential for all animals. Several different forms of vitamin A occur in animals; of these transretinol is considered to have 100 percent vitamin A activity. Other forms in animals and carotenoids in plants have varying degrees of vitamin A activity. Commercial catfish feeds are supplemented with vitamin A.

Vitamin A and beta-carotene (from plants) are absorbed from the intestine with dietary lipids. Beta-carotene is converted to vitamin A in the intestinal mucosa. Vitamin A is transported via the lymphatic system to the liver, which is the primary storage site. It is required for the formation of rhodopsin (visual purple) in the eye, where it functions in vision. The vitamin is also needed for maintenance of epithelial cells, reproduction, embryonic development, and bone development.

Catfish fed vitamin A-free diets for extended periods develop deficiency signs, including exophthalmia, edema, and kidney hemorrhages. The minimum dietary level needed for maximal weight gain and prevention of deficiency is between 1,000 and 2,000 international units/kilogram (IU/kg) of diet. Catfish can utilize beta-carotene as a source of vitamin A if it is included in the diet at levels exceeding 2,000 international units/kilogram. Vitamin A should be added to commercial catfish feeds at a level of at least 2,000 international units/kilogram and perhaps more if losses are anticipated during feed processing.

Vitamin D. Vitamin D is a general term used to describe several fat-soluble sterol derivatives that have antirachitic activity. Of these compounds vitamin D_2 (ergocalciferol) and vitamin D_3 (cholecalciferol) are the most important nutritionally. Ultraviolet irradiation of the provitamins ergosterol (found in plants) and 7-dehydrocholesterol (found in animals) yields vitamins D_2 and D_3, respectively. Animals that are not exposed to sunlight require a dietary source of vitamin D. Catfish feeds are supplemented with vitamin D_3.

Dietary vitamin D is absorbed from the intestine in the same manner as lipids. Vitamin D_3 from the diet or from ultraviolet irradiation is converted to 25-hydroxycalciferol in the liver and then converted to the physiologically active form, 1,25-dihydroxy calciferol in the kidneys of most animals. It is presumed that this is also the case with catfish. The hydroxy-metabolites have been identified in catfish. Storage of vitamin D is primarily in the liver and excretion is in the bile. In animals, the primary function of vitamin D is as a precursor of 1,25-dihydroxy calciferol, which facilitates absorption of calcium from the intestine by functioning in the synthesis of calcium-binding protein. Vitamin D is essential for catfish and certain other fish, but since fish may meet their calcium requirement by absorption of calcium from water via the gills vitamin D may not be essential for calcium homeostasis as in other animals. Vitamin D is needed for the utilization of inorganic phosphate and for the function of alkaline phosphatase.

Catfish fed vitamin D-deficient diets show reduced weight gain as well as decreased body ash, phosphorus, and calcium. A precise dietary requirement for vitamin D has not been established; however, in terms of weight gain and feed efficiency the requirement does not appear to be greater than 250 international units/kilogram of diet. It has been suggested that the dietary vitamin D requirement might be as great as 2,000 international units/kilogram of diet if serum levels of the hydroxylated metabolites are used as an indicator of the requirement. Vitamin D levels of 1,000 to 2,000 international units/kilogram of diet are necessary to maintain peak levels of serum hydroxylated metabolites. Catfish feeds should contain about 2,000 international units of vitamin D/kilogram of diet, and the most efficacious form to use is D_3. Vitamin D_2 appears to be utilized as well as vitamin D_3 at dietary levels up to 1,500 international units/kilogram, but higher levels depress growth.

Vitamin E. Vitamin E is a group of compounds, tocopherols, which are found in plants. The vitamin is considered to be an essential nutrient for all animals. It is added to catfish feeds in the form of alpha-tocopherol acetate. Alpha-tocopherol is the most active form and the acetate ester helps to stabilize the vitamin during feed processing and storage. Vitamin E is easily oxidized.

Ingested vitamin E is absorbed from the intestine with lipids. It is transported in the lipoprotein fraction of the blood to various tissues, where it serves as a biological antioxidant. Vitamin E is stored throughout the body; the liver contains the highest concentration. Apparently, the primary functions of vitamin E are related to its antioxidant activity. It protects highly unsaturated fatty acids found in biological membranes from oxidation. Other functions of vitamin E are not clearly defined, although it has been implicated as having a role in blood clotting, electron transport, disease resistance, phosphorylation, synthesis of vitamin C and ubiquinone, and metabolism of sulfur amino acids and vitamin B_{12}.

One important aspect of vitamin E function is its interrelationship with selenium. Selenium can prevent certain signs of vitamin E deficiency. Both vitamin E and selenium protect biological membranes from oxidation but by differing mechanisms. Vitamin E protects against free-radical formation in membranes, whereas selenium destroys peroxides in the cytoplasm. Although selenium can alleviate some vitamin E deficiency signs, both are essential nutrients.

Catfish fed vitamin E-deficient diets show typical signs of vitamin E deficiency, including exudative diathesis, muscle dystrophy, erythrocyte hemolysis, and splenic and pancreatic hemosiderosis as well as skin depigmentation. Most vitamin E studies required that the diets contain high levels of polyunsaturated fatty acids to demonstrate deficiency signs. However, vitamin E deficiency can occur in catfish fed diets containing low levels of polyunsaturated fatty acids. The amount of dietary vitamin E required for normal growth and prevention of all deficiency signs in catfish is 50 ppm of diet. Higher levels, five to six times the requirement, may be beneficial in improving storage quality of frozen processed catfish. Work to assess the effects of vitamin E-fortified diets on product quality during frozen storage is continuing.

Vitamin K. Vitamin K comprises a group of quinones that have antihemorrhagic activity. Several naphthoquinone compounds that have vitamin K activity have been isolated or synthesized. The most important and the form added to commercial catfish feeds is a synthetic vitamin K, menadione or K_3.

Apparently natural forms of vitamin K are absorbed from the intestine with the lipids, but synthetic forms are somewhat water-soluble and therefore may be absorbed freely. Vitamin K is metabolized in the liver and is required for the synthesis of several proteins involved in blood clotting. Vitamin K may also be involved in other reactions in which carboxylase enzyme systems are functional.

Catfish fed vitamin K-deficient diets may exhibit skin hemorrhages. No dietary requirement has been established even in fish fed vitamin K-deficient diets for up to 30 weeks. Apparently catfish can meet their vitamin K requirement from sources other than diet. Synthesis of vitamin K in the intestine is a important source of the vitamin in some animals. Intestinal synthesis of vitamin K in catfish has not been determined.

Minerals

The same minerals required for metabolism and skeletal structure of other animals are presumably required by catfish. Fish also require minerals for osmotic balance between body fluids and their environment, some of which can be taken from the water. The ability to regulate mineral balance via the gills and problems with formulating mineral-free diets make it difficult to determine quantitative requirements for fish. Mineral requirements for catfish have generally been based on weight gain, feed efficiency, and tissue mineralization of fish fed purified diets limiting in a particular mineral to which graded levels of the limiting mineral is added. Enzyme activity and tissue morphology have also been used to establish mineral requirements of catfish.

Minerals are generally classified into macrominerals or microminerals on the basis of the amount required in the diet. Macrominerals are required in relatively large quantities and microminerals are required in trace quantities. Fourteen minerals are thought to be required by catfish. Deficiency signs and quantitative requirements of catfish for macro- and microminerals are summarized in Table 5.11.

TABLE 5.11.
Mineral Deficiency Signs and Minimum Dietary Levels Required to Prevent Deficiency Signs in Catfish

Mineral	Deficiency Signs	Requirements
Macrominerals		
Calcium[a]	Reduced bone ash	<0.1%, 0.45%
Phosphorus[b]	Reduced bone mineralization	0.45%
Magnesium	Muscle flaccidity, sluggishness, reduced bone, serum, and whole body magnesium	0.04%
Sodium, potassium, and chloride	Not determined	Not determined
Sulfur	Not determined	Not determined
Microminerals		
Cobalt	Not determined	Not determined
Iodine	Not determined	Not determined
Zinc[c]	Reduced serum zinc and serum alkaline phosphatase activity, reduced bone zinc and calcium concentrations	20 ppm
Selenium	Reduced liver and plasma selenium-dependent glutathione peroxidase activities	0.25 ppm
Manganese[c]	None	2.4 ppm
Iron	Reduced hemoglobin, hematocrit, erythrocyte count, reduced serum iron and transferrin saturation levels	20 ppm
Copper	Reduced hepatic copper-zinc superoxide dismutase, reduced heart cytochrome coxidase activities	4.8 ppm

[a]Deficiency cannot be demonstrated in catfish reared in water containing sufficient calcium.
[b]Requirement expressed on an available basis.
[c]Requirement increases in presence of phytic acid.
Note: Anorexia, reduced weight gain, and mortality are not listed as deficiency signs since they are common deficiency signs of several minerals. Minerals listed as not determined are assumed to be required.
Source: Adapted from Robinson (1989).

Macrominerals

Calcium. Calcium is an essential mineral for all animals. Calcium is needed for bone formation, blood clotting, muscle and nerve function, and osmoregulation and as an enzymatic cofactor. Calcium is found in catfish in higher concentrations than other minerals. A 1-ounce (30-gram) catfish contains about 2.5 percent calcium. Catfish fingerlings contain about 21 percent bone calcium (expressed on a fat-free, dry weight basis).

Calcium deficiency, which is characterized by reduced weight gain and bone ash, can be produced in catfish reared in calcium-free water. Catfish reared in water containing as little as 14 ppm calcium and fed a diet essentially void in calcium did not exhibit signs of deficiency. Apparently, catfish can meet their calcium requirement from the water. Because the water used in commercial catfish culture usually contains ample levels of calcium and the feed ingredients contain considerable calcium, it is not necessary to supplement catfish feeds with calcium.

Phosphorus. Phosphorus is found in all tissues, but the majority is associated with calcium in bone and is required for normal bone formation. Organic phosphates serve functions in phospholipids, coenzymes, deoxyribonucleic and ribonucleic acids, high-energy compounds, and various metabolic intermediates. Inorganic phosphates also serve as biological buffers. As a result of these functions, phosphorus has a role in protein, lipid, and carbohydrate metabolism and in the metabolic processes in various tissues.

Catfish fed phosphorus-deficient diets exhibit decreased weight gain, feed efficiency, and bone mineralization. Catfish cannot meet their phosphorus needs from the water because phosphorus concentrations in natural waters are generally low. The dietary requirement of catfish for phosphorus is approximately 0.4 to 0.5 percent available phosphorus. Since one-third to two-thirds of the phosphorus from natural feed ingredients used in commercial catfish feeds is biologically unavailable, a phosphorus supplement is used in catfish feeds.

Calcium/phosphorus ratio. The dietary calcium/phosphorus ratio (Ca:P) does not appear to be critical in catfish diets as it is in diets of certain terrestrial animals. The optimum Ca:P ratio for chickens is near 1:1. Catfish have been fed diets varying widely in Ca:P ratio without any detrimental effects. It appears that fish can regulate calcium via the gills and thus the Ca:P ratio may be unimportant in fish feeds, provided calcium in the water is adequate.

Magnesium. Magnesium is found throughout the body; about 60 percent is in bone and the remainder in various tissues. Magnesium is essential for proper muscle and nerve function and osmoregulation and is a cofactor for various enzymes.

Catfish fed magnesium-deficient diets exhibit decreased weight gain, anorexia, sluggishness, muscle flaccidity, increased mortality, and lower concentrations of whole body, serum, and bone magnesium. About 0.04 percent dietary magnesium is needed for normal growth and prevention of deficiency in catfish. Magnesium is abundant in most feed ingredients (particularly plant sources); thus a dietary supplement is not generally needed in commercial catfish feeds.

Sodium, potassium, and chloride. Sodium, potassium, and chloride are the most abundant electrolytes found in body fluids and soft tissues. Sodium and chloride are the primary ions found in extracellular fluids, whereas potassium is a major ion found in the cells. These ions are important in maintaining osmotic pressure and acid-base balance as well as water balance.

A dietary requirement has not been established for sodium, potassium, or chloride for catfish, but they are presumed to be required. Dietary levels of NaCl up to 2 percent do not affect catfish growth. Since these ions are generally abundant in natural waters and in feedstuffs, it is not necessary to supplement commercial catfish feeds with either of them.

Sulfur. Sulfur is largely found in organic compounds in the body, such as proteins, glutathione, heparin, taurine, and chondroitin sulfates. Some animals can utilize inorganic sulfate to meet part of their sulfur amino acid requirement, but catfish do not. A dietary requirement for sulfur has not been determined for catfish. Sulfur supplements are not added to commercial catfish feeds because catfish apparently meet their needs from sulfur inherent in feed ingredients.

Microminerals

Cobalt. Cobalt is an essential part of vitamin B_{12}. Catfish can synthesize vitamin B_{12} in the intestine if cobalt is present. No other function of cobalt in catfish has been identified. It is added to commercial catfish feeds to ensure adequacy, although no quantitative dietary requirement has been established.

Iodine. Iodine is part of the thyroid hormones thyroxine and triiodothyronine, which regulate metabolic rate. Iodine deficiency results in thyroid hyperplasia (goiter) in fish. Although the iodine requirement for catfish has not been determined, it is added to commercial catfish feeds. Fish may be able to take iodine from the water, but concentrations in natural waters are generally rather low.

Zinc. Zinc exerts its physiological effects as a component of numerous metalloenzymes. Over 80 zinc-containing enzymes and proteins have been identified. Thus zinc has many functions, including a role in protein, nucleic acid, lipid, carbohydrate, and mucopolysaccarede metabolism. Zinc also functions in carbon dioxide transport and hydrochloric acid secretion in the stomach as a component of carbonic anhydrase. It also appears to have a role in the immune response.

Catfish fed zinc-deficient diets exhibit reduced weight gain, appetite, serum zinc levels, serum alkaline phosphatase activity, and zinc and calcium levels in bone. The amount of dietary zinc needed for normal growth and prevention of deficiency signs in catfish is 20 ppm. In diets that contain considerable levels of phytic acid, such as practical catfish feeds, a higher level (150 ppm) of supplemental zinc is needed, because phytic acid reduces the bioavailability of zinc.

Selenium. The chief function of selenium is as part of the enzyme glutathione peroxidase, which, in conjunction with vitamin E, serves as an antioxidant to protect cells and membranes from oxidation.

Selenium deficiency in catfish causes a reduction in weight gain and activity of plasma selenium-dependent glutathione peroxidase. The dietary selenium requirement for catfish is 0.25 ppm. Levels of selenium above 15 ppm are toxic to catfish. Commercial catfish feeds should contain a selenium supplement.

Manganese. Manganese serves as a cofactor in several enzyme systems. Thus it is involved in the synthesis of urea from ammonia, oxidation of glucose, and metabolism of amino acid and fatty acid.

Manganese deficiency has not been demonstrated in catfish. A dietary level of 2.4 ppm manganese is adequate for normal growth of catfish. Commercial catfish feeds are generally supplemented with manganese at a level exceeding the suggested requirement because phytic acid reduces manganese availability.

Iron. Iron is found in greatest quantities in hemoglobin; smaller amounts are found in myoglobin and cytochrome enzymes as well as various other enzymes. Thus iron is essential for oxygen transfer and electron transport.

Iron deficiency signs in catfish include a decrease in the following: weight gain, feed efficiency, hemoglobin, hematocrit, erythrocyte count, and serum iron and transferrin saturation values. Approximately 20 ppm dietary iron is necessary to prevent deficiency signs. Iron may be taken from the water by fish, but concentrations in water are usually rather low because iron is readily precipitated. Iron is found in feedstuffs, but its bioavailability is not known for catfish. Catfish feeds should be supplemented with iron.

Copper. Copper is a part of a large number of enzyme systems, such as cytochrome oxidase, involved in oxidation and reduction reactions. Copper has a role in iron absorption, maturation of red blood cells, bone and connective tissue formation, pigmentation, reproduction, and function of the heart.

Gross deficiency signs are not apparent in catfish fed diets containing copper at 0.8 ppm. However, 4.8 ppm dietary copper (4.0 ppm supplemental plus that in the feed ingredients) is needed to prevent a reduction in the activities of the enzymes hepatic copper-zinc superoxide dismutase and heart cytochrome C oxidase. Commercial catfish feeds should be supplemented with copper.

DIGESTION

Digestion is generally thought of as a series of processes taking place in the gastrointestinal tract that prepare ingested food for absorption. These processes involve mechanical reduction of particle size and solubilization of food particles by enzymes, pH, or emulsification. Once digestion has occurred, absorption (the uptake of small molecules from the gastrointestinal tract into the blood or lymph) may occur by diffusion, active transport, or pinocytosis (engulfment).

Specific digestive processes have not been extensively studied in catfish, but digestion in catfish is presumed to be similar to that of other simple-stomach animals. The digestive tract of catfish is similar to that of other simple-stomach animals, divided into the mouth, pharynx, esophagus, stomach, and intestine as well as the accessory digestive organs, the pancreas, liver, and gall bladder. The pH of the stomach and intestine of catfish ranges from 2 to 4 and 7 to 9, respectively. Digestive enzymes such as trypsin, chymotrysin, lipase, and amylase have been identified in catfish intestine.

A brief description of the digestive processes as well as methods to determine digestibility of feeds in catfish is presented. If more detail is desired, any book on animal nutrition should provide ample information on digestion in simple-stomach animals. Also, the anatomy and histology of the digestive system of the channel catfish have been described by Grizzle and Rogers (1976).

Digestive Organs

Mouth, pharynx, and esophagus. The mouth of the catfish is large and is used for ingesting food. It is distinct from the pharynx and is bound laterally by gill slits. The mouth and pharynx contain numerous small toothlike abrasive pads that serve to reduce the size of food particles. There are no salivary glands. The esophagus extends from the pharynx to the stomach and is separated from the stomach by a cardiac sphincter. The esophagus is a passage for food to reach the stomach. A mucus is secreted from glands found in the esophagus that aids the passage of food. The cardiac sphincter prevents food from backing out of the stomach as well as separating water from food in the stomach.

Stomach. Catfish have what is termed a true stomach in that they secrete hydrochloric acid and pepsinogen. Digestion of various nutrients begins in the stomach. The stomach is J-shaped and the ascending limb terminates at the pyloric sphincter, which separates the stomach from the intestine. The pyloric sphincter keeps stomach contents from entering the intestine until the proper time as well as preventing intestinal contents from backing into the intestine. The stomach is divided into two regions, the fundic and pyloric. The fundic region contains gastric glands that secrete hydrochloric acid.

Intestine. The intestine is not divided into a small and large intestine as is that of other simple-stomach animals. The intestine is shorter than the length of the catfish's body and begins at the pyloric sphincter. Although there are no distinct divisions of the catfish intestine, the following divisions have been suggested: pyloric intestine, which is the first loop posterior to the stomach; middle intestine, the coiled portion; and rectal intestine, the portion posterior to the last loop. An intestinal sphincter separates the middle and rectal intestines. Most digestion and absorption of nutrients occur in the intestine. Digestive enzymes and bile enter the pyloric intestine via the common bile and pancreatic ducts.

Accessory digestive organs. The liver, pancreas, and gall bladder are involved in the digestive process. The liver produces bile that is stored in the gall bladder. The pancreas provides digestive enzymes that are released into the intestine. The pancreas is not a discrete organ;

rather it is diffuse and is scattered in the mesenteries and on the intestinal surface as well as in the liver and spleen.

Digestion and absorption of nutrients. Pepsinogen is converted to the proteolytic enzyme pepsin in the presence of hydrochloric acid in the stomach. This process initiates protein digestion. Proteins are reduced to polypeptides that are further digested in the intestine. Proteolytic enzymes in the intestine—trypsin, chymotrypsin, carboxypeptidases, aminopeptidases, tripeptidases, and dipeptidases—hydrolyze protein to short peptides and free amino acids for absorption. Trypsinogen and chymotrypsinogen are secreted into the intestine from the pancreas, where they are converted into active proteolytic enzymes. Other proteolytic enzymes are released from the intestinal mucosa.

Other enzymes released from the pancreas, pancreatic amylase and lipase, function in the digestion of carbohydrates and lipids, respectively. Carbohydrates are broken down to monosaccharides for absorption. Lipase breaks down lipids to their component fatty acids, monoglycerides, or glycerol. Proper digestion and absorption of lipid are dependent on bile salts for emulsification. Bile is produced in the liver, stored in the gall bladder, and released into the intestine.

Water-soluble vitamins are absorbed from the intestine. Fat-soluble vitamins are also absorbed, along with the lipids. Minerals are absorbed from the intestine either free, as mineral complexes, or bound to proteins. Bile salts, certain minerals, and water are apparently absorbed from the posterior intestine in catfish.

Determination of nutrient digestibility. Knowledge of the digestibility of nutrients enables the nutritionist to formulate feeds on an available nutrient basis, which allows for more precision in meeting nutrient needs on a cost-effective basis. Digestion coefficients provide insight into the usefulness of feedstuffs and finished feeds. Digestion coefficients are easily determined for terrestrial animals, but determining nutrient digestion for aquatic animals poses unique problems. Nutrients or indicators can be solubilized and lost from feeds or fecal material collected from the water. Also, since the catfish's intestine is not well differentiated, absorption occurs along most of its length, and this poses a problem in selecting the correct site for collecting fecal material. The rectal intestine is the section from which the fecal material should be taken for determination of digestion coefficients.

Nutrient digestibility can be determined by direct or indirect quantitative measurement of the amount of nutrient consumed and then eliminated in the feces. Endogenous nitrogenous materials eliminated via the feces are usually not accounted for in digestion trials with catfish, because a nitrogen-free diet would have to be fed to correct for endogenous nitrogen and such diets are not readily consumed by catfish. Some researchers use a low-nitrogen diet to estimate true digestibility. However, most digestion coefficients determined for catfish are apparent digestibility coefficients; that is, no correction has been made for endogenous nutrients eliminated in the fecal material. True digestibility coefficients offer few advantages over apparent digestibility coefficients because endogenous contributions to fecal matter are associated with consumption of food and thus are proportional to feed intake.

Total collection of fish fecal material is difficult; thus digestibility trials with catfish have generally used an indirect method that normally involves feeding of diets containing an inert indicator such as chromic oxide. In such studies, the fish are fed a diet containing the indicator and the fecal material is collected several hours later either from the water or directly from the rectal intestine by catheterization and aspiration of the material or by sacrificing and dissection. The following formula is used to calculate apparent nutrient digestibility:

$$100 - 100 \times \frac{\% \text{ chromic oxide in feed}}{\% \text{ chromic oxide in feces}} \times \frac{\% \text{ nutrient in feces}}{\% \text{ nutrient in feed}} \quad (5.4)$$

In some cases, catfish will not consume the test diet containing the ingredient under study. Thus the fish are force-fed or a standard reference diet is used. Force-feeding catfish appears to work well for determining protein digestion or amino acid availabilities, but the stress of force-feeding reduces energy digestion. To determine energy digestion coefficients, a standard reference diet (Table 5.12) that is voluntarily consumed should be used. Test diets using 70 percent of the standard reference diet and 30 percent of the test ingredient are fed and digestion coefficients calculated by the following formula:

$$\frac{\text{Digestion}}{\text{coefficient}} = \frac{100}{30} \frac{\text{digestion coefficient of test diet}}{} - \frac{70}{100} \frac{\text{digestion coefficient of reference diet}}{} \quad (5.5)$$

Digestion Coefficients

Protein. Catfish protein digestibility coefficients (Table 5.13) are fairly high, particularly for good-quality protein sources. Some variation in reported values exists because of differences in experimental methodology and the impact of processing on protein quality of various feedstuffs. Different sites along the intestinal tract and different methods for fecal collection have been used. Also, single- or mixed-protein feeds have been used. Processing of feed ingredients affects protein digestibility. For example, the protein digestibility of soybean meal and fish meal is reduced if it is improperly cooked.

TABLE 5.12.
Standard Reference Diet Used in Digestibility Trials with Catfish

Ingredient	International Feed Number	Percentage
Soybean meal	5-04-604	48.5
Corn grain	4-02-935	38.0
Menhaden fish meal	5-02-009	9.1
Wheat grain	4-05-268	3.2
Vitamin mix[a]		0.1
Mineral mix[a]		0.1
Chromium oxide		1.0

[a]Provides all recommended vitamins and minerals.
Source: Adapted from Wilson and Poe (1985).

TABLE 5.13.
Average Apparent Protein Digestibility Coefficients

Feedstuff	International Feed Number	Percentage Digestibility			
		(1)	(2)	(3)	(4)
Alfalfa meal (17%)	1-00-023	12	13		
Blood meal	5-00-381	23			74
Corn, grain	4-02-935		60	97	
Corn, cooked			66		
Corn, dist. sol.	5-28-241	67			
Corn, gluten meal	5-04-900	80			92
Cottonseed meal	5-01-621	76	81	83	
Fish, anchovy meal	5-01-985	85	90		
Fish, menhaden meal	5-02-009	74	87	85	70, 86
Meat meal	5-00-385	40			
Meat meal with bone	5-00-388		75	61	82
Peanut meal	5-03-649			74	86
Poultry, by-product meal	5-04-798	27			65
Poultry, feather meal	5-03-795	63	74		
Rice, bran	4-03-928	71		73	
Rice, mill feed	1-03-941			63	
Soybean meal (44%)	5-04-604		77		
Soybean meal (48%)	5-04-612	72	84	97	85
Wheat, bran	4-05-190		82		
Wheat, grain	4-05-268		84	92	
Wheat, shorts	4-05-201		72		

Source: (1) Hasting (1966); (2) Cruz (1975); (3) Wilson and Poe (1985); (4) Brown et al. (1985).

TABLE 5.14.

Percentage Average Apparent Amino Acid Availabilities for Various Feedstuffs

Amino Acid	Peanut[a] Meal	Soybean[a] Meal	Meat and Bone[a] Meal	Menhaden Fish[a] Meal	Corn[b]	Cottonseed[a] Meal	Rice[b] Bran	Wheat[b] Middlings
Ala	88.9 ± 0.5	79.0 ± 2.8	70.9 ± 3.0	87.3 ± 1.5	78.2 ± 1.0	70.4 ± 1.3	82.0 ± 0.9	84.9 ± 0.9
Arg	96.6 ± 0.2	95.4 ± 0.7	86.1 ± 3.4	89.2 ± 0.7	74.2 ± 0.2	89.6 ± 0.2	91.0 ± 1.1	91.7 ± 0.5
Asp	88.0 ± 0.4	79.3 ± 1.4	57.3 ± 0.5	74.1 ± 1.8	53.9 ± 3.1	79.3 ± 0.5	82.4 ± 0.7	82.8 ± 2.7
Glu	90.3 ± 1.0	81.9 ± 1.0	72.6 ± 3.8	82.6 ± 0.1	81.4 ± 1.6	84.1 ± 0.3	88.8 ± 0.4	92.3 ± 0.5
Gly	78.4 ± 0.3	71.9 ± 2.8	65.6 ± 4.7	83.1 ± 1.2	53.1 ± 3.2	73.5 ± 0.6	80.0 ± 0.9	85.2 ± 0.4
His	83.0 ± 0.6	83.6 ± 1.2	74.8 ± 2.0	79.3 ± 2.2	78.4 ± 0.6	77.2 ± 2.0	70.4 ± 2.1	87.4 ± 1.2
Ile	89.7 ± 0.2	77.5 ± 4.0	77.0 ± 5.2	84.8 ± 1.0	57.3 ± 3.4	68.9 ± 0.6	81.4 ± 0.9	81.8 ± 1.9
Leu	91.9 ± 0.1	81.0 ± 3.4	79.4 ± 3.1	86.2 ± 0.6	81.8 ± 1.0	73.5 ± 0.7	84.1 ± 0.9	84.6 ± 1.3
Lys	85.9 ± 0.5	90.9 ± 1.3	81.6 ± 2.6	82.5 ± 1.2	69.1 ± 4.8	66.2 ± 1.2	81.3 ± 0.3	85.9 ± 2.1
Met	84.8 ± 0.2	80.4 ± 2.1	76.4 ± 3.7	80.8 ± 0.3	61.7 ± 4.9	72.5 ± 0.9	81.9 ± 0.8	76.7 ± 2.4
Phe	93.2 ± 0.3	81.3 ± 4.5	82.2 ± 3.0	84.1 ± 1.1	73.1 ± 7.2	81.4 ± 0.4	82.9 ± 3.5	87.2 ± 1.1
Pro	88.0 ± 1.3	77.1 ± 2.1	76.1 ± 4.0	80.0 ± 0.6	78.4 ± 1.0	73.4 ± 0.3	79.5 ± 1.4	88.3 ± 0.5
Ser	87.3 ± 0.4	85.0 ± 0.5	63.7 ± 0.1	80.7 ± 1.9	63.9 ± 1.4	77.4 ± 1.4	82.0 ± 0.6	83.0 ± 2.3
Thr	86.6 ± 0.5	77.5 ± 1.3	69.9 ± 3.2	83.3 ± 1.7	53.9 ± 3.9	71.8 ± 0.4	77.3 ± 3.6	78.8 ± 3.2
Tyr	91.4 ± 0.3	78.7 ± 2.6	77.6 ± 3.7	84.8 ± 1.4	68.7 ± 5.0	69.2 ± 2.6	86.7 ± 3.2	83.0 ± 2.0
Val	89.6 ± 0.2	75.5 ± 3.7	77.5 ± 2.9	84.0 ± 0.6	64.9 ± 4.6	73.2 ± 0.3	83.2 ± 0.6	84.5 ± 0.7
Ave	88.4	81.0	74.3	82.9	68.3	75.1	82.2	84.9

[a]Determined after ad libitum feeding test diets.
[b]Determined after force-feeding test diets.
Note: Mean ± SEM.
Source: Wilson et al. (1984).

TABLE 5.15.

Average Apparent Digestion Coefficients of Lipids and Carbohydrates

Feed Sources	International Feed Number	Percentage Digestibility	
		Lipid	Carbohydrate
Fish, oil		97	
Fish, anchovy meal	5-01-985	97	
Meat meal with bone	5-00-388	77	
Poultry, feather meal	5-03-795	83	
Soybean meal (44%)	5-04-604	81	
Cottonseed meal, solvent extracted	4-01-621	81	17
Wheat, grain	4-05-268	96	59
Uncooked corn (30% of diet)	4-02-935	76	66
Uncooked corn (60% of diet)			59
Cooked corn (30% of diet)		96	78
Cooked corn (60% of diet)			62
Glucose (30% of diet)			88
Glucose (60% of diet)			92
Dextrin (30% of diet)			73
Dextrin (60% of diet)			48

Source: Cruz (1975).

Amino acids. Amino acid availability coefficients (Table 5.14) indicate that the availability of certain amino acids is variable within and among the various proteins tested. To use protein digestibility coefficients to formulate catfish feeds, it must be assumed that all amino acids are equally available. It is obvious that this is not the case; thus it would be more proper to use amino acid availability data to formulate catfish feeds. However, there is a paucity of amino acid availability data, and those presented in Table 5.14 are based on relatively few samples and should be considered as preliminary.

Lipids. Lipid digestion coefficients (Table 5.15) demonstrate that lipids are highly digestible by catfish. Lipid sources containing appreciable amounts of unsaturated lipids appear to be more highly digested than sources containing predominantly saturated lipids. Also, highly saturated lipids (beef tallow) may be poorly digested at low water temperatures.

Carbohydrates. Carbohydrates are not as digestible by catfish as lipids (Table 5.15). Glucose is more highly digestible by catfish than

TABLE 5.16.

Average Percentage Apparent Digestible Energy Coefficients

Feedstuff	International Feed Number	Percentage Digestibility (1)	(2)
Alfalfa meal	1-00-023	16	
Corn, grain	4-02-935	26	57
Corn (cooked)		59	
Cottonseed meal	5-01-621	56	80
Meat meal with bone	5-00-388	81	76
Fish, menhaden meal	5-02-009	85	92
Peanut meal	5-03-650		76
Poultry, feather meal	5-03-795	67	
Rice, bran	4-03-928		50
Rice, mill feed	1-03-941		14
Soybean meal (44%)	5-04-604	56	
Soybean meal (48%)	5-04-612		72
Wheat, bran	4-05-190	56	
Wheat	4-05-268	60	63

Source: (1) Cruz (1979); (2) Wilson and Poe (1985).

dextrin or starch. Dietary carbohydrate level also affects carbohydrate digestion. Starch and dextrin digestion decrease as the dietary level increases from 30 to 60 percent. Cooking improves the digestibility of corn starch.

Energy. Fish meal appears to be a good source of digestible energy for catfish (Table 5.16). Oilseed meals are also relatively good sources of energy for catfish. Different findings for digestibility of the same ingredients in various studies are due in part to differences in experimental methods and to processing of feed ingredients as well as other unexplained factors.

Minerals. Phosphorus availability data (Table 5.17) indicate that inorganic sources of phosphorus are highly digestible by catfish. In general, phosphorus availabilities from plant and animal sources are 25 to 30 and 40 percent, respectively. Most of the phosphorus in plants is in the form of phytic acid, which requires phytase enzymes to free the phosphorus. Phytase enzymes are not present in catfish tissues.

TABLE 5.17.

Average Percentage Apparent Availability of Phosphorus

Source	International Feed Number	Availability
Phosphates		
Sodium phosphate, mono basic	6-04-288	90
Calcium phosphate, mono basic	6-01-082	94
dibasic	6-01-080	65–80
Fish meals		
Anchovy	5-01-985	40
Menhaden	5-02-009	39
Purified protein sources		
Egg albumin		71
Casein	5-01-162	90
Plant sources		
Wheat, middlings	4-05-205	28
Corn, grain	4-02-935	25
Soybean meal, with hulls	5-04-604	50
Soybean meal, dehulled	5-04-612	29–54

Source: All values except those for purified protein sources and dehulled soybean meal (Lovell 1978); values for purified sources and dehulled soybean meal (Wilson et al. 1982).

REFERENCES

Brown, P. B., R. J. Strange, and K. R. Robbins. 1985. Protein digestion coefficients for yearling channel catfish fed high protein feedstuffs. *Progressive Fish-Culturist* 47:94–97.

Church, D. C., and W. G. Pond. 1982. *Basic Animal Nutrition and Feeding.* New York: John Wiley & Sons.

Cruz, E. M. 1975. Determination of nutrient digestibility in various classes of natural and purified feed materials for channel catfish. Ph.D. Dissertation, Auburn University, Alabama.

Grizzle, J., and W. Rogers. 1976. *Anatomy and Histology of the Channel Catfish.* Auburn: Alabama Agricultural Experiment Station.

Halver, J. E. 1989. *Fish Nutrition.* San Diego: Academic Press, Inc.

Hastings, W. H. 1966. Progress in sport fisheries research, 1966: Feed formulation; physical quality of pelleted feed; digestibility. U.S. Bureau of Sport Fisheries and Wildlife, Washington D.C. Publ. 39:137–141.

Lovell, R. T. 1978. Dietary phosphorus requirement of channel catfish (*Ictalurus punctatus*). *Transactions of the American Fisheries Society* 197:617–621.

Lovell, R. T. 1989. *Nutrition and Feeding of Fish.* New York: Van Nostrand Reinhold.

References

Maynard, L., J. Loosli, H. Hintz, and R. Warner. 1979. *Animal Nutrition.* New York: McGraw-Hill.
National Research Council. 1977. *Nutrient Requirements of Warmwater Fishes.* Washington D.C.: National Academy of Sciences.
National Research Council. 1983. *Nutrient Requirements of Warmwater Fishes and Shellfishes.* Washington D.C.: National Academy of Sciences.
Robinson, E. H. 1989. Channel catfish nutrition. *Reviews in Aquatic Sciences* 1:365–391.
Robinson, E. H., and R. P. Wilson. 1985. Nutrition and feeding. In *Channel Catfish Culture,* ed. C. S. Tucker, pp. 323–404. Amsterdam: Elsevier.
Santiago, J. B. 1985. Amino acid requirements of Nile tilapia. Ph.D. Dissertation, Auburn University, Alabama.
Wilson, R. P., and W. E. Poe. 1985. Apparent digestible protein and energy coefficients of common feed ingredients for channel catfish. *Progressive Fish-Culturist* 47:154–158.
Wilson, R. P., and E. H. Robinson. 1982. *Protein and amino acid nutrition for channel catfish,* Information Bulletin 25. Mississippi State, Mississippi: Mississippi Agricultural and Forestry Experiment Station.
Wilson, R. P., E. H. Robinson, D. M. Gatlin III, and W. E. Poe. 1982. Dietary phosphorus requirement of channel catfish. *Journal of Nutrition* 112:1197–1202.

PART II
CULTURAL PRACTICES

The next eight chapters constitute the "how-to" of catfish farming. It will become evident as these chapters are read that seldom does only one correct technique or practice exist. Catfish farmers have developed specific practices to fit their own individual needs, as a result there are myriad "accepted practices." Also, cultural practices continually change as the industry evolves toward a large-scale national agribusiness. Practices that are common today may be outdated in the near future. We have tried to offer concrete suggestions wherever possible, but quite often we have opted to discuss the alternatives and have left the final decision to the farmer. Most of the discussion that follows is focused on pond cultural practices because growing fish in ponds is by far the most common way catfish are produced. Chapter 13 addresses the culture of channel catfish in systems other than ponds.

CHAPTER 6
Breeding

The performance of modern livestock and poultry far exceeds that of their ancestors. Improved production efficiencies are largely attributable to breeding programs based on genetic theory. Commercial channel catfish culture lags far behind most other farm animal culture in using genetic improvement programs to increase productivity, despite indications from research (summarized by Dunham and Smitherman 1987) that considerable progress can be made. Failure to implement long-term genetic improvement programs is due in large part to the relatively recent development of the channel catfish farming industry. Catfish producers and researchers have focused primarily on short-term programs in nutrition, water quality management, and disease control to provide immediate increases in production. As the industry has matured and profit margins have decreased, the benefits of genetic improvement have become more obvious. Production efficiency cannot be optimized unless the biological potential of the fish is optimized.

THE NEED FOR PLANNED BREEDING PROGRAMS

Selection for certain phenotypes (traits) occurs whenever brood stock are chosen from a population. If the choice of stock is made according to a plan, progress can be made in improving the performance of subsequent generations. If little thought is given to the choice of brood stock, the culturist may unintentionally select for undesirable traits.

Unintentional selection is common in commercial catfish farming because many catfish fingerling producers have no planned breeding program. An excellent example of unintentional selection is the manner in which many commercial channel catfish fingerling producers obtain brood stock. Brood fish commonly come from food fish growout ponds that contain large fish. These fish are available in quantity and at relatively low prices. The culturist has the impression that the fish will be good brood stock because they are large, appealing fish when compared to the rest of the pond population. However, most growout ponds in the lower Mississippi River valley are operated for years without draining, and after several cycles of harvesting and restocking, it is impossible to know the age, origin, or history of the individual fish in the pond. The larger fish are usually older fish that evaded capture rather than fast-growing fish. When these fish are chosen as brood stock, the culturist may well be unintentionally selecting for slow-growing fish that are difficult to harvest.

Unintentional selection can even occur when the culturist attempts to select fast-growing fish from a population of fish of the same year class. Fry from spawns produced over several weeks are usually stocked into the same nursery pond; differences in size after several months of growth may result from the slight differences in age and not differences in genetic potential for growth. Choosing the larger fish may inadvertently select for slower-growing fish. Also, if fry from only a few spawns are used to stock a nursery pond, the larger fish may be mostly sibs (brothers and sisters) because they came predominantly from the oldest spawn. Selection of siblings for brood stock could result in inbreeding or unintentional selection for undesirable traits.

The results of unintentional selection for undesirable traits can be insidious; decreased productivity is difficult to detect in large-scale commercial culture because actual fish growth rates are almost never measured. Realized on-farm yield is affected by market constraints (being able to harvest and sell fish only when the processor wants the fish), disease losses, bird predation, and other factors. Furthermore, maintaining accurate fish inventory records is difficult for most commercial ponds. Consequently, the farmer has no reliable measure of the population's performance.

Unintentional selection can have severe consequences for brood stock management in hatcheries raising game fish. Fish raised in these hatcheries may be selected for good performance in the hatchery with the unintentional consequence of culling fish that perform well in the wild. Because it is difficult to know what phenotypes should be selected in hatchery stock to assure good performance in the wild, it is important to establish a program of no selection to maintain the average genetic potential of the population. A program of no selection does not mean that no effort is made to obtain proper brood stock; it is a conscious, highly planned effort to prevent unintentional selection for potentially undesirable traits. Brood fish must be chosen to represent as much of the phenotypic variation of the population as possible, and records must be maintained to verify that inadvertent selection has not occurred.

Although expenditure of resources on a program of no selection is often warranted when managing hatchery populations of game fish, effort is better spent on a positive breeding program to improve productivity when managing populations of channel catfish raised as food. Also, the best way to prevent inbreeding and other problems associated with unintentional selection is through the use of an orderly, planned breeding program.

Although unplanned breeding programs are to be avoided, unintentional selection is not necessarily bad in channel catfish populations. Through several generations, fish that are poor at spawning, inordinately susceptible to disease, or intolerant of poor environmental conditions are eliminated from the population. Brood fish are thus inevitably selected from fish that perform relatively well under food fish culture conditions. This process is an important part of domestication. Domesticated strains of channel catfish generally have better survival and growth than wild fish under commercial culture conditions because selection for these traits is almost unavoidable.

GUIDELINES FOR A MINIMAL BREEDING PROGRAM

Channel catfish farming is a laborious, time-consuming activity, and many farmers feel that the added effort and facilities required to implement a planned breeding program of selection or crossbreeding are not justified. However, farmers unwilling to develop even a simple planned breeding program should realize that imprudent selection of brood stock can cause great harm to long-term productivity and profits. A little extra effort and use of common sense can help the farmer avoid the pitfalls of unintentional selection. As a minimal breeding program, the following guidelines can be used:

1. Choose brood stock from domestic stock. Wild fish often are

unreliable spawners in captivity, and the progeny may be susceptible to diseases or grow slowly in culture environments.
2. Select brood fish from stocks that are known to perform well under commercial culture conditions. This is more difficult than it sounds because few field trials have been conducted.
3. Do not mistake large fish for fast-growing fish. Try to select brood fish from fish of known age; however, selection of the largest fish even from a pond containing a single year class of fish could inadvertently result in selection of mostly sibs.
4. Prevent inbreeding by obtaining fish that will become brood fish from as many different spawns as possible. Initial stock should be obtained from several different ponds or, preferably, from unrelated stocks from different locations.
5. If replacement brood stock come from progeny produced on the farm, they should come from at least fifty random matings. If this is not possible because the breeding population is small (fewer than seventy-five to one hundred pair of fish), enrich bloodlines by adding unrelated stock as part of the brood fish replacement program.
6. Keep accurate records of spawning success, egg hatchability, fry survival, and, if possible, growth rate of fingerlings and food-sized fish. A trend of decreasing performance over time may indicate inbreeding or other problems related to disorderly selection.

BREEDING FOR QUALITATIVE TRAITS

Qualitative traits exhibit an "all-or-nothing" expression of phenotype. At present, the only potentially valuable qualitative trait of channel catfish is albinism. The flesh of albino channel catfish is lighter in color than that of normally pigmented fish and is preferred by some consumers (Heaton et al. 1973). The genetics of albinism in channel catfish is discussed in the section Qualitative Traits. Albinism is rare in wild channel catfish. Most albinism in commercial stocks appears to be related to the Marion strain of channel catfish, which has been widely distributed throughout the southeastern United States (Dunham and Smitherman 1984).

A few studies have compared the performance of albino and normal channel catfish (reviewed by Dunham and Smitherman 1987), but the results are contradictory. The most thorough evaluation of the relative performance of albino and normal channel catfish was conducted by Bondari (1984). His results showed that albino × albino brood fish spawn at a lower rate and produce smaller egg masses with smaller eggs than normal × normal fish of the same strain. Eggs of the albino × albino matings hatched at a lower rate and fry had lower survival than those produced by normal parents. Normally pigmented fish grew faster and had higher survival rates than albino fish when grown to about 0.25 pound in ponds. Survival and growth to food-fish size were not determined.

The ancestry of the Tifton strain of channel catfish used by Bondari (1984) indicates that the albino parents used to make the albino × albino matings were more likely to be inbred than the normal parents, and this may have affected the performance of the all-albino families (Dunham and Smitherman 1987). More recent studies (C. A. Goudie, personal communication) indicate that albino catfish may have stringent temperature requirements for spawning, but reproductive performance and growth are similar to those of normally pigmented fish of the same strain.

A breeding program to produce all-albino progeny is simple because fish with the recessive phenotype (albinism) breed true: matings between albino parents produce all-albino progeny. Thus, albino fish can be selected from a mixed color population and used in any convenient breeding program as long as only albino × albino matings are allowed. Performance of subsequent populations can then be improved through a planned program of selection. Estimates of heritability (h^2; see the section Quantitative Traits) made by Bondari (1984) for body weight and length were moderate to high for both albino and normal channel catfish, indicating that selection can be effective at increasing growth rates regardless of pigmentation.

BREEDING FOR QUANTITATIVE TRAITS

Quantitative traits (phenotypes) do not segregate individuals into discrete groups such as albino and normal pigmentation. Quantitative traits exhibit a range of expression and are measured rather than described qualitatively. Most production traits, such as body weight, feed conversion efficiency, body fat content, egg production, tolerance to environmental stressors, disease resistance, and seinability, are quantitative traits. The genetics of quantitative phenotypes are described in the section Quantitative Traits.

Catfish farmers can improve performance of their fish in three ways: (1) obtain and raise superior fish, (2) use selective breeding to develop superior fish, or (3) crossbreed strains or species to produce hybrids that often outperform either parental type. The most desirable program or combination of programs depends on the traits to be improved, the mechanism by which these traits are inherited, and the resources available to the breeder.

Obtaining Superior Strains

A *strain* is a group of fish sharing a similar history and possessing unique characteristics. Strains may come from a particular river system or may be produced in a hatchery breeding program. Domesticated strains are catfish grown on farms or hatcheries that are at least two breeding generations removed from the wild strain. Domestic strains generally outperform wild fish when raised in food-fish culture systems. Acquisition of a superior domestic strain is the fastest way to improve productivity.

In practice, however, obtaining a superior strain can be an uncertain procedure. Although several performance trials have been conducted, only a few strains have been evaluated and most of the strains presently are not easy to obtain in large numbers by commercial catfish farmers. Also, performance trials have been conducted at few locations and the relative performance of the strains may vary under different environmental conditions. Most of the commercial stocks used by catfish farmers have not been subjected to performance trials, and between-farm comparisons of performance of different commercial stocks are not valid because many factors other than genetic potential affect farm productivity. Until more research is conducted at different locations and an effective system is developed to maintain founder stocks and distribute sufficient seed stock to industry, acquisition of strains to improve productivity will be a hit-or-miss proposition. The establishment of a system of yield trial centers (Tave 1989) would be beneficial by providing farmers with the opportunity to assess the performance of commercial stocks.

Dunham and Smitherman (1984) describe many of the strains and stocks used in research and commercial production and summarize results of strain evaluations. The information in Table 6.1 illustrates differences for four quantitative traits among four of the more extensively tested strains of channel catfish (Auburn, Marion, Kansas, and Rio Grande). In addition to the traits listed in Table 6.1, differences among strains exist for resistance to bacterial and parasitic diseases, seinability, age of sexual maturity, reproductive performance, time of spawn-

TABLE 6.1.

Some Quantitative Phenotypes of Four Channel Catfish Strains

	Weight[a] (lb)	Feed Conversion Efficiency[b]	Dress-out[c] (%)	Susceptibility to Channel Catfish Virus[d] (Percentage Mortality)
Auburn	1.06	1.36	63	NT[e]
Marion	1.36	1.26	59	33
Kansas	1.53	1.26	59	NT
Rio Grande	0.97	1.42	64	72

[a]Weight after second summer of growth in ponds stocked at 3,000 fish/acre. *Source:* Chappell (1979).
[b]Pounds feed offered ÷ pounds weight gain. *Source:* Chappell (1979).
[c]Weight dressed fish (deheaded, skinned, and eviscerated) ÷ weight whole fish. *Source:* Dunham et al. (1983).
[d]Average percentage mortality of fingerling channel catfish experimentally infected with channel catfish virus. *Source:* Plumb et al. (1975).
[e]NT = not tested for this strain.

ing, and variation in length. Tolerance to environmental stressors probably also varies among strains.

Of the limited number of channel catfish strains evaluated thus far, the Kansas strain grows fastest and is most disease-resistant; the Rio Grande strain is the least disease-resistant. Kansas and Marion strain fish convert feed to flesh most efficiently, and the Marion strain is the easiest to harvest from ponds by seining. Auburn and Rio Grande strains have high dress-out percentages, due mainly to the relatively small head of fish of these strains. Fish of the Minnesota strain spawn earliest; the Rio Grande strain spawns latest. The Rio Grande strain matures sexually at 2 years and the Kansas strain at 4 years. Most other strains generally become sexually mature at 3 years.

Selection

Selection is a breeding program in which superior individuals are identified and used as future brood stock to produce progeny that are also superior. Two types of selection programs have been used to improve channel catfish populations: mass selection (also called individual selection) and family selection. In *mass selection*, the performance of all individuals is compared and selection is based on the performance of each regardless of parentage. In *family selection*, average performance of families is compared and whole families are selected or culled.

Selection programs are successful only if the genetic component responsible for the variation in phenotypes is passed from parents to offspring in a reliable manner. Only additive genetic variance (see the section Quantitative Traits) is transmitted to progeny in a reliable way and heritability (h^2) is an estimate of the proportionate contribution of additive genetic variance to the total phenotypic variance. Values for h^2 range from 0 to 1.0; large values for h^2 indicate that additive genetic variance is responsible for much of the phenotypic variation seen in the population and selection will be an efficient way to improve performance. It is generally assumed that phenotypes will be difficult to improve by selection if h^2 for that phenotype is less than about 0.15.

Heritability estimates for a particular phenotype are valid only for the particular population for which the estimate is made and only under the specific culture condition used to raise that population. Heritability estimates for growth rate, feed conversion, dress-out percentage, body fat content, and morphometrics (body shape) have been made for few populations or strains of channel catfish (summarized by Tave 1986). These heritability estimates can be used to infer response to selection for other populations of the same strain, but the culturist must realize that response to selection, and thus success in improving subsequent generations, will vary from population to population.

Generally, heritabilities are moderate to high for channel catfish growth rate, and selection to improve weight gain has been successful in every reported trial (Dunham and Smitherman 1987). As an example, one generation of mass selection resulted in increases in body weight of 17 percent for Rio Grande, 18 percent for Marion, and 12 percent for Kansas strains after 18 months of growth. Realized heritabilities for increased body weight were as follows: Rio Grande, 0.24; Marion, 0.50; Kansas, 0.33 (Dunham and Smitherman 1983a).

Heritability estimates also have been made for other traits, but selection programs have not been attempted. Low to moderate heritabilities for feed conversion efficiency (Burch 1986) and dress-out percentage (El-Ibiary and Joyce 1978) indicate that these traits will be more difficult to improve by selection than growth rate. A relatively high estimated heritability (0.61) for fat content (Reagan 1979) indicates that selection for leaner fish could be successful. Heritability estimates are moderate (0.15 to 0.40) for body conformation phenotypes such as body depth, girth, and head weight (El-Ibiary and Joyce 1978).

Correlated responses to selection. If a trait responds to selection, other traits may also change or they may remain the same. A positive correlation between two traits means that selection for one trait increases the other; a negative correlation means that selection for one decreases the other. When selecting for one trait, it is important to monitor other economically important traits to ensure that none of the correlated responses are undesirable. For example, selection for increased growth rate may not be desirable if a correlated response is decreased resistance to disease.

Selection for body weight in three strains of channel catfish increased fecundity (Kansas and Rio Grande strains) and increased disease resistance (Marion and Rio Grande strains). Selection did not affect dress-out percentage, seinability, spawning date, spawning rate, hatchability of eggs, or fry survival (Dunham 1981). Increased body weight is positively correlated with feed consumption and negatively correlated with feed conversion values (Dunham 1981; Burch 1986). Selected fish grow faster primarily because they eat more, but they also convert the feed to flesh more efficiently.

Direct selection for feed conversion efficiency would probably be inefficient because heritability estimates are low. However, feed conversion efficiency can be selected indirectly by selecting for weight gain because the two traits are correlated. Similarly, heritabilities for dress-out percentage are low, but better dress-out percentage could possibly be selected indirectly by selecting for body conformation phenotypes that have high heritabilities.

An important correlation exists for body weights at different ages (Dunham and Smitherman 1983a). Select lines grow faster than control lines both through fingerling and food-fish growout. This means that if the aim of the program is to select for weight at 18 months (a typical period for growth to 1 to 1.5 pounds), one can make the initial selection from fingerlings rather than waiting to select only from 18-month-old fish. This reduces considerably the number of culture units (ponds, tanks, and so on) necessary to conduct a selection program.

A mass selection program for improved growth. The efficiency of a breeding program decreases if more than one trait is under selection. Mass selection for one trait, growth, is relatively simple and appears to improve several other commercially important traits. Although improved performance is likely through mass selection, actual results of selection cannot be known without controlled experimentation, but this is not practical for most catfish farmers.

A simple program to select for improved growth is illustrated in Figure 6.1. The initial population (the one to be improved) should consist of a good strain of fish with a broad gene pool. Spawn as many fish as possible to prevent a bottleneck, which could reduce genetic variance and increase inbreeding. If you wish to make the initial selection from 100,000 fingerlings, it is much better to obtain a random sample of 1,000 fish from 100 different spawns than 10,000 fish from only 10

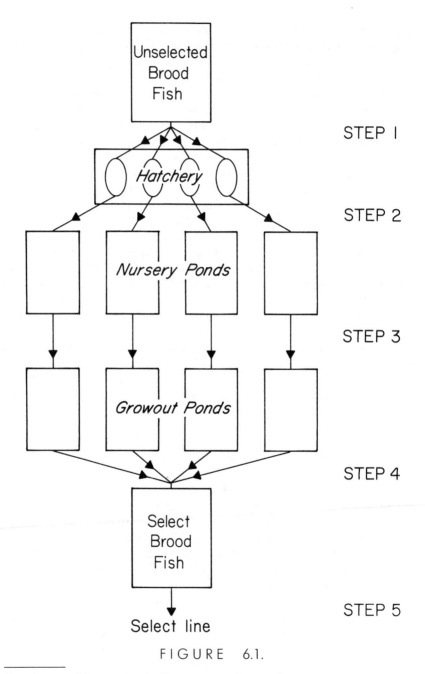

FIGURE 6.1.

Flow diagram of the steps involved in a program of mass selection. Fish are segregated by age group through the hatchery, nursery, and growout phases. See text for details.

spawns. A portion of each spawn can be saved and used in the selection program; the remaining eggs can be hatched and the resulting fingerlings sold or grown to food-fish size as unselected fish.

Spawns or portions of spawns are segregated in the hatchery by age (Step 1 in Figure 6.1). Each group should consist of spawns that differ in age by no more than 4 days. Eggs are hatched and fry are reared in the hatchery as discrete groups by age. Segregation by age is important because the initial selection must be based on genetic differences and not age-related growth advantages. Stock each age group into separate nursery ponds (Step 2). Small nursery ponds (0.25 to 2.0 acres) are most convenient because several different age groups can be raised without committing an inordinate amount of pond space to the selection program. For instance, six 0.5-acre ponds can be stocked with 50,000 fry/acre (each pond with 25,000 fry of one age group) to produce a total of 150,000 fingerlings from which to make the initial selection. Stocking density and management practices should be similar for all ponds and should reflect the practices that will be used when the select line is raised; for example, do not select fish for fast growth at very low fish densities if their progeny will be raised at high densities. You may unintentionally select for fish that perform well only in ponds stocked at low densities. Also, if fish are being selected for growth in raceways or other intensive culture systems, conduct the selection process in that system, rather than in ponds as described here. Fish selected for growth in ponds could perform differently in other systems.

After the first season of growth, select the largest 25 to 35 percent of the fingerlings in each pond and transfer them to separate ponds by age group for growout (Step 3). Depending on the size of the nursery pond, selection can be conducted by using grader boxes or a seine or live car with a mesh size that retains only the largest fish (see the section Fingerling Production). Some trial and error will be necessary to find the right grader bar spacings or seine mesh size to retain only the larger fish. Again, small ponds are convenient for growout, and stocking rates and culture practices should be similar to those normally used for growout.

At the end of the second summer, fish should weigh 1 to 2 pounds. The largest 10 to 20 percent of males and females from each age group are selected (Step 4) and saved for future brood stock (Step 5). Fish will probably have to be hand-selected and sexed to ensure that adequate numbers of both sexes are selected. Selected fish from all ponds can now be mixed. One more year of growth will be needed to reach prime spawning size and age.

Family selection. Family selection differs from mass selection in that selection is based on performance of families rather than individuals. Specific pairs of brood fish are mated and the progeny are reared separately as individual family groups. The performance characteristic of interest is determined as a family average and whole families are selected or culled. Family selection is more efficient than mass selection when additive genetic variance exists for a phenotype, but heritability is low. Family selection is also useful when selecting for traits that are difficult to measure on individual fish, such as feed conversion efficiency. Finally, family selection is necessary for improving traits such as dress-out percentage that cannot be measured without sacrificing the fish. In such instances, dress-out percentage can be measured for a random sample of family members; the remaining family members are retained and used for future brood stock if dress-out is superior to that of samples from other families.

Family selection is effective at improving weight gain in channel catfish (Bondari 1983), and there is evidence that family selection for growth in fish may be more efficient than mass selection (Kincaid, Bridges, and Von Limback 1977). However, family selection requires more labor and facilities than mass selection and is not practical for most commercial catfish farmers.

Hybridization

If additive genetic variance for a trait contributes little to total phenotypic variance (heritability is low), selection will be an inefficient way to improve the phenotype. To improve the trait, dominance genetic variance must be exploited by creating new combinations of alleles. The new combination of alleles is created by mating fish that have different genetic backgrounds. This process is called *crossbreeding* or *hybridization*, and the progeny are hybrids. Parents can be different strains within the same species (intraspecific hybridization) or different species (interspecific hybridization).

The production of superior hybrids is largely a matter of luck. Some crosses result in superior fish; some do not. If the cross produces superior fish, it is due to interactions between fortuitous combinations of alleles. Combinations of alleles are disrupted during production of gametes, so hybrids cannot transmit their superiority to their progeny.

Hybridization is the only practical way to improve performance when heritabilities for a phenotype are small. But exploitation of dominance genetic variance is independent of additive genetic variance, so hybridization can be used to improve a phenotype even if heritability for the phenotype is large. The independence of dominance and additive genetic variance also means that both hybridization and selection can be incorporated into a single breeding program to produce superior

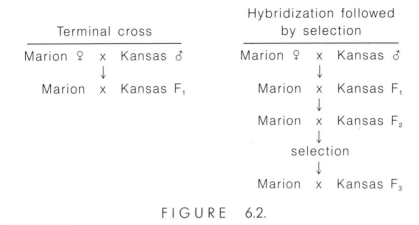

FIGURE 6.2.

Comparison of two uses of crossbreeding. The two breed terminal cross (left) is used to improve productivity through hybrid vigor. Crossbreeding can also be used to produce a new strain that can then be used in a selection program for improved productivity.

fish. For example, two strains known to produce good hybrids can undergo selection to produce brood fish and hybridization can then be used to produce fingerlings for growout.

Production of hybrids also prevents inbreeding. If the parental lines remain pure they will not be related, and since only females from one line are mated with males from the other, the progeny cannot be inbred, even if the parental lines are inbred.

Intraspecific hybridization. Intraspecific hybridization is usually used to improve performance in one of two ways: by producing a new strain that can then undergo selection or by conducting a terminal two-breed cross to improve performance through hybrid vigor. These two programs are compared in Figure 6.2.

Creation of a new strain involves crossing strains to form a new gene pool and then using mass selection to improve performance. Selection is not begun until after the second generation in the breeding program (the F_2 generation) because superiority of individuals in the first generation (the F_1 generation) is due to dominance genetic variance, which cannot be passed to offspring. In effect, the new gene pool created by crossing two strains provides a new pool of additive genetic variance that can be exploited by selection. Strain crossing followed by selection can dramatically improve performance of channel catfish. For example, Dunham and Smitherman (1987) crossed female Marion strain and male Kansas strain fish to produce an F_1 population of Marion × Kansas hybrids.* The F_1 hybrids were mated to produce an F_2 population, which was then selected for fast growth. The resulting F_3 generation grew faster, had a higher spawning rate, and produced more fingerlings per pound of female than the best select pure strain (Kansas).

Some terminal two-breed crosses result in improved growth, disease resistance, survival, and reproductive performance. Success depends on mating the correct strains and using the best of the reciprocal hybrids (*reciprocal hybrids* are the two possible hybrids produced by mating two groups of fish; i.e., the progeny of A × B and B × A).

Not all between-strain crosses produce superior hybrids. When eleven different crosses were made among ten domestic and wild strains of channel catfish, hybrids superior to both parent lines were produced in six of the crosses (Dunham and Smitherman 1983b). Crosses between domestic strains produced superior fish more often (80 percent) than domestic × wild crosses (33 percent). Domestic × wild hybrids usually had growth intermediate to that of parent strains.

Reciprocal hybrids usually grow at different rates. For instance, after one summer of growth, Marion × Kansas hybrids weighed 20 percent more than Kansas × Marion hybrids, 30 percent more than Kansas × Kansas, and 32 percent more than Marion × Marion (Dunham and Smitherman 1983b). The amount of hybrid vigor expressed decreased with age, and the Marion × Kansas hybrids averaged less than 10 percent heavier than the best parental strain after a second summer of growth.

Strains of channel catfish spawn at different times in ponds, probably because of different temperature requirements for induction of spawning. Strain of the female largely determines the spawning date of between-strain pairings, and this may limit success in intraspecific hybridization. Smitherman et al. (1984) found lower spawning rates for between-strain pairings of Marion and Kansas strains (33 percent) than for within-strain pairings (52 percent). Even poorer spawning success can be anticipated for between-strain matings of strains that differ widely in spawning date, such as Minnesota (which spawns early) and Rio Grande (which spawns late).

Although the success of intraspecific hybridization is not as predictable as selection for improved growth, a hybridization program can be the simplest and quickest way to improve performance. Selection requires at least one generation to produce brood fish; if two good strains are obtained, a simple two-breed terminal cross can produce

---

*In designating the parents of hybrids the female parent is listed first and the male parent second.

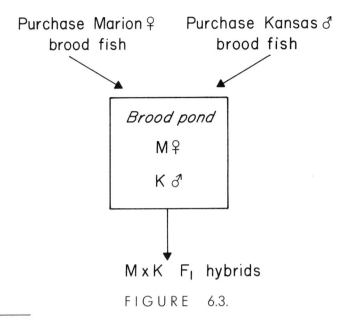

FIGURE 6.3.

A simple way to make a two breed cross: purchase female brood fish of one strain, male brood fish of another strain, and let them mate.

immediate improvement. A major limitation to the success of intraspecific hybridization is the lack of thorough and widespread performance evaluations of hybrids. Only a few intraspecific hybrids have been evaluated, and the evaluations have been conducted almost entirely at one location, Auburn University in Alabama. Since any breeding program requires allocation of time and resources, the farmer should be certain that any hybrid produced is in fact a superior fish. Research has indicated that chances are good (about three in four) that crosses between domestic strains will produce a superior hybrid. However, the only hybrid identified so far whose performance justifies the effort involved in making the cross is the Marion × Kansas hybrid. This cross will be used to illustrate three variations of programs involving two-breed terminal crosses. If other superior hybrids are identified in the future, they can be used in similar programs.

The simplest way (Fig. 6.3) to produce Marion × Kansas hybrids is to purchase pure strain female Marion and male Kansas brood fish from a public research institution or reputable breeder. The fish must be accurately sexed to ensure against within-strain matings. The appropriate number of each sex (strain) is stocked into brood ponds and only Marion × Kansas hybrids will result. Handling of spawns, fry, and fingerlings and growout to food fish are then conducted as for any fish (Chapters 8 and 9). Replacement brood fish are purchased from outside sources as the need arises.

This program is well suited to small farms where a relatively small number of brood fish are needed. The cross eliminates inbreeding, which can be a problem on farms with a small breeding population. Also, a minimal number of brood ponds are needed, another benefit for small farms. Success in this program depends on obtaining genuine pure strain fish and performing accurate sexing. The major limitations of this program are the availability and cost of brood fish. To overcome this limitation, fry or fingerling pure strain Marion and Kansas fish can be purchased and grown separately to food-fish size. Female Marion and male Kansas fish are then obtained from the two populations and moved to other ponds for additional growth or directly to brood ponds (Fig. 6.4).

When large numbers of brood fish are required it may be necessary to maintain isolated, self-reproducing populations of Marion and Kansas strains to produce replacement brood fish (Fig. 6.5). It is important to maintain relatively large brood fish populations within each strain so that inbreeding within strains will not reduce reproductive efficiency or have other deleterious effects. This program requires more labor and pond space than the simple programs outlined previously.

The most effective program combines selection within strains with a terminal cross (Fig. 6.6). Selection is made within strains as outlined previously. The select female Marion and male Kansas brood fish are then combined to produce the hybrid. Although this program is highly effective, it requires a large commitment of pond space and labor. Fingerling producers would have to receive a premium price for fingerlings to justify the added expenses incurred in production.

Interspecific hybridization. The channel catfish possesses a combination of qualities that make it the most desirable species for commercial culture; however, other species have certain qualities that, individually, are superior to those of channel catfish. For instance, the white catfish tolerates low concentrations of dissolved oxygen better than the channel catfish. The blue catfish is more seinable and shows less size variation among individuals during growout. Attempts have been made to take advantage of these characteristics by producing interspecific hybrids. About thirty interspecific hybrids have been produced from seven species of catfish. Many of the hybrids are difficult to produce or the crosses result in a high proportion of nonviable or abnormal progeny. However, hybridization between the female channel catfish and male blue catfish produces a promising F_1 hybrid. The reciprocal blue × channel hybrid is not as good.

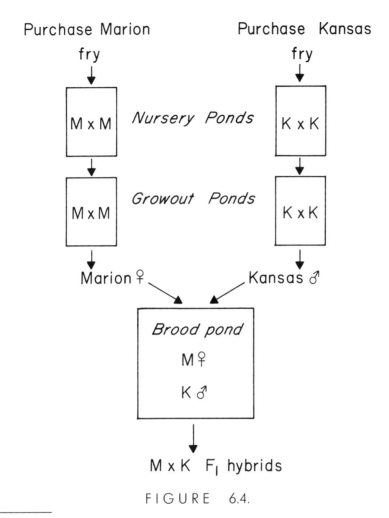

FIGURE 6.4.

A two breed cross made by purchasing fry of two strains, growing them to adults, then selecting females of one strain, males of the other, and letting them mate.

Research evidence indicates that the channel × blue hybrid grows faster than both parental lines and is more disease-resistant, more uniform in size, more easily captured by seining, and more tolerant of low dissolved oxygen than the channel catfish parental line (Dunham and Smitherman 1987). The major obstacle to commercial production of the hybrid is the poor hybridization rate: fingerlings cannot be produced in sufficient numbers for commercial application. Apparently, behavioral differences interfere with reproductive success.

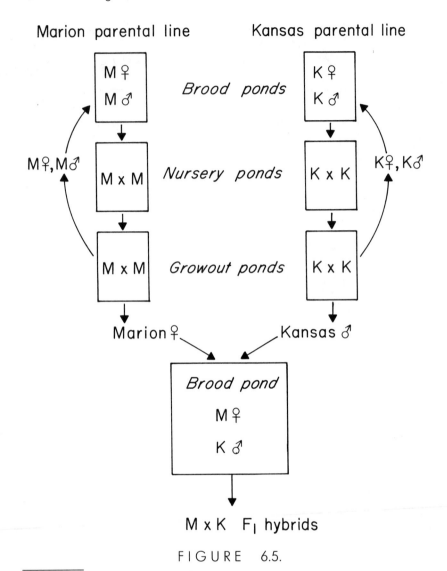

FIGURE 6.5.

A two breed cross made between males and females of two closed populations. A self-reproducing line of each strain is maintained on the farm.

Breeding for Quantitative Traits **129**

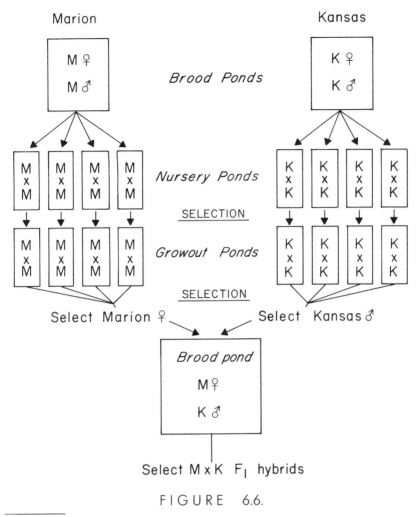

FIGURE 6.6.

A breeding program consisting of mass selection within two strains followed by a two breed cross between select brood fish.

Spawning rates in ponds are much lower than for channel × channel or blue × blue matings. Hatchability of hybrid eggs and survival of fry, on the other hand, are as good as those of parental species. If future research improves spawning success, breeding programs to produce the channel × blue hybrid would be similar to those described for intraspecific hybridization. In one respect, the channel × blue hybrid would be easier to produce because parental lines are easy to differenti-

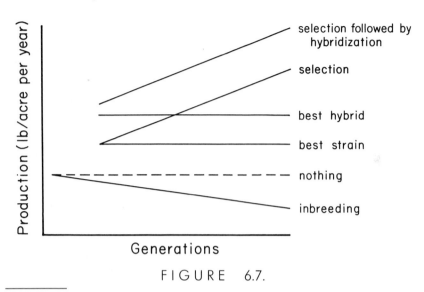

FIGURE 6.7.

Relative changes in fish production possible using various breeding programs (Tave 1986).

ate visually and there is little or no chance that parental lines would become mixed on the farm.

Comparison of Breeding Programs

Recommendations on breeding programs are difficult to offer because few yield trials have been conducted under commercial culture conditions. Also, selection of a breeding program will depend on production goals, farm size, and the farmer's willingness to devote resources to the program. Tave (1986) discusses the relative merits of different breeding programs in detail and offers a schematic representation of how they affect productivity (Fig. 6.7).

Disorderly or unplanned selection of brood fish can cause long-term decreases in productivity. As a minimum, commercial breeding programs should be conducted to maintain some level of productivity by minimizing inbreeding or other negative consequences of unintentional selection.

Productivity on many farms could be improved by obtaining a superior strain of fish and then properly managing that population. Productivity could further be improved by using a simple two-breed terminal cross to produce a hybrid. The best intraspecific hybrid (Marion × Kansas) grows at about the same rate as the best interspecific hybrid (channel × blue). The channel × blue hybrid is, however, more tolerant of low dissolved oxygen concentrations, more seinable, and more uniform in size, but inconsistent mating success between the two species limits its commercial application. Hybridization may also require more pond space and effort than most farmers are willing to allocate.

Mass selection probably has the greatest commercial potential for improving performance. Mass selection is relatively simple and does not necessarily require a lot of pond space (although it does require several ponds, they can be relatively small). Much of the labor involved in the actual selection process can be conducted during the fall and winter, which are the less labor-demanding times on most commercial farms. However, mass selection programs must be conducted under a fairly rigid set of conditions to ensure against unintentional selection or inbreeding.

The most effective breeding program (but not necessarily the most efficient in terms of use of resources) is a combined program of selection followed by hybridization. Strains that produce good hybrids must be obtained and resources allocated to both selection and hybridization. This requires considerable pond space and labor.

NONTRADITIONAL GENETIC IMPROVEMENT PROGRAMS

Presently the only practical means of genetically improving performance of farm-raised channel catfish is through the traditional breeding programs described previously. However, other approaches to genetic improvement are being investigated and may hold some promise for the future.

Polyploidization

Polyploids are individuals with more than the normal $2N$ number of chromosome sets. Polyploidy can be induced in fish by temperature or pressure shocks applied to eggs soon after fertilization (see the section Basic Genetics). Triploid ($3N$ chromosomes) and tetraploid ($4N$ chromosomes) channel catfish have been produced (Wolters, Libey, and Chrisman 1982; Bidwell, Chrisman, and Libey 1985). Tetraploid channel catfish should be fertile, but the growth rate of tetraploids relative to diploids is not known. Triploid channel catfish appear to be sterile and grow faster than normal, diploid fish. The reason for improved growth of triploids is not clear. Possibly, metabolic energy normally used in sexual maturation is directed to growth. Also, for some organisms, polyploids have larger cell size than diploids, and this translates to increased body size. Triploid channel catfish also have a better dress-out percentage than diploid fish as a result of lack of gonadal development (Chrisman, Wolters, and Libey 1983).

Commercial application of triploidy is presently constrained by the difficulties in collecting, fertilizing, and shocking large numbers of channel catfish eggs. Some egg-shocking treatments also result in low survival. An alternative approach that may have some commercial application is the development of tetraploid brood fish that can be mated with normal diploid brood fish by using traditional pond spawning methods to produce triploid fish for growout.

Production of All-Male Populations

Male channel catfish generally grow faster than females in mixed populations. If the faster growth rate is genetically determined in male catfish and not just due to behavioral or social interactions in mixed-sex populations, then productivity could be improved by raising all-male populations.

Hormone-induced sex reversal has been used to produce monosex populations of several fish species. Hormones are usually applied to fertilized eggs or fed to sexually undifferentiated fry. Depending on the hormone used, the functional (phenotypic) sex of half the population reverses, but the genetic sex remains unchanged. Male sex hormones (androgens) have been used to produce all-male populations of tilapia to control reproduction during growout. All-female salmon populations have been produced by using female sex hormones (estrogens).

Attempts to produce all-male, sex-reversed populations have failed. Oral administration of the androgen 17-α-ethynyltestosterone to swim-up fry results in paradoxical feminization and production of 100 percent females (Goudie et al. 1983). Administration of estrogens produces all females, as expected. The cause of paradoxical feminization of androgen-fed channel catfish is not known but may be related to the presence of enzymes in channel catfish that convert androgens to estrogenlike compounds. Even if a process is developed to hormone-induce sex reversal in channel catfish, the United States Food and Drug Administration would have to approve the process before the fish could be produced and sold as human food.

A more promising approach to creation of all-male populations is to use sex-reversed female fish to produce brood fish that will then produce all-male progeny: fry are fed estrogens and the normal XY male fry become functional females but retain XY sex chromosomes. These XY functional females are identified and then mated with normal XY males to produce XX normal females, XY normal males, and YY "supermales." The YY supermales are identified and mated with XX normal females to produce 100 percent XY males. This process may take up to 10 years and requires considerable labor to identify which fish are sex-reversed XY females and YY supermales. These fish cannot be differentiated from normal fish of the same functional sex except through progeny testing. Progeny testing involves mating fish with an unknown genotype to fish with a known genotype and then examining the phenotypic ratios of progeny to decipher the unknown genotype. Production of YY supermales is impractical on commercial farms because of the time and labor involved; if this process proves to be valuable, YY supermale brood fish will probably be produced by specialized breeders and then sold to commercial fingerling producers.

Genetic Engineering

Genetic engineering is a process in which an individual gene from one organism is transferred into the genetic makeup of another organism. The process is successful if the gene is expressed in the organism and transmitted to offspring. Fish possessing the new gene are called *transgenic*.

Genetic engineering research in fish has primarily been limited to development of experimental procedures. In the future, it may be possible to transfer genes for improved growth or disease resistance to channel catfish. Although the future contributions of genetic engineering to channel catfish farming are likely to be significant, progress will be slow. Genetic engineering is a relatively new science and experimental procedures are still being developed.

REFERENCES

Bidwell, C. A., C. L. Chrisman, and G. S. Libey. 1985. Polyploidy induced by heat shock in channel catfish. *Aquaculture* 51:25–32.
Bondari, K. 1983. Response to bidirectional selection for body weight in channel catfish. *Aquaculture* 33:73–81.
Bondari, K. 1984. Comparative performance of albino and normally pigmented channel catfish in tanks, cages, and ponds. *Aquaculture* 37:293–301.
Burch, E. P. 1986. Heritabilities for body weight, feed consumption, and feed

conversion, and correlations among these traits in channel catfish, *Ictalurus punctatus*. Master's Thesis, Auburn University, Alabama.

Chappell, J. A. 1979. An evaluation of twelve genetic groups of catfish for suitability in commercial production. Ph.D. Dissertation, Auburn University, Alabama.

Chrisman, C. L., W. R. Wolters, and G. S. Libey. 1983. Triploidy in channel catfish. *Journal of the World Mariculture Society* 14:279–293.

Dunham, R. A. 1981. Responses to selection and realized heritability for body weight in three strains of channel catfish grown in earthen ponds. Ph.D. Dissertation, Auburn University, Alabama.

Dunham, R. A., and R. O. Smitherman. 1983a. Response to selection and realized heritability for body weight in three strains of channel catfish, *Ictalurus punctatus*, grown in earthen ponds. *Aquaculture* 33:89–96.

Dunham, R. A., and R. O. Smitherman. 1983b. Crossbreeding channel catfish for improvement of body weight in earthen ponds. *Growth* 47:97–103.

Dunham, R. A., and R. O. Smitherman. 1984. *Ancestry and Breeding of Catfish in the United States*, Circular 273. Auburn: Alabama Agricultural Experiment Station.

Dunham, R. A., and R. O. Smitherman, eds. 1987. *Genetics and Breeding of Catfish*. Southern Cooperative Series Bulletin 325. Auburn: Alabama Agricultural Experiment Station.

Dunham, R. A., M. Benchakan, R. O. Smitherman, and J. A. Chappell. 1983. Correlations among morphometric traits of 11-month-old blue, channel, white, and hybrid catfishes and relationship to dressing percentage at 18 months of age. *Journal of the World Mariculture Society* 14:668–675.

El-Ibiary, H. M., and J. A. Joyce. 1978. Heritability of body size traits, dressing weight, and lipid content in channel catfish. *Journal of Animal Science* 47:82–88.

Goudie, C. A., B. D. Redner, B. A. Simco, and R. A. Davis. 1983. Feminization of channel catfish by oral administration of steroid sex hormones. *Transactions of the American Fisheries Society* 112:670–672.

Heaton, E. K., T. S. Boggess, R. E. Worthington, and T. K. Hill. 1973. Quality comparisons of albino and regular (gray) channel catfish. *Journal of Food Science* 38:1194–1196.

Kincaid, H. L., W. R. Bridges, and B. Von Limbach. 1977. Three generations of selection for growth rate in fall spawning rainbow trout. *Transactions of the American Fisheries Society* 106:621–625.

Plumb, J. A., O. L. Green, R. O. Smitherman, and G. B. Pardue. 1975. Channel catfish virus experiments with different strains of channel catfish. *Transactions of the American Fisheries Society* 104:140–143.

Reagan, R. E. 1979. Heritabilities and genetic correlations of desirable commercial traits in channel catfish. Mississippi Agricultural and Forestry Experiment Station Research Report Vol. 4, No. 5. Mississippi State University, Mississippi.

Smitherman, R. O., R. A. Dunham, T. O. Bice, and J. L. Horn. 1984. Reproductive efficiency in reciprocal pairings of two strains of channel catfish. *Progressive Fish-Culturist* 46:106–110.

Tave, D. 1986. *Genetics for Fish Hatchery Managers*. Westport, Conn.: AVI Publishing.

Tave, D. 1989. Channel catfish yield trial centers: an idea whose time has come. *Aquaculture Magazine* 15(1):50–52.

Wolters, W. G., G. S. Libey, and C. L. Chrisman. 1982. Effect of triploidy on growth and gonadal development of channel catfish. *Transactions of the American Fisheries Society* 111:102–105.

CHAPTER 7
Egg and Fry Production

Two important considerations when selecting a fish species for culture are whether it will reproduce in captivity and whether the reproductive phase can be controlled or manipulated. These are desirable characteristics in a cultured fish because natural supplies of seed stock cannot be depended on since their availability and quality are inconsistent. Also, unwanted reproduction in food-fish production ponds is undesirable because the competition for space and nutrients of new recruits will reduce yield of marketable fish. Thus a reliable source of high-quality seed stock at an economical cost is essential to large-scale commercial fish culture.

It was originally thought that the channel catfish required flowing water to spawn because it is primarily a river fish. However, channel catfish are easily spawned in static ponds. Spawning can be induced in brooders and prohibited in food-fish production ponds by habitat manipulation, which primarily means stocking at appropriate densities and providing a suitable artificial spawning site. Hormones can be used to induce ovulation in ripe channel catfish, but using hormones to produce early egg maturation has generally not been successful.

Advances in nutrition, water quality management, disease control, and genetics made over the last three decades have resulted in increased survival and improved quality of catfish fingerlings. Large numbers of high-quality channel catfish fingerlings can now be produced at an economical price. Although improved technologies have been developed in many areas of catfish fry and fingerling production, the procedure used to spawn catfish developed years ago (providing spawning containers to brood fish in ponds) remains basically the same.

Although spawning of catfish has changed very little over the years and there is a general protocol to be used, there is no single "best" method to produce catfish seed stock. That is, since fish culture is as much an art as a science, many catfish hatchery managers may use techniques that vary somewhat from the general protocol discussed in the following sections. In addition to the material presented herein, the reader is referred to the Alabama Cooperative Extension Service Circular ARN-327 (Jensen, Dunham, and Flynn 1983).

BROOD FISH MANAGEMENT

The population (or populations) from which brood stock are acquired is determined by the type of breeding program chosen by the fingerling producer. From that population, the number of brood fish selected and the manner in which they are selected must be such that the chances of inbreeding are minimized. The genetic principles of brood stock selection were discussed in Chapter 6. Brood fish must also be selected and then managed to maximize reproductive success. It does little good to obtain fish with a desirable genetic background if spawning success is poor.

Selecting Brood Fish

It is difficult to inspect individual brood fish on large farms, but every effort should be made to select healthy brood fish of the proper size with well-developed secondary sex characteristics. Channel catfish can mature sexually at 2 years of age and at weights as low as 0.75 pounds but for reliable spawning fish should be at least 3 years old and weigh at least 3 pounds. Channel catfish of 4 to 6 years of age weighing between 4 and 8 pounds are prime spawners. Older and larger fish produce fewer eggs per pound body weight, are more difficult to handle, and may have difficulty entering certain containers used as spawning sites. Fish over 10 pounds should be culled and replaced with younger, smaller fish.

If possible, avoid using fish with a history of channel catfish virus disease and enteric septicemia of catfish. There is some evidence that these diseases are transmitted from parent to progeny. Fish with deformities or other defects should also be avoided or culled from existing brood stock.

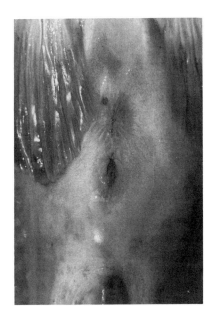 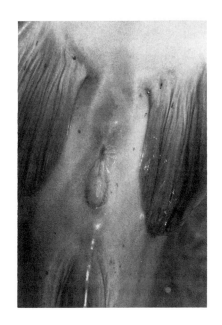

FIGURE 7.1.

Comparison of the urogenital areas of a sexually mature male (right) and female (left) channel catfish.

Sex of brood fish must be determined so that females and males can be stocked into brood ponds in the desired ratio. The development of robust secondary sex characteristics also provides some indication of potential reproductive success. The urogenital areas of both sexes are on the ventral side, posterior to the anus and anterior to the anal fin (Fig. 7.1). The male releases sperm and urine through a common opening called the *urogenital pore*, which is located on the urogenital papilla, a small fleshy nipple just posterior to the anus. The female genital opening is separate from the urinary opening and lies in a groove covered by folds of skin. The groove and surrounding folds of skin form a distinct slit posterior to the anus. The slit of the female readily contrasts to the papilla of the male, especially during spawning season in fish over 3 pounds. The sex of smaller fish can be confirmed with the use of a probe to determine whether the genital and urinary openings are separate or united.

Mature channel catfish also develop secondary sex characteristics

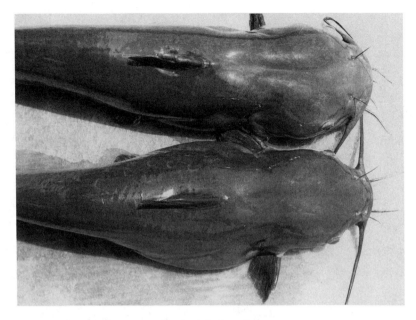

FIGURE 7.2.

Secondary sex characteristics of sexually mature male (top) and female (bottom) channel catfish near spawning time. The male has a wide, muscular head and the female has a soft, full abdomen.

that are useful for selecting fish (Fig. 7.2). These characteristics are most evident near spawning time. The male has a broad, muscular head; thickened lips; and dark mottled pigmentation on the underside of the jaw and abdomen. The female has a slender head with no distinct muscular pads and undersides that are lighter in color. Near spawning time, the female has a soft, full abdomen caused by gonad development, and the urogenital area is swollen and inflamed.

Brood Fish Care

Brood fish must be managed to optimize spawning success. Reproduction can be inhibited by poor water quality, overcrowding, or poor nutrition. Water temperature is the most important factor influencing reproductive success, but water temperature cannot be controlled or modified in large ponds typically used to hold brood fish. Otherwise, good egg production is ensured by maintaining an appropriate standing crop and providing an adequate food supply.

Brood fish standing crops should not exceed about 1,200 pounds of fish/acre. A relatively low standing crop is necessary in brood ponds to maintain optimal environmental conditions and to minimize suppression of spawning by overcrowding. Spawning success is relatively poor when brood fish standing crops exceed about 1,500 to 2,000 pounds/acre. Initially stock brood ponds with 600 to 800 pounds of fish/acre to allow for about 50 percent weight gain by brood fish per year. Brood fish should be seined from ponds in late fall or winter to cull large or otherwise undesirable fish. Large fish are replaced with younger fish to reestablish the initial standing crop.

Spawning success can be improved by moving brood fish each year to ponds that have been drained and recently reflooded. Fish are moved in the late winter when water temperatures are low to minimize handling stress. Water quality is good throughout the spawning season in the recently flooded pond. Additionally, moving fish to a new pond provides an excellent opportunity to inventory the brood stock, adjust sex ratios, and inspect and cull unwanted fish. Moving brood stock each year may not be feasible on large farms, but every effort should be made to rotate brood fish to new ponds at least every other year.

Brood fish must be provided adequate food throughout the year. Inadequate diet can result in poor egg quality and quantity. Also, if both sexes of brood fish are held in the same pond, an insufficient food supply can result in poor-quality female brood fish because the larger, more aggressive males consume most of the limited ration. When water temperatures exceed 70°F, feed a nutritionally complete supplemental feed of at least 32 percent crude protein at about 2 percent of the body weight daily. At water temperatures of 55 to 70°F, feed at about 1 percent of body weight on alternate days. Do not feed when water temperatures are below 55°F.

Some producers stock forage fish into brood ponds to ensure that adequate feed is available at all times. The most commonly used forage fish are fathead minnows. These minnows seldom exceed a length of 3 inches, do not compete with brood fish for feed, and do not prey upon catfish eggs. Mature (1-year-old) fathead minnows are stocked in the late winter or early spring at 1,000 to 2,000 fish/acre. Natural reproduction then supplies an abundance of forage. Species of *Tilapia* (see the section Polyculture Systems) also make excellent brood fish forage, although the production of *Tilapia* is banned or restricted in many states. Ten to twenty pair of adult *Tilapia* per acre are stocked in late spring or summer. The fish reproduce rapidly and establish a high standing crop of small (1 to 2 inches long) fish by fall. *Tilapia* are intolerant of cold temperatures and become lethargic in the fall. The slow-moving *Tilapia* are easily captured by channel catfish brood fish. Brood catfish continue to eat dead *Tilapia* after they are killed by cold water temperatures. The total kill of *Tilapia* each winter in ponds over most of the southeastern United States is an advantage because there is no chance that fish populations will become established in ponds where they are not wanted. *Tilapia* also help control aquatic weeds that sometimes become established in catfish brood ponds.

Sex Ratios

Female channel catfish spawn once a year, but males can spawn two or three times and consequently can mate with more than one female. Fry production per total number of brood fish in a breeding population is thus increased if there are more females than males. Bondari (1983) found equal spawning success in ponds stocked at male to female ratios of 1:1, 1:2, 1:3, and 1:4. The fewest brood fish needed to achieve a given fingerling production goal would be obtained by using the 1:4 male/female ratio. However, the use of highly skewed sex ratios can be detrimental if the progeny are to be used as future brood fish. Skewed sex ratios increase the chance that individuals obtained from random matings have at least a male parent in common. The probability of inbreeding is therefore increased when randomly breeding populations have skewed sex ratios (Tave 1986). The probability of inbreeding is greatly increased if the breeding population also is small. If fewer than about 200 brood fish are used in a closed* brood population, a sex ratio of 1:1 should be used to minimize inbreeding. Most large-scale commercial fingerling producers stock spawning ponds of male/female ratios of 2:3 or 1:2.

Estimating Brood Fish Requirements

Fingerling producers usually set an annual production goal. If producers grow fingerlings for their own use, the production goal is based on the expected number of fingerlings needed to replace food fish that will be harvested and sold in the following year. If the fingerlings are to be sold to other farmers for growout, production goals are based on the number of fingerlings that must be sold to achieve a certain income within the limitations of available pond space.

Once a fingerling production goal is set, the number of brood fish required to produce that number can be estimated from a series of assumptions regarding egg production, survival of egg to fry in the hatchery, and survival of fry to fingerling in the nursery.

The pounds of female brood fish needed to produce a specific number of fingerlings is calculated as follows:

---

*A *closed population* is a population in which only progeny become future brood stock (no brood stock are imported from outside sources).

$$\text{Pounds } \female \text{ brood fish} = \frac{\text{number of fingerlings required}}{(E \times FS \times H \times N)} \quad (7.1)$$

where E = number of eggs produced per pound of female brood fish

FS = decimal fraction of females in the brood population that will spawn in a given spawning season

H = decimal fraction of eggs that hatch and survive to swim-up fry in the hatchery

N = decimal fraction of swim-up fry that survive to fingerling in the nursery pond

The parameters involved in this calculation vary among brood populations and from year to year. Egg production by channel catfish varies from over 4,000 eggs/pound body weight for young females to fewer than 2,500 eggs/pound body weight for older, larger females. Spawning success (percentage of females spawning) ranges from less than 40 to over 80 percent, depending on water temperatures during the spawning season and condition of brood stock. Hatchery survival is extremely variable, but in a well-managed hatchery with a good water supply, survival rate should exceed 80 percent. Likewise, survival in nursery ponds varies considerably but should exceed 70 percent. Individual producers can use estimates for these parameters derived from actual experience; however, the following assumptions can be used as a conservative, but realistic, starting point:

where E = 3,000 eggs produced/pound body weight

FS = 0.40 females spawning/total females

H = 0.80 fry surviving/eggs produced

N = 0.70 fingerling surviving/fry stocked

Therefore:

$$\text{Pounds } \female \text{ brood fish} = \frac{\text{number fingerlings required}}{(3{,}000 \times 0.4 \times 0.8 \times 0.7)}$$

$$= \frac{\text{number fingerlings required}}{672} \quad (7.2)$$

As an example, assume that the production goal is 5 million fingerlings per year. The estimated weight of female brood fish required to produce this many fingerlings is as follows:

$$\text{Pounds } \female \text{ brood fish} = \frac{5{,}000{,}000}{672}$$

$$= 7{,}440 \text{ pounds} \quad (7.3)$$

142   Egg and Fry Production

The required weight of males is calculated by first determining the number of female brood fish that will be stocked. The number of females is calculated by dividing the weight of females required by the average weight of the female brood fish. The number of males needed is then determined by the male/female ratio to be stocked into the brood pond:

$$\text{Number of } \male = (\text{number of } \female) \times (\text{males/females}) \qquad (7.4)$$

The weight of males needed is calculated by multiplying the number of males needed by the average weight of the male brood fish. For example, assume that 7,440 pounds of females is needed, the females average 4.5 pounds each, a 2:3 male/female ratio is desired, and the males average 6.5 pounds each.

$$\begin{aligned}\text{Number of } \female &= 7{,}440 \text{ pounds}/4.5 \text{ pounds per fish} \\ &= 1{,}653 \; \female \text{ fish}\end{aligned} \qquad (7.5)$$

$$\begin{aligned}\text{Number of } \male &= 1{,}653 \text{ fish } (2 \; \male/3 \; \female) \\ &= 1{,}102 \; \male \text{ fish}\end{aligned} \qquad (7.6)$$

$$\begin{aligned}\text{Weight of } \male &= 2{,}479 \text{ fish } (6.5 \text{ pounds/fish}) \\ &= 7{,}163 \text{ pounds}\end{aligned} \qquad (7.7)$$

In this example, a total of 7,440 + 7,163 = 14,603 pounds of brood fish are required to produce ultimately about 5 million fingerlings. If the brood fish were stocked at 800 pounds/acre, about 18 acres of water would be required. The entire brood population could be stocked into one 18-acre pond. This is risky, however, because a single catastrophe could kill the entire population. It would be prudent to stock the fish into two 10-acre ponds or several even smaller ponds. Small (<10 acres) brood ponds are also easier to manage in general. If the brood population is stocked into two ponds of the same size, simply divide the female and male weights into two equal weights. In the preceding example, 7,440 pounds ÷ 2 = 3,720 pounds of females are stocked into each pond along with 7,163 ÷ 2 = 3,582 pounds of males.

METHODS OF PROPAGATION

Open Pond Method

Channel catfish are most commonly spawned by using the open pond technique in which brood fish held in ponds are provided with spawn-

Methods of Propagation **143**

FIGURE 7.3.

Three types of spawning containers.

ing containers and allowed to select their mates and spawn naturally. This method is more likely to produce a spawn than pen or aquarium spawning methods, particularly if the brood fish are of marginal quality. The pond method requires minimal facilities and the least amount of skill of the culturist. Large numbers of fish can be spawned without the absolute necessity of sexing each fish or checking gonadal development at stocking as are required by more sophisticated methods; late-maturing fish can continue to develop and spawn when ready. Although the pond spawning method has a number of advantages, it does not allow for certain genetic manipulations through selective pairing of brood fish as does the pen or aquarium spawning methods.

Spawning containers are used to provide a nesting site for pond-spawned catfish. A variety of containers similar to those pictured in Figure 7.3 have been used successfully. These include nail kegs, tile, wooden boxes, earthenware crocks, ammunition cans, milk cans, plastic buckets, and plastic containers made specifically for spawning catfish. There is evidence that catfish prefer certain types of containers when given a choice (Busch 1983; Steeby 1987). However, if the fish are not given a choice, spawning success is about the same regardless of the type of spawning container (Busch 1983). Any of the containers shown in Figure 7.3, or containers similar in size to those, should be adequate for channel catfish weighing up to about 10 pounds.

The selection of spawning containers more often depends on cost, availability, durability, and ease of handling than on fish preference. For example, the metal can shown in Figure 7.3 is very durable, but it weighs about 35 pounds and can be cumbersome. On the other hand, 5-gallon plastic buckets fastened together make suitable lightweight containers, but they will not last as long as the metal containers. Some of the commercially available molded plastic containers are lightweight and durable but are more expensive. Wooden boxes, although suitable for spawning fish, are not recommended because when waterlogged they become heavy and tend to break apart when moved. Containers that have a lip below the opening, such as those depicted in Figure 7.3, may help keep the egg mass inside the container if it becomes detached from the container wall. Because almost any container that provides adequate space for the adults to enter and spawn can be used, choose a container that is most suited to your particular situation. Do not use recycled pesticide containers that might contain toxic chemical residues or transformer cases that may have contained polychlorinated biphenyls (PCBs) because of possible contamination.

The containers should be placed into the brood fish ponds several days before anticipated spawning to allow time for the fish to clean and prepare the spawning site. However, the containers should generally not be put into the pond before the water temperature reaches 75°F to discourage early spawning. Containers should be placed along the pond bank with the opening toward the pond center at intervals of 10 to 30 feet at a water depth of about 2 to 3 feet. Steeby (1987) reported that if dissolved oxygen were above 4 ppm, spawning container depth had no effect on spawning success. Containers in deeper waters are more difficult to check and may be exposed to lower dissolved oxygen concentrations. All containers should be marked with a float or stake so they can be easily located. Since all of the fish will not spawn at the same time, a container for each pair of brood fish is not needed. Various ratios of containers to female brood fish have been used, but ratios of 1:2, 2:3, or 3:4 are commonly used.

After the brood fish spawn, the fish culturist has a choice of how to handle either the eggs or the hatched fry (Fig. 7.4). The simplest alternative is to allow the eggs to hatch in the spawning container and let the fry remain in the pond. This approach (sometimes called "wild spawning") requires the least skill, labor, and facilities, but it is unreliable and should not be used for commercial catfish fingerling production. Under this method it is difficult to estimate the number of fry produced and fry survival is generally poor. Thus the number of fingerlings available for stocking growout ponds is uncertain. Additionally, fry are produced over the entire spawning season and those that begin feeding first have an initial size advantage that is magnified as the fry grow. Fingerlings from "wild-spawn" ponds usually vary greatly in size. If this production strategy is used, it is recommended that the brood fish be removed from the pond after spawning is complete to

Methods of Propagation **145**

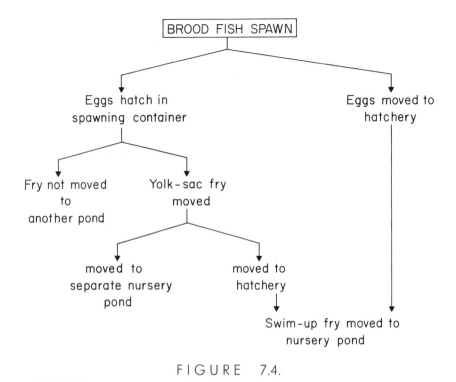

FIGURE 7.4.

Options for handling eggs and fry.

reduce chances of disease transfer from adults to fry and to prevent cannibalism. Also, the culturist should attempt to stock the number of brood fish that will theoretically yield a suitable stocking density of fry.

One method that can be used to eliminate some of the problems encountered when brood fish ponds are used as nursery ponds is to transfer yolk-sac fry to previously prepared nursery ponds or troughs. This method will improve fry survival and allow for more accurate estimates of fingerling inventories. This method also eliminates or reduces the incidence of disease transfer from adult to young fish and prevents cannibalism.

To evaluate egg development or retrieve eggs the spawning containers must be checked periodically. The frequency at which the containers need to be checked for spawns depends on the number of brood fish in the pond and on the rate spawning is progressing. As a rule, they should be checked at least every 3 or 4 days; some culturists check them daily. Since spawning generally occurs at night or early morning, it is better to check for spawns in the late morning, when it is less likely

146 Egg and Fry Production

FIGURE 7.5.

Checking a spawning container for the presence of a spawn.

that spawning activity will be interrupted. To check a container for spawns, gently raise it to the surface and tilt it to drain water to improve visibility (Fig. 7.5). If the male remains in the spawning container use caution when the container is lifted to prevent being bitten by an attacking male. The male should be allowed to swim out before examining or retrieving the egg mass.

If the fry are to be transferred, check spawning containers about every 3 days. When an egg mass is found, remove a few eggs and determine their age by using Table 7.1 to predict the hatching date. The fry should be removed the day after the predicted hatching date. Transfer the fry to a bucket containing pond water by pouring them from the spawning container or leave the fry in the spawning container and

TABLE 7.1.

Estimated Age and Time to Hatch for Channel Catfish Eggs at Various Stages of Development

Egg Description	Estimated Age (78°F Water)	Estimated Days to Hatching
No pulsation	Less than 24 hours	7–8
Pulsating motion	1–2 days	6–7
Bloody streak	2–3 days	5–6
Blood throughout egg	3–4 days	4–5
Eyes visible	4–5 days	3–4
Eyes visible, embryo turns inside shell	5–6 days	2–3
Complete fish visible, no bloody streak	6–7 days	1–2
Hatching begins	7–8 days	0–1

Note: For every 2°F above or below 78°F subtract or add 1 day, respectively.

transfer the container, which will act as a shelter once placed into a nursery pond or a trough. If the temperatures of the brood pond water and the water into which the fry are being transferred differ, the fry must be slowly acclimated to the new temperature. The main advantages of using a trough instead of a pond for rearing sac fry are that survival to swim-up stage is higher in trough-reared sac fry and swim-up fry are more easily trained to accept artificial feeds when reared in troughs. After the swim-up fry are feeding they can be moved from the troughs to nursery ponds.

The most productive and reliable method currently used to produce catfish eggs is to transfer the egg mass to a hatchery. The fish are allowed to spawn in containers in the pond, but the fertilized eggs are removed immediately after they are found and allowed to incubate in the hatchery. The egg mass (Fig. 7.6), which sticks to the floor of the container, can be removed by scraping it from the container using a hand, plastic card, spatula, or similar device. The eggs are then placed into an insulated ice chest or bucket containing pond water for transfer to the hatchery. The eggs should be shaded and should not remain in the chest or bucket for more than 15 minutes unless aerated. The older the egg mass the more oxygen it requires. An egg mass near hatching should be taken to the hatchery immediately. This method of rearing fry offers the best opportunity for the development, survival, and growth of fry because the eggs can be incubated and the fry grown under controlled, optimal conditions. Egg incubation and fry rearing in hatcheries are discussed in the section Hatchery Practices.

FIGURE 7.6.

A typical channel catfish spawn (egg mass).

Pen Method

The pen method is similar to the open pond spawning methods described previously except that pens are constructed in the pond and used to confine the brood fish. Pen spawning is used primary for mating selected pairs of brood fish. Pens can be constructed next to the pond bank or with all four sides in the pond. Adjacent pens can have common sides to reduce cost (Fig. 7.7). The pens should be constructed by using plastic-coated wire attached to a wooden frame. Other materials such as wooden slats or concrete blocks can be used, but the wire pens are more desirable. The mesh of the wire should be as large as possible to allow for good water circulation but not allow the fish to escape. Mesh sizes of 1/2 to 2 inches are satisfactory. Pens are commonly about 4 to 5 feet wide and 9 to 10 feet long, but a 4-by-6-foot pen is adequate. Pen walls should be at least 6 inches below the pond bottom and extend about 12 inches above the water level. The water depth in the pen should be maintained between 2 and 3 feet.

In the pen method the brood fish must be accurately sexed and the spawning condition evaluated by the culturist before placing the pair into the pen. A spawning container is placed into each pen with

Methods of Propagation **149**

FIGURE 7.7.

Pens used to confine catfish for spawning.

seasonally ripe brood fish. The male and female fish should be about equal in size; however, some culturists prefer a male that is slightly larger than the female to prevent the female from injuring the male or eating the eggs after spawning is complete. Only one female and one male should be placed into a pen. The female should be removed after spawning to prevent the aggressive male from injuring her. If another female is to be put into the pen with a male that has already spawned, the eggs or fry (if the eggs are allowed to hatch in the pen) should be removed before placing the new female into the pen. Otherwise, the male may continue to care for the initial spawn and not spawn with the new female. Eggs and fry are cared for in the same manner as described in the section Open Pond Method.

The primary advantage to using the pen method over the open pond method is that the culturist can pair brood fish to make whatever crosses are desired. The pen method also allows the spent female to be removed, offers some protection for the brood fish, and provides control over the time of spawning. Because the fish are captive, injection of hormones can also be used to increase spawning success.

Aquarium Method

The aquarium method, which involves the pairing of brood fish in an aquarium or tank and using hormones to induce ovulation, is not used on commercial catfish farms because of the general ease of obtaining adequate numbers of fry by the open pond spawning method. The aquarium method requires that the culturist select fish that are ready to spawn and inject the female with some type of hormonal material. Fish that are not in good shape and ready to spawn will not generally spawn or will only partially spawn even when injected with hormones; thus the fish selected must be mature and in good physical condition. The culturist must be experienced and able to determine fish condition accurately by external examination because channel catfish oocytes are opaque and identification of stages of oocyte maturation by microscopic examination of catheterized oocytes is not possible. Aquarium spawning of catfish is generally restricted to use in teaching and research, in which observation of spawning is desirable, and when hand-stripping of eggs and artificial fertilization are used for hybridization or triploids are to be produced.

Flowing-water aquaria fitted with tops are generally used to spawn channel catfish. Various sizes of aquaria have been used successfully to spawn channel catfish. Fish up to about 2.5, 6.5, and 12 pounds have been spawned in 10-, 20-, and 30-gallon aquaria, respectively. However, it is generally best to use larger aquaria, of 30 to 50 gallons. Mature, seasonally ripe female catfish are injected with a fish pituitary extract or human chorionic gonadotropin (HCG) and placed into the aquarium several hours before the male is introduced. The male is usually not injected. After the male is placed into the aquarium with the female, the fish are observed for prespawning behavior and spawning activity. Three to five injections of hormone may be needed to induce the female to ovulate, and if after these injections spawning does not occur the female should be replaced. After spawning is completed the female should be removed from the tank; the male may remain to care for the eggs, but it is best to remove the egg mass to the hatchery for incubation.

Intraperitoneal injections at the base of the pelvic fin are generally used to deliver the hormone (Fig. 7.8). Pituitary glands obtained from several different species of fish and then acetone-dried have been used to induce ovulation in gravid channel catfish. Common carp pituitary, which is available in a powdered form, is most commonly used. Most female channel catfish require a total of about 6 milligrams (mg) of pituitary/pound of body weight. Dosages of 2 milligrams of pituitary/pound of body weight administered at 24-hour intervals until ovulation occurs (generally after three to four injections) or for up to 10 days is recommended. The recommended total dosage of HCG for inducing ovulation in female channel catfish is about 800 IU/pound of body weight, which is given in one or several injections until ovulation oc-

Hatchery Design **151**

FIGURE 7.8.

Delivering an intraperitoneal injection of a gonadotropic hormone to a seasonally ripe, mature female channel catfish.

curs. The luteinizing hormone-releasing hormone (LH-RH) analog has also been used to induce ovulation in female channel catfish. The dosage of LH-RH analog is about 0.45 milligram/pound of body weight given as a single injection, or as a primer injection of about 0.005 milligram/pound of body weight and then 12 hours later in a second injection of 0.04 milligram/pound of body weight.

The use of hormones in channel catfish has primarily been to induce mature, seasonally ripe females to ovulate. Although some culturists suggest that catfish can be spawned at any time of the year by using hormones, consistent use of hormones to spawn channel catfish by control of oocyte maturation and ovulation has not been accomplished. Brauhn (1971) was successful in spawning channel catfish in August and November by using temperature and photoperiod manipulation in conjunction with hormone injections.

HATCHERY DESIGN

Elaborate facilities are not essential to producing channel catfish fry, but a well-designed hatchery is beneficial as it allows for more efficient and economical production of fry. Although it is desirable to have hatching facilities that are dedicated to this purpose, a catfish hatchery can be set up on a temporary basis and dismantled after hatching and the facility used for other purposes. Almost any type of available structure could be turned into a hatchery. For example, successful catfish hatcheries have been built on cotton trailers and in barns, sheds, and so on.

Water Sources

Although hatchery designs vary greatly, the one requirement that cannot be overlooked is a good water supply. Whether the hatchery is permanent or temporary, an adequate supply of good-quality water is essential.

Groundwater. Groundwater is considered the best source of water for use in catfish hatcheries. Groundwaters are generally free of suspended matter, pollutants, and fish disease organisms, and the temperature and chemical composition of a specific body of groundwater are relatively constant. In regions with abundant groundwater, the supply is constant and dependable. Obtain a chemical analysis of the intended groundwater supply to ensure that it is suitable for use. Analyses can sometimes be obtained from well logs kept by local water-well drilling companies or state water resource agencies. Alternatively, a test well can be drilled and a sample sent to a laboratory for analysis.

Some groundwaters may have to be treated to make them suitable for hatchery use. Depending on the water, treatments may include aeration to increase dissolved oxygen concentrations and reduce total gas pressure and carbon dioxide and hydrogen sulfide concentrations, temperature regulation using water heaters or mixing of waters of different temperatures, sedimentation and filtration to remove iron, and addition of calcium to waters of low hardness.

Surface water. Surface water can be from streams and rivers, ponds, lakes, or reservoirs. A major constraint to using surface water for hatcheries is the potential for contamination by fish disease organisms or water-borne predators, such as wild fish, insects, and other invertebrates that could enter a hatchery and cause losses of fry. Surface waters in areas with intensive agriculture may be polluted with pesticide residues or turbid from suspended clay particles. All surface waters should be carefully evaluated before use as a hatchery water supply. Because of the potential variability in volume and quality of surface water, a historical record is required to ensure that the volume

Hatchery Design **153**

and quality of water are adequate at all times. If surface water is used, methods to prevent entry of unwanted organisms should be implemented. For example, screens and filters may use an inlet water supply to assist in eliminating organisms and reducing turbidity.

Water Quality

Desirable characteristics of channel catfish hatchery water supplies as well as common water quality problems are summarized in Table 7.2. A water flow of about 2 to 5 gallons a minute is needed for a typical 100-gallon hatching trough, or approximately one complete water change every 45 to 60 minutes to maintain proper water quality for hatching eggs and rearing fry. In addition to the specific quality requirements presented in Table 7.2, the water should also be free of pesticides, solvents, petroleum products, and other toxicants; free of fish disease organisms; and relatively constant in quality and availability.

Temperature. The optimal temperature range for development of eggs and rearing of fry is between 78 and 82°F. If the temperature is too low, egg hatching and fry development are prolonged, and fungi, which thrive in cool waters, may invade the egg mass. At higher water temperatures, embryos develop too fast and there may be a high incidence of malformed or nonviable fry. Bacterial diseases of eggs or fry and channel catfish virus disease of fry also are more common if the water temperature is greater than 82°F.

Heating and cooling of water are only feasible on small scale because of the cost; therefore, water entering the hatchery should be near 80°F. Groundwater from depths of 500 to 1,000 feet should be warm enough for use in catfish hatcheries without treatment. Water from wells of less than 300 feet and some surface waters are too cool for direct use. Cool groundwaters can be pumped into a reservoir pond where solar heating can raise the temperature somewhat; however, the temperature of reservoir pond water and surface water vary with local weather conditions. If the water is too cool, an in-line, demand water heater or a heat exchanger can be used to raise the inlet water temperature. When surface waters or groundwater pumped into a pond are used, it is possible that late in the spawning season the hatchery water supply will become too warm and must be cooled. The most reasonable method to cool water is to blend the warm water with a cool water source, such as shallow well water.

Total gas pressure. Gas bubble trauma is a condition that affects fish living in water supersaturated with gases (also see the section

TABLE 7.2.
Common Water Quality Problems in Channel Catfish Hatchery Water Supplies

Variable	Desired Level	Problem	Solution
Temperature	78–82°F	Too low	Reservoir pond for solar heating or use water heaters
		Too high	Blend with cooler water
Total dissolved gases	<105% TGP[a]	Too high	Vigorous aeration
Dissolved oxygen	6 ppm to saturation	Too low	Vigorous aeration of incoming water and supplemental aeration in vats or troughs
Carbon dioxide	<10 ppm	Too high	Vigorous aeration of incoming water and supplemental aeration in vats or troughs
Calcium hardness	>20 ppm as $CaCO_3$	Too low	Addition of calcium chloride to water supply
Ammonia (un-ionized)	<0.05 ppm NH_3-N	Too high	Remove from incoming water with zeolite filter; decrease fry density or increase water flow
Iron	<0.5 ppm total iron	Too high	Aeration followed by precipitation or sand filtration
Hydrogen sulfide	<0.005 ppm H_2S-S	Too high	Vigorous aeration of incoming water

[a]TGP = total gas pressure.
Source: Tucker (1988).

Total Gas Pressure). Fish living in gas-supersaturated water may be killed when dissolved gases within their tissues and vascular system come out of solution and form bubbles that restrict circulation. Fish in hatcheries are particularly susceptible to gas bubble trauma because of the reduced water depth, long exposure times, and crowding. Any total gas pressure over about 105 percent of total gas saturation is undesirable in catfish hatcheries. Ideally, the total gas pressure should be 100 percent of saturation or less. Supersaturation can occur in both ground and surface waters. If supersaturation occurs, the most effective and economical method for reducing it is vigorous aeration of the incoming water. Total gas pressure is measured with instruments called *saturometers*, which are not commonly available to hatchery managers. However, problems with gas supersaturation in hatcheries are easy to identify. The most obvious clinical sign (symptom) is development of gas bubbles in the yolk sac of yolk-sac fry. The bubbles cause fish to float upside down at the surface.

Dissolved oxygen. Dissolved oxygen concentrations should not fall below 5 to 6 ppm in water used in egg hatching troughs or fry rearing vats, because eggs and fry have a high metabolic rate and thus a high dissolved oxygen requirement. The hatchery water should contain an adequate concentration of dissolved oxygen before it enters the hatching troughs or rearing vats. Groundwaters usually contain no dissolved oxygen and should be aerated in a tank or packed-column aerator before use. Dissolved oxygen concentrations should be maintained above 5 ppm in the hatchery tanks by use of mechanical aerators. Rotating paddles typically used in egg hatching troughs also serve to aerate the water and to ensure that oxygenated water is circulated around the egg masses.

Carbon dioxide. Carbon dioxide interferes with oxygen uptake and use by eggs and fry. Ideally, there should be no carbon dioxide in hatchery water, but concentrations less that 10 ppm are tolerated. Groundwaters that are high in carbon dioxide should be vigorously aerated to eliminate it.

Salinity. Egg hatchability and fry development are not hindered in waters with salinities up to at least 8 ppt.

Hardness. Hardness is the amount of calcium and magnesium in water. It is expressed as ppm equivalent calcium carbonate ($CaCO_3$). Environmental calcium is required for "hardening" of eggs and for normal bone and tissue development of fry. A deficiency of calcium in the water can result in poor egg hatchability as well as slow development, lack of vigor, decreased resistance to stress, and low survival of sac fry. A minimum of 5 ppm calcium hardness is required for adequate egg hatchability and development and vigor of sac fry. At higher concentrations calcium also protects fry from ammonia and metal toxicosis. It is recommended that hatchery water supplies contain at least 20 ppm calcium hardness. Calcium can easily be added in solution to the incoming water by using an inexpensive "drip system" or a chemical metering pump.

Alkalinity. Alkalinity, which is expressed as ppm equivalent $CaCO_3$, is a measure of the capability of a water to neutralize acids. Generally, the predominant bases in natural waters are bicarbonate and carbonate. Although catfish eggs and fry do well in waters of a wide range of alkalinity, waters with an alkalinity of less than 20 ppm as $CaCO_3$ should not be used in the hatchery, if possible. Low-alkalinity waters are poorly buffered and pH can fluctuate drastically with small additions of acid or base. Also, dissolved metals such as copper and zinc are very toxic to fry in waters of low alkalinity. These metals may leach from metal pipes used to supply water.

Ammonia. The water in rearing troughs should be free of ammonia for optimal health and growth of fry. The maximum concentration of un-ionized ammonia that should be allowed is about 0.05 ppm NH_3-N, because at higher concentrations eggs and fry develop more slowly and are under chronic stress, which makes them more susceptible to disease. If ammonia must be removed from the incoming hatchery water, a zeolite filter can be used (Liao and Lin 1981). However, fish are the primary source of ammonia in the system since ammonia is a product of fish metabolism. In rearing troughs stocked at high densities ammonia production can be significant. To reduce ammonia levels in rearing troughs, the fry density can be reduced or the water flow can be increased. In a typical 100-gallon fry-rearing tank, fry density should not exceed 50,000 fry/gallon/minute of water flow. For example, with a water flow of 2 gallons/minute, a maximum of 100,000 fry can be held until about a week past swim-up without an accumulation of ammonia. Fry density must be decreased if fish are held longer than 7 to 10 days in the hatchery because ammonia production increases as more feed is given in response to fry growth.

Iron. Most surface waters contain very low concentrations of iron. Some groundwaters contain over 10 ppm iron, which is in a dissolved ferrous (Fe^{2+}) form. However, when the water is aerated the iron is oxidized to a rust-colored precipitate of iron oxide. Dissolved iron is not particularly toxic to aquatic animals, but precipitates of iron oxide can coat the gills of fry and interfere with respiration or coat fish eggs and interfere with gas exchange and suffocate the eggs. Total iron concentrations should be less than about 0.5 ppm for hatchery water supplies.

The simplest system used to remove iron is to pump the water into a 1- to 2-acre pond where the iron is oxidized and some of the

Hatchery Design

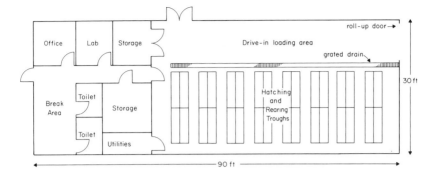

FIGURE 7.9.

Floor plan for a hatchery capable of producing about 10 million fry/year. This facility has an office and laboratory space and would also serve as headquarters for the entire farm.

iron oxide precipitate settles out. The remainder of the precipitate is removed by using sand filters. An alternate method is to oxidize the iron by vigorously aerating the water in a tank before sand filtration. The residence time of the water within the tank should be at least 15 minutes to allow sufficient time for oxidation. Complete removal of iron is difficult regardless of the system used.

Hydrogen sulfide. Hydrogen sulfide gives water a "rotten-egg" odor and is very toxic to channel catfish fry. As little as 0.005 ppm unionized hydrogen sulfide is toxic to sac fry. Hydrogen sulfide should be removed from water before it enters the rearing vat. Removal can be accomplished by vigorous aeration, which volatilizes the hydrogen sulfide and oxidizes some of it to sulfate, which is nontoxic.

Facilities

A floor plan representative of a commercial channel catfish hatchery is illustrated in Figure 7.9. There are many variations of the plan that would be functional, but several factors should be considered. There should be ample space for egg hatching troughs and fry rearing troughs as well as room for office, storage, and laboratory space. There should be easy assess between and around the troughs to facilitate egg and fry handling and trough cleaning. The hatchery floor should be sloped to drains to prevent collection of water. Drain pipes should be embedded in the floor or placed in covered, recessed channels to avoid the nui-

158 Egg and Fry Production

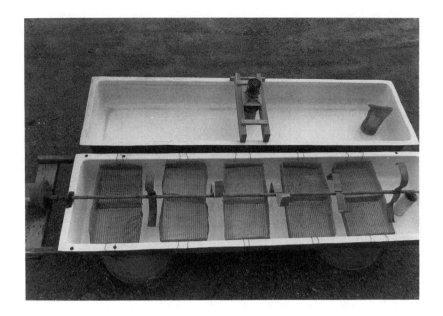

FIGURE 7.10.

A fry rearing trough (top) with a small electric aerator and an egg hatching trough (bottom) with egg baskets and paddles.

sance of having to step over the drain pipes when working in the hatchery. The drain leaving the hatchery should have a trap to prevent fry from escaping from the facility. Water and electric lines should be located overhead for convenience. Electrical outlets and wiring should be installed in a manner that will make them safe to work around in a wet environment. Plastic pipe is preferred for plumbing because it is nontoxic and easy to install. A backup electrical generator that turns on automatically during electrical power failures is desirable. Other emergency systems that are less expensive can be used. For example, a generator that works off the power takeoff of a tractor combined with a telephone alarm system that will notify personnel of electrical power failures can be used.

Troughs used to incubate and hatch eggs are generally flat-bottomed wooden, Fiberglas, or metal structures about 8 feet long, 2 feet wide, and 10 inches deep (Fig. 7.10). These troughs hold approximately 100 gallons of water. Flat-bottomed troughs reduce the chances of localized overcrowding and suffocation of fry on the bottom. The troughs have a water inlet at one end and a drain at the opposite end.

Hatchery Design **159**

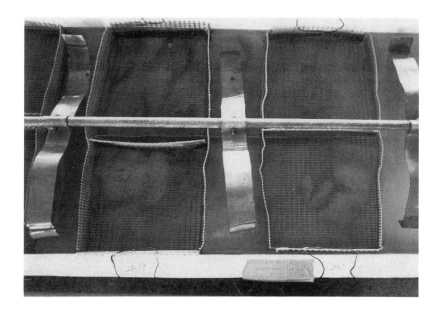

FIGURE 7.11.

A channel catfish spawn in an egg basket suspended in a hatching trough.

A series of paddles attached to a shaft is suspended along the length of the trough with enough space between paddles to allow wire-mesh baskets that hold the egg masses to fit between them (Fig. 7.11). Paddles, which are constructed of thin sheet metal so that they can readily be shaped to obtain optimal egg movement and water circulation, are generally 2 to 5 inches wide. The paddles should be positioned so that they reach about halfway to the bottom of the trough and should extend below the bottoms of the baskets so that their rotation will cause water movement through the egg mass similar to that produced by the male catfish. A paddle speed of 30 revolutions per minute (rpm) is maintained by a gearmotor or belt and pulley reduction system. A 1/20-horsepower, 30-rpm gearmotor is sufficient to power paddle wheels for two egg hatching troughs.

Baskets used to hatch catfish eggs are generally constructed from 1/4-inch galvanized or rubber-coated hardware cloth. Egg basket dimensions are about 12 inches wide, 24 inches long, and 4 inches deep. The baskets should be suspended just beneath the water surface, and five to six baskets can be placed into each trough. Each basket can hold two or three egg masses. As the eggs hatch the fry fall or swim out into the trough and tend to clump in the corners. They are then easily siphoned for transfer into fry rearing troughs.

Two fry rearing troughs should be available for each egg hatching trough. A single hatching trough could yield up to 200,000 eggs, and a similar number of swim-up fry could be handled in a fry trough of the same size. However, more fry rearing space is needed as the fish grow. Only about 100,000 10-day-old fry can be held in a similar tank. Some culturists use hatching and rearing troughs; others raise the fry in the egg hatching trough. In the latter case, the egg baskets and paddles are removed and aerators are placed into the trough. Fry troughs should be equipped with one or two 1/20-horsepower aerators, which must be covered with small-mesh window screening to prevent injury to the fry. The drain in the fry trough should also be screened to prevent escape of fry. Fry troughs can be constructed from wood, metal, plastic, or Fiberglas and are typically 8 to 12 feet long, 12 to 24 inches wide, and 10 to 12 inches deep. Water level should be maintained at about 8 inches by using a standpipe. In large tanks, use a fry holding box, which is a wooden box about 2 feet square and 1 foot deep. The bottom should be made of 1/16-inch window screen. Each box can hold 20,000 to 30,000 fry. If a fry box is used, water should spray directly into the box; otherwise adequate oxygen levels may be difficult to maintain within the box.

HATCHERY PRACTICES

Catfish eggs and fry are very susceptible to disease and should be handled with care. When the eggs are brought into the hatchery they should be tempered to the hatchery water temperature. As a rule of thumb, when the water temperatures differ by about 5°F, eggs and fry should be tempered before transfer. Bacterial diseases and fungal infections are threats in the hatchery. Disease problems usually are related to stress caused by crowding, poor sanitation, and deviation from optimal water temperatures. Thus the best disease control is prevention. A continuous supply of good-quality water, scrubbing and disinfection of troughs and equipment, and maintenance of the proper water temperature are essential. Two commonly occurring disease problems in catfish hatcheries are bacterial egg rot and fungus. Bacterial egg rot generally occurs when water temperatures are higher than 82°F and may be recognized as a milky white dead patch of eggs normally on the underside or center of the egg mass. Fungus grows on infertile or dead eggs when the temperature of hatchery water is below 78°F and appears as a white or brown cottonlike growth. Good sanitation and maintenance of proper water temperature for hatching help prevent these problems. A more detailed discussion of catfish diseases is presented in Chapter 11.

Hatching troughs should be stocked with egg masses of similar age if possible, and the egg masses should not be crowded into the baskets. Large egg masses should be broken apart to prevent problems associated with crowding, such as poor water circulation within the egg mass, which could result in the death of eggs inside the mass. After the eggs are in the hatchery and distributed properly, they should be checked several times daily for development of embryos and signs of disease. When the egg mass is checked, it should be shaken briskly to expel debris and turned over to change its position in the basket. The water in the hatching trough should not be murky, but should this occur, reduce the number of egg masses in the trough, clean the trough, and increase the water flow through the trough.

The incubation time for catfish eggs varies from 5 to 8 days, depending on water temperature. As the eggs mature they change from yellow to brownish before hatching. At hatching the sac fry are golden and fall through the egg basket and school in a tight cluster on the bottom of the tank. Sac fry do not need to be fed because they receive nourishment from the attached yolk sac. If a separate fry-rearing trough is used, the sac fry are siphoned into a bucket by using a 1/2-inch hose and then transferred to the rearing trough.

The yolk sac is absorbed by the fry in about 3 to 5 days, depending on water temperature. At this time, the fry turn black and begin to swim to the surface seeking food. They must be fed several times a day for good survival and growth. Though fry can be fed finely chopped liver and egg yolk, commercial feeds are generally used to feed fry. Fry feeds are dry, finely ground, nutritionally complete feeds that contain at least 50 percent protein. Most of the protein in the feed should be supplied by fish meal. Fry feeds are discussed in detail in Chapter 10. Fry should be fed all they will consume every 2 to 4 hours, 24 hours a day. Fry are fed for 2 to 10 days before being moved into nursery ponds. The amount of time they stay in the hatchery varies with the personal sentiment of the culturist and the production schedule of the hatchery. Fry should be checked daily and debris and uneaten feed removed.

Egg hatching and fry rearing troughs should be cleaned and sterilized before restocking with new egg masses or fry. Troughs are commonly cleaned with a dilute solution of household chlorine bleach. Chlorine is extremely toxic to eggs and fry, so troughs must be thoroughly rinsed after they are cleaned.

FRY INVENTORY METHODS

Using a hatchery provides an opportunity to inventory the fry before stocking in nursery ponds, allowing an estimate of the number of fingerlings that will be available for stocking production ponds. Numbers of hatched fry can be estimated from the weight of brood fish stocked. This is an extremely crude estimate and should not be used unless it is not possible to use more accurate methods. Egg numbers can be estimated by weighing the egg mass when they are brought into the hatchery, but egg numbers derived from weight of an egg mass are highly variable because the mass contains water and the amount of water varies with size, shape, and age of the egg mass. Egg numbers per weight of egg mass may differ as much as 35 to 40 percent using this method.

Two acceptable methods for estimating numbers of catfish fry are the volumetric and weight comparison methods. To use the volumetric method, count 300 fry into a graduated cylinder or cup containing a premeasured quantity of water, taking care not to add any extra water with the fry. Record the displacement of water in the container. Three to five samples should be taken and the results averaged. Then add the entire group of fry to a measuring container and record their water displacement. The number of fry can be calculated from the following equation:

$$\text{Number of fry} = 300 \times \frac{\text{water displaced by all fry}}{\text{water displaced by 300 fry}} \quad (7.8)$$

For example, three samples of 300 fry each raise the water level in a 100-milliliter cylinder by 11, 13, and 15 milliliters. On average the 300 fry displace 13 milliliters. The sample of fry to be inventoried raises the water level in a 1,000-milliliter cylinder by 450 milliliters. The number of fry in that sample is

$$\text{Number of fry} = 300 \times \frac{450}{13}$$

$$= 10{,}384 \quad \text{(round to 10,400)} \quad (7.9)$$

To estimate fry numbers by weight, weigh a sample of water (about 8 fluid ounces or 250 milliliters) and record the weight. Add 300 fry to the water, taking care to add as little extra water as possible, and record the new weight. Subtract the weight of water from the weight of water plus fry. Repeat this process at least three times. Then add the entire sample to a suitable volume of weighed water and record the weight. Subtract the weight of water from the weight of water plus fry. The number of fry is calculated as follows:

$$\text{Number of fry} = 300 \times \frac{\text{weight of all fry}}{\text{weight of 300 fry}} \quad (7.10)$$

For example, the weights of three groups of 300 fry each are 15, 17, and 18 grams. The average weight is then 16.7 grams. The sample of fry to be inventoried weighed 3,200 grams. The number of fry in the sample is

$$\text{Number of fry} = 300 \times \frac{3{,}200}{16.7}$$

$$= 57{,}844 \quad \text{(round to 57,800)} \quad (7.11)$$

Groups of fry vary in size, depending on the size of egg at hatching, fry density in the rearing trough, feed consumption, and number of days fed. Thus each distinct group of fry stocked should be inventoried.

REFERENCES

Bondari, K. 1983. Efficiency of male reproduction in channel catfish. *Aquaculture* 35:79–82.

Brauhn, J. L. 1971. Fall spawning of channel catfish. *Progressive Fish-Culturist* 33:150–152.

Busch, R. L. 1983. Evaluation of three spawning containers for channel catfish. *Progressive Fish-Culturist* 45:97–99.

Busch, R. L. 1985. Channel catfish culture in ponds. In *Channel Catfish Culture*, ed. C. S. Tucker, pp. 13–84. Amsterdam: Elsevier.

Jensen, J., R. A. Dunham, and J. Flynn. 1983. *Producing Channel Catfish Fingerlings*, Circular ANR-327. Auburn: Alabama Cooperative Extension Service.

Liao, P. B., and S. S. Lin. 1981. Ion exchange systems for water recirculation. *Journal of the World Mariculture Society* 12:32–39.

Steeby, J. A. 1987. Effects of spawning container type and placement depth on channel catfish spawning success in ponds. *Progressive Fish-Culturist* 49:308–310.

Tave, D. 1986. *Genetics for Fish Hatchery Managers*. Westport, Conn.: AVI Publishing.

Tucker, C. S. 1988. Water quality requirements for channel catfish hatcheries. In *Proceedings of the Louisiana Aquaculture Conference 1988*, ed. R. Reigh, pp. 60–66. Baton Rouge: Louisiana State University Agricultural Center.

CHAPTER 8

Fingerling and Food-Fish Production in Ponds

Growing fish in earthen ponds is an ancient practice, and ponds continue to be the most common fish culture system used worldwide. Over 95 percent of the catfish produced in the United States are grown in ponds. Nevertheless, pond culture of channel catfish is profitable only when the proper combination of resources is available. Water temperature cannot be controlled in ponds, and commercial pond culture is feasible only in regions that provide a growing season long enough to produce a 1-pound fish from egg in less than 18 months. Pond culture also requires large tracts of relatively inexpensive land. The land must be of the correct topography and soil type for economical construction and operation of ponds. Pond culture is water-intensive, and large volumes of high-quality water must also be readily available. Even when these resources are available, production of fish in ponds will be uneconomical if there is no market for the fish or if variable operating costs (particularly feed cost) are high. The proper industry infrastructure is necessary for profitable production of channel catfish.

When the proper combination of resources is available, production costs are generally lower for catfish grown in ponds than for any other culture system. This is due in large part to natural processes in ponds acting to maintain an adequate environment for fish growth. Photosynthesis and diffusion supply the oxygen required by fish, and the wastes produced by fish are processed by naturally occurring microorganisms. The capacity of the pond biological community to maintain adequate water quality for fish growth is limited, but rather high yields of fish are possible with limited technological intervention. Water quality management in more intensive culture systems generally requires more energy to move water, provide oxygen, and process and remove wastes. Production of catfish in systems other than ponds is economical only when some unique circumstance exists, such as the opportunity to sell fish at an exceptional price or the availability of an unusual resource (e.g., geothermal artesian groundwater; see the section Raceway Culture).

At first glance, the production of catfish in ponds appears to be a simple process. Fry are stocked into nursery ponds, fed daily, and harvested as fingerlings. These fingerlings are then stocked into food-fish production ponds, fed daily, and harvested when they reach a size desired for processing. This simple scenario is complicated by a number of management decisions that must be made to optimize the production strategy for each farm. The complex biology of pond culture systems also results in many day-to-day decisions regarding feeding practices, water quality management, and disease control.

This chapter discusses general features of pond culture systems, management strategies for fingerling and food-fish production, water use in ponds, economics of growing catfish in ponds, and culture of secondary fish species in catfish ponds. The daily management of ponds, including water quality management, feeding, and fish disease control, is discussed in later chapters.

POND CULTURE SYSTEMS

Ponds used to culture channel catfish are classified as either levee ponds or watershed ponds, although in some situations the classification is indistinct. Levee ponds (also called dike ponds) are built by removing dirt from the area that will be the pond bottom and using that dirt to form levees around the pond perimeter. There is no watershed, so a source of pumped surface water or groundwater must be available. Watershed ponds (also called hill ponds) are made by building a dam across a watercourse. The main water supply is runoff from the watershed above the dam. Runoff is sometimes supplemented with pumped water from a well.

Both types of ponds are successfully used to grow channel catfish, but levee ponds are preferred because they can be built adjacent to one another, making management practices such as feeding and water quality monitoring more convenient. The regular features and relatively flat

Pond Culture Systems   **167**

bottoms of levee ponds also make fish harvest easier. Fish can easily escape around or under seines during harvest of watershed ponds, which often have irregular bottom features. Precipitation is sporadic, and dependence on runoff as the primary water source can limit the success of watershed ponds that are too deep or too irregular to harvest by seining unless the pond water level is dropped. If the pond is drained during dry periods, considerable time may elapse before the pond refills and can be restocked with fish. Valuable production time is lost. Also, during prolonged droughts the pond volume may decrease to such an extent that fish densities become excessively high and water quality deteriorates. Most watershed ponds are best suited for use when channel catfish production is a supplemental source of income. Under these conditions, fish stocking rates are lower and management is easier.

The following sections outline the general features of channel catfish culture ponds. Complete accounts of surveying and construction methods are presented by Anonymous (1982) and Tisdale (1982).

Levee Ponds

Adequate water supplies, flat land, and soils that hold water are the basic requirements for levee pond sites. Soils should also be free of harmful pesticide residues. The area should be large enough to accommodate the desired number of ponds along with areas for feed and equipment storage. The site must be accessible to an all-weather road and should be near a source of electricity.

Water source. Groundwater is the preferred source for levee ponds. The supply of groundwater is usually dependable in the short run, and it is free of wild fish and is less often polluted with wastes or pesticides than surface waters. The water supply should be of proper salinity, total alkalinity, and hardness (Chapter 4). The water should also be free of pesticides or other harmful substances. Most groundwaters are devoid of oxygen, but the water is rapidly oxygenated after it is added to the pond. High concentrations of dissolved iron are occasionally present in some groundwaters, but this poses no problem in pond culture systems. The iron is rapidly oxidized and precipitated as iron oxide, which settles to the bottom. The suitability of a groundwater supply can be assessed from chemical analyses of the supply obtained from local water-well drilling companies or state water resource agencies, or from analysis of a test well. If there are other fish farmers in the area, use their experiences to assess potential problems with a water supply. If no other fish farms are located nearby and the quality

TABLE 8.1.

Water Volumes Pumped in 1 Day at Different Pumping Rates

Flow Rate (Gallons/Minute)	Acre-feet/Day	Flow Rate (Gallons/Minute)	Acre-feet/Day
50	0.22	1,000	4.42
100	0.44	2,000	8.84
250	1.10	3,000	13.26
500	2.21	4,000	17.68
750	3.31	5,000	22.09

Note: Volume in acre-feet. To determine the approximate time (in days) required to fill a pond, divide the pond volume (in acre-feet) by the acre-feet pumped in a day.

of the water is suspect, a pilot-scale facility can be constructed to ensure that the water is suitable for use.

Surface water supplies include streams, rivers, lakes, and reservoirs. Evaluate surface waters carefully before use because they are subject to contamination and often have high silt loads and their availability and quality may vary dramatically with time. Surface waters also contain wild fish, fish disease organisms, and waterborne predators. Water inlets often are fitted with fine-mesh screens to prevent introduction of wild fish from surface water supplies. These screens are not practical for debris-laden waters or systems using high flow rates.

The supply should provide enough water to replace losses to evaporation and seepage and to fill ponds in less than about 2 weeks. If ponds fill slowly, production time can be lost and weed problems are more likely. The absolute minimum water requirement for levee ponds is 15 gallons/minute/acre of water. This is about twice the maximum daily evaporation rate from ponds in the southeastern United States; it allows maintenance of water levels during drought periods with some excess supply available to meet moderate seepage losses. Typically, one well supplies four levee ponds, which average about 17 acres each. This well must supply about 1,000 gallons/minute to meet the minimum requirement of 15 gallons/minute/acre. If each 17-acre pond is 4 feet deep, about 15 days of constant pumping is required to fill one pond (Table 8.1). Ideally, at least 30 gallons/minute/acre of water should be available. This would necessitate a well supplying about 2,000 gallons/minute to four 17-acre ponds; 8 days would be required to fill each pond.

The availability of water must be assessed before building ponds. Depth to water and the geology of the water-bearing formation determine well construction and pumping costs, so shallow aquifers (depth to water less than 250 feet) are preferred. Contact local water-well drillers about water availability. If information is limited, have a test well drilled to assess the depth to water, available flow rates, and water quality.

Permits or registration is required in many states before a water well can be drilled. Personnel with the local office of the United States Soil Conservation Service should be able to supply information regarding permit procedures.

Land and soils. Restrictions may apply to the use of certain lands for agricultural purposes, including building ponds for catfish farming. A permit is required from the United States Army Corps of Engineers before clearing or building on lands classified as wetlands. The definition of wetlands and the permit procedure are in a state of flux at the time of this writing. Some states may also require additional permits before construction of ponds. Check with the local Soil Conservation Service office for information before purchasing land or initiating any pond construction, even if you already own the land.

The lay of the land determines the amount of dirt that has to be moved during construction. Less dirt moving is required on flat land; hence construction costs are lower than on rolling or hilly land. The size of individual ponds is virtually unlimited on level land, but only small levee ponds can be built on rolling land.

The site must allow drainage by gravity at any season. Avoid sites that are seasonally flooded. High water could limit access to ponds and make it impossible to harvest or feed fish until waters recede. Fish will be lost if flood waters rise above the levees.

Check with local utility companies to make sure the proposed pond site does not lie over pipeline or under power line right-of-ways. Also make sure that ponds will not block drainage from nearby lands or cause any damage off the property.

Soils in the proposed pond site should contain at least 20 percent clay to minimize water lost to seepage. Soils can be initially assessed by this simple test: make a ball out of slightly moistened soil and drop it to the ground; a good soil for ponds will remain intact. Take a series of soil borings over the entire site to assure that sand or gravel beds do not underlie the site. Borings should extend at least 3 feet beneath the surface. Local Soil Conservation Service personnel can offer guidelines and assistance. Ponds located on pervious soils can be lined with plastic or a "blanket" of clay to reduce seepage, but such measures are too expensive to be used under most commercial situations.

Chlorinated hydrocarbon insecticides such as toxaphene, dieldrin, and endrin were widely used on row crops before 1975. Although the use of these pesticides is now prohibited, they are long-lasting and concentrations are still high in some soils. High concentrations of pesticides in soils used to construct ponds are undesirable because the resulting residues in fish may exceed United States Food and Drug Administration tolerance limits or they may directly affect fish health. The recommended maximum concentration of toxaphene in soil to be used for ponds is 0.5 ppm. The combined concentrations of dieldrin and endrin should not exceed 0.1 ppm. These are general guidelines and have no regulatory status.

Chlorinated hydrocarbon pesticides are generally found only in the top few inches of soil, so samples for analysis should be taken to a depth not to exceed 6 inches. When taking samples, pay particular attention to low areas where runoff collects, any site previously used for pesticide storage or disposal, and areas where aerial or ground pesticide application equipment has been loaded or washed. Pesticide analyses can be conducted by private laboratories or, in some states, by government chemical laboratories.

If the top layer of soil is contaminated, but the soil beneath is pesticide-free, move the top layer and incorporate it into the core or outside slope of the levee. Do not use contaminated soils for outside levees if future expansion of ponds will use that levee as an inside slope.

Pond morphometry. The ideal pond size and shape are a compromise among construction costs, operating costs, management ease, topography of the land, and property lines. Small ponds are generally easiest to manage. There is also some indication that fish production per acre is somewhat greater in small ponds than in large ponds. However, construction costs increase as pond size decreases and small ponds have less water area on a given amount of land because more area consists of levees. As a compromise, most ponds in current use are about 17 acres of water, built on 20 acres of land. The actual number and size of ponds depend on production goals and type of farm. For instance, some farmers prefer to have several small ponds (5 to 10 acres) for brood fish or fingerlings and remaining ponds 15 to 20 acres for food fish.

Most ponds are rectangular and should be no more than 700 feet wide to facilitate harvesting. Actual shape depends on topography and property lines. Irregularly shaped ponds with peninsulas, bays, or islands are extremely difficult to manage and harvest and should be avoided.

Minimum pond depth should be about 3 feet; maximum depth should not exceed 5 feet. The height of the levee above normal water level (freeboard) should be 1 to 2 feet. Greater freeboard increases con-

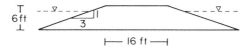

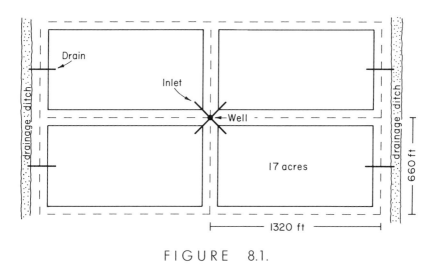

FIGURE 8.1.

Cross section of a pond levee (top) and typical layout for four 17-acre levee ponds (bottom).

struction costs and makes it difficult to get equipment in and out of the pond. Levees with less than 1 foot of freeboard are quickly eroded. The bottom of the pond should have a slope to the drain of 0.05 to 0.1 percent to facilitate draining.

Levees should be at least 16 feet wide at the top to accommodate feeding and harvesting equipment. Wider levees are more convenient and are more resistant to wave damage but are more expensive to build. Narrower levees are quickly eroded and more hazardous in wet weather. The slope on the levee depends on soil type, but a 3:1 slope is usually satisfactory (Fig. 8.1). More gentle slopes are less erodable but more costly to build and encourage the growth of aquatic weeds. Levee tops on at least one side of each pond should be graveled to permit all-weather access. The side slopes should be seeded with a perennial grass to reduce erosion.

Drains are extended through the levee at the lowest end of the pond. The elevation of the pond bottom at the drain must be 1 or 2 feet above the water level in the drainage ditch. Two types of drains are commonly used: inside swivel drains and outside valve drains (Fig.

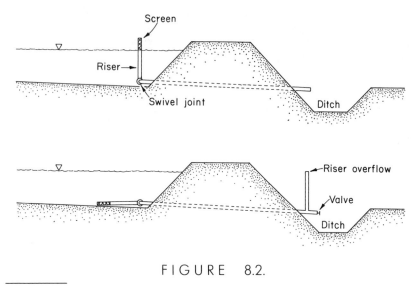

FIGURE 8.2.

Two types of drains used in levee ponds: inside swivel drain (top) and outside drain with valve (bottom).

8.2). Outside valve drains are easier to maintain and use. Drainpipes for ponds larger than 10 acres should be about 12 inches in diameter to facilitate rapid draining. Drain lines should have at least a 1 percent slope for good flow.

The well is usually drilled after levees are constructed. The number of ponds served from one well depends on pond size and well pumping capacity. As pointed out earlier, one 2,000 gallon/minute well can service four 17-acre ponds. If possible, locate the well at the intersection of the four ponds to reduce the length of pipe required (Fig. 8.1).

Watershed Ponds

Site selection for watershed ponds is more difficult than for levee ponds because the topography is irregular and watershed requirements must be considered. Ponds for commercial catfish farming preferably should be built in gently sloping, shallow valleys rather than deep valleys with steep slopes. Ponds built in shallow valleys can be cut and filled to make the basin of relatively uniform shallow depth (Fig. 8.3) to facilitate harvest. Deep ponds will have to be partially drained for harvest or harvested by trapping fish. Water quality is also more difficult to manage in deep ponds because the water tends to stratify during

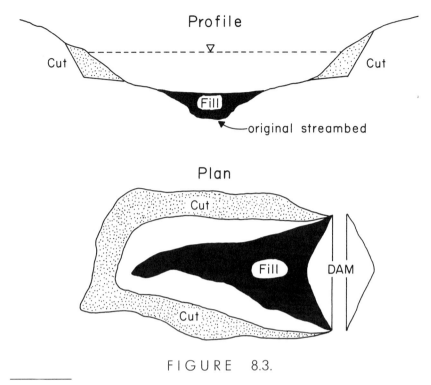

FIGURE 8.3.

By cutting and filling areas in a watershed pond the bottom profile can be made uniform to allow harvest by seining.

hot weather with cool, oxygen-depleted water on the bottom. This foul water may mix with surface waters during extreme weather events and kill fish (see the section Dissolved Oxygen and Aeration).

When surface runoff is the main source of water, the watershed must be large enough to maintain water in the pond during dry seasons but not so large that elaborate overflow structures are needed to handle excessive runoff during storms. The amount of watershed required per acre-foot of storage varies with climate, soils, and topography. In the southeastern United States, from 2 to 5 acres of watershed is required for each acre-foot of pond storage (Anonymous 1982). The watershed should not receive drainage from feedlots, industrial sites, orchards, or row-crop fields to prevent pollution of the pond water. Runoff must also be relatively free of suspended matter to prevent excessive turbidity in the pond and to prevent rapid filling with silt. Land under cover of trees or grasses is the most desirable drainage area.

Most of the site selection criteria mentioned for levee ponds are also applicable to watershed ponds. Soils should be relatively impervious and pesticide-free; the site should be accessible all year and should be near a source of electricity. Be sure to check with the Soil Conservation Service about possible permit requirements before building any ponds.

A supplemental water source is highly desirable even in areas with abundant annual rainfall. A small well of 10 to 15 gallons/minute/water acre will supply enough water to prevent dangerous drops in water level during drought periods. Water from the well can also be used to supplement runoff to hasten refilling after draining for fish harvest or repair. This will reduce lost production time.

Design and construction of watershed ponds vary greatly, depending on topography. Consult Anonymous (1982) for specific information.

FINGERLING PRODUCTION

After their brief stay in a hatchery, fry are moved to a nursery pond for further growth. To ensure good fry survival, ponds must be properly prepared before stocking. Fry must be stocked at a density that will result in production of large numbers of fingerlings of the desired size. Proper feeding practices and water quality management are then required for rapid growth and healthy fingerlings.

Pond Preparation

Before stocking of fry, the pond must be free of wild fish and aquatic insects that feed on fry. Both the water inlet and drain structures of the nursery pond should be constructed in a manner to prevent the entrance of "trash" fish such as the green sunfish. Predation by green sunfish can completely destroy the fry population. Undesirable fish can be eliminated by drying the pond before stocking. If wild fish are present and the pond cannot be drained or if fish remain in potholes after draining, they can be destroyed before stocking the fry by treating the water with rotenone or chlorine. Rotenone is available in both a powder and an emulsifiable 5 percent rotenone liquid; the liquid is preferred. The recommended dosage for the 5 percent rotenone is to provide a concentration of about 1.0 ppm of the product; the actual concentration of rotenone is only about 0.05 ppm. This is equivalent to adding about 1 gallon of the 5 percent product to a 1-acre pond with an average depth of 3 feet. Dosages in excess of 2.0 ppm may be required to control certain species, such as bullheads and mosquito fish. Rotenone is more toxic at higher temperatures and should be applied at temperatures greater than 60°F. Rotenone toxicity may persist for a week or two in warm waters and even longer if applied at cool water temperatures; thus care should be used to assure that the water is no longer toxic before stocking fry. Potassium permanganate can be used to detoxify rotenone if necessary; 2 ppm detoxifies about 0.05 ppm of rotenone (1 ppm of the 5 percent product).

Chlorine can be used to kill unwanted fish in a nursery pond before stocking. Calcium hypochlorite (70 percent available chlorine) kills all fish at a concentration of 10.0 ppm active ingredient; this is equivalent to adding about 39 pounds of calcium hypochlorite (70 percent active) to 1 acre-foot of water. Chlorine deteriorates rapidly and generally is not toxic after a few days. Sodium thiosulfate or sodium sulfite can be used to neutralize chlorine. Chlorine will kill most organisms in the pond as well as fish.

Antimycin A (Fintrol®) is a selective fish toxicant approved for use in ponds used to grow food fish. It is effective against scaled fish and can be used in catfish ponds without harming the catfish. To use antimycin A, follow the label instructions closely because its activity is dependent on water chemistry and temperature.

Two types of insects can cause problems in catfish fry ponds by preying on the fry. They may be classed, on the basis of their method of respiration, as either "air breathers," which must come to the surface, or "gill breathers," which respire under water. The air breathers, such as the back swimmers (order Hemiptera, family Notonectidae), can be controlled by applying oil to the pond to form a film on the pond surface. The film prevents insects from respiring at the surface. A variety of oils have been used, but a common practice is to spread a mixture of 3 to 5 gallons of diesel fuel or kerosene mixed with 1 quart of motor oil per acre over the pond surface 2 days before stocking or at 4-day intervals for 2 to 3 weeks after the fry have been stocked. The oil mixture should be applied on a calm day when there is just enough breeze to spread the film over the pond. Supplemental feeding should be discontinued until the film has dissipated.

Problems with insects that are gill breathers, such as aquatic nymphs (naiads) of the dragonflies (order Odonata), can be prevented by stocking fry as the pond is being filled. Aquatic nymphs cannot migrate as easily as certain other aquatic insects, and more time is required for their colonization of a new pond. However, if the pond is not stocked soon after being filled, the nymphs can increase in number and cause a serious reduction in fry survival. Certain chemicals can be used to control both types of insects before stocking of fry; however, they have not been approved for use in fish reared for food in the United States.

Catfish fry nursery ponds should be fertilized to establish a phytoplankton bloom that will help prevent weed problems and provide natural food for the fry. Liquid fertilizers are particularly effective at promoting phytoplankton growth. Apply 2 to 4 pounds/acre of liquid fertilizer (10-34-0 or 13-38-0) every other day for 8 to 14 days or until a noticeable phytoplankton bloom develops.

Stocking the Pond

After the nursery pond is properly prepared, the fry can be transferred from the hatchery to the pond. Fry should be stocked in early morning while it is cool. Gently transport the fry in buckets or other suitable containers and slowly condition them to the pond by gradually adding small quantities of pond water to the fry container. Shelters such as spawning containers or frames covered with black plastic can be used to improve survival rate and to provide areas for fry to congregate, aiding in feeding. Although yolk-sac fry can be stocked directly into the nursery pond after hatching, survival rate is improved if the fry are reared in the hatchery to swim-up stage. If possible, fry of similar age and size should be stocked together to reduce size variation at harvest. The fry should be stocked at a density to provide the size of fingerlings desired at harvest. The relationship between fingerling size and fry stocking rate is given in Table 8.2; the production indicated in the table can be achieved with good management. Survival of catfish fry to fingerlings in excess of 80 percent is good. As indicated, anticipated production is dependent on good management, as well as initial stocking density.

Feeding the Fry and Fingerlings

Catfish fry are usually fed on the basis of their standing crop weight. Feed should be distributed along the shoreline in amounts to ensure that it is available to the fry. Fry and small fingerlings should be fed frequently, at least two or three times daily for the first 2 weeks. A finely ground high-protein feed (45 to 50 percent protein with about 60 percent of the protein from fish meal) should be given. Feeding rates up to 50 percent of the fry standing crop weight have been used. It is better to overfeed initially than to underfeed. Overfeeding is not critical during the first few weeks because the biomass is low and any excess feed will serve to fertilize the pond; however, as the fish grow, care must be used not to feed enough to cause water quality problems. Feeds

TABLE 8.2.

Estimated Fingerling Size after 120- to 150-Day Growing Season at Different Stocking Densities

	Fingerling Size	
Fry Stocking Density (Fish/Acre)	Average Length (Inches)	Approximate Weight (Pounds/1,000 Fish)
10,000	7–10	150
30,000	6–8	95
53,000	5–7	60
73,000	4–6	40
95,000	3–5	30
120,000	3–5	20
140,000	3–4	10
200,000	2–3	5
300,000	1–2	3
500,000	1	1

Source: Jensen, Dunham, and Flynn (1983).

for feeding fry and fingerlings, and feeding techniques are discussed in more detail in Chapter 10.

The fry are usually not seen after stocking into nursery ponds for several weeks. Eventually they will be observed feeding on the supplemental feed. At that stage they are classified as small fingerlings. They typically school and swim near the pond surface in the late afternoon. They appear as a large black mass. This activity is normal and should not be interpreted as a sign of stress.

Water Quality Management in Nursery Ponds

The principles of water quality management in nursery ponds are fundamentally similar to those in food-fish growout ponds (Chapter 9). However, nursery ponds are managed somewhat differently than growout ponds, and minor differences do exist in water quality management practices.

Nursery ponds are drained between crops and are filled with high-quality water just before stocking with fry. Amounts of feed added to the pond initially are low, so water quality is good for 1 or 2 months after stocking. This sometimes lulls new producers into a false sense of well-being. Fingerlings begin to grow rapidly after they reach 2 to 3 inches in length, and feed allowances must rapidly increase in response to growth. When feed allowances exceed 30 to 50 pounds/acre/day, supplemental aeration is required to keep fish alive. By 5 or 6 months after stocking, the total weight of fish in the pond can be quite high and large amounts of feed are necessary to meet the requirements of the fish. At that time, water quality problems are as common in nursery ponds as in food-fish growout ponds. The producer should be aware of this and monitor water quality from the day of stocking until the pond is completely harvested.

Low dissolved oxygen concentrations are particularly troublesome in nursery ponds. Fry or small fingerlings may not yet be trained to seek oxygenated water near an aerator. Also, small fish are relatively weak swimmers and cannot travel long distances through oxygen-poor water to get to an aerator. Aeration devices should be turned on well in advance of anticipated problems, so that the fish have plenty of time to find the zone of oxygenated water before dissolved oxygen concentrations fall to stress-provoking levels. Small fish consume much more oxygen per pound than large fish, so aerators in nursery ponds should be sized to provide more dissolved oxygen per acre of water than aerators in food-fish growout ponds. The use of smaller ponds (<10 acres) for nursery ponds helps to alleviate these problems. Small ponds are easier to aerate, and the fish have less distance to swim to find oxygenated water near an aerator.

Harvesting Fingerlings

Fingerlings are harvested and transported in basically the same manner as food-sized fish. These principles are discussed in Chapter 12. There is, however, a fundamental difference between fingerling and food-fish harvesting: fingerlings are harvested and moved to another pond for further growth; food fish are usually harvested for slaughter. Fingerlings must be handled in a manner that minimizes stress and maximizes survival after restocking. If fingerlings are mishandled, particularly in warm water, they will be susceptible to secondary, stress-related infectious diseases.

Fingerlings should not have an established infectious disease when harvested. The stress of handling may kill many of the fish already weakened by the disease. The disease also may be spread to other ponds as the fingerlings are stocked. Do not seine fingerlings immediately after feeding or those from ponds with stress-causing levels of low dissolved oxygen, un-ionized ammonia, or nitrite. Again, the added stress of handling could result in excessive mortality rate.

Every effort should be made to complete the harvest as quickly

Fingerling Production **179**

FIGURE 8.4.

A fingerling grader box being used to sort fingerlings by size.

and efficiently as possible. Load the fingerlings onto transport trucks immediately after crowding, unless fish are to be graded in the pond. Position live cars used for grading so that oxygenated water from an aerator or well flows through the live car. However, the live car should not be placed in heavy currents that force fish to swim vigorously.

Fingerlings in most large commercial nursery ponds vary considerably in size. Fry stocked into the pond just a week or two apart may differ in length by 2 or 3 inches after 150 days of growth. The smaller fish resulting from the later stocking may not be of a size desired for stocking into food-fish growout ponds. If fingerlings vary considerably in size, they should be graded to assure that most of those harvested will be large enough to grow to food-fish size within the desired period. Generally, fish larger than 6 inches are preferred for growout; fish of this size can grow to 1 pound or more in an additional 150 to 200 days.

Small numbers of fingerlings can be graded through grader boxes (Fig. 8.4) or through bar graders in a tank. This is generally too slow and labor-intensive for the large numbers of fingerlings harvested from commercial ponds. Most fingerling producers who grade fish do so with live cars or seines with a mesh size that selectively retains fish of a certain minimum size (Table 8.3). A popular seine square mesh size

TABLE 8.3.
Net Mesh Sizes and Box Grader Bar Spacings Required to Grade Fingerlings

Approximate Fingerling Size Retained		Square Mesh Size (Inches)	Spacing between Bars (Inches)
(Inches)	(Pounds/1,000 Fish)		
8	140	1	1
7	95	3/4	56/64
6	60	5/8	48/64
5	35	1/2	40/64
4	20	3/8	32/64
3	7	—	27/64
1–2	2	1/4	—

for grading is 5/8 inch; this mesh size retains fingerlings of about 6 inches (60 pounds/1,000 fish) and up. Preliminary seining and grading with a 1-inch square mesh seine or live car is sometimes beneficial if size variation is great. This harvest will consist of large fingerlings of 8 inches (140 pounds/1,000 fish) or more. Fish of this size are highly desirable in food-fish ponds managed under the single-batch production system because they can be grown to "harvest size" in one growing season.

Grading occurs when fish are crowded in the seine or live car after seining the pond. Grading efficiency depends on the degree of crowding, water temperature, variation in fish size, and grading time. Inspection of samples taken during grading will indicate the progress of grading.

FOOD-FISH PRODUCTION

When fingerlings reach some desired size (usually at least 5 to 6 inches long), they are harvested from the nursery pond and restocked into growout ponds. Fingerlings are stocked into growout ponds at roughly one-tenth to one-twentieth the density of nursery ponds because fish will be ten to twenty times heavier when harvested as food-sized fish. Channel catfish grow best when water temperatures are above 70°F, but there is no well-defined production schedule on most commercial catfish farms. Food-sized fish are harvested and fingerlings are stocked year-round. Ponds on a given farm contain fish at various stages of growout throughout the year.

A variety of cropping systems and stocking rates are used in growout ponds; the optimal strategy differs for each farm situation. Farm managers must assess the advantages and disadvantages of different production systems and decide which one will best achieve production goals given the resources available. Each pond on the farm and each new production cycle will present a different set of factors to be considered, so it is important to be flexible and not become locked in to a single production strategy.

Size of Fingerlings to Stock

A large fingerling will reach harvestable size faster than a small fingerling, but more time and pond space are required to produce a large fingerling; consequently, they cost more to grow or purchase. The best fingerling size to stock is thus a compromise that depends on the type of cropping system (discussed later), fish stocking density, and length of growing season.

In ponds managed under the *single-batch* cropping system on a 1-year production cycle, relatively large fingerlings are needed because small fingerlings will not grow to harvestable size in 1 year. Ideally, the fingerling should reach 1 to 1.5 pounds in a 150- to 200-day growout period or less. Fingerlings smaller than about 5 inches will not grow to an average weight of 1.25 pounds in a single 200-day growing season in ponds stocked at 4,000 to about 6,000 fish/acre. This does not mean that fingerlings should average 5 inches in length. Pond-raised fingerlings are extremely variable in size and, unless graded, a group of fish that averages 5 inches in length will contain many fish that are less than 4 inches and will not achieve market size in a single growing season. If possible, the group of fingerlings stocked should contain few fish under 5 inches. For pond-run (ungraded) fingerlings, this means that the average length should be at least 6 to 7 inches. Fish of this size weigh about 60 to 90 pounds/1,000 fish. Fingerlings of this size often are not available and are expensive. Food-fish producers often are forced to use smaller fingerlings, many of which will not grow to market size in a single season. If only small fingerlings are available, it may be advisable to stock at a higher density (up to 8,000 fish/acre) and manage the pond on a 2-year production cycle. As the fish grow, larger fish are cropped off (without restocking) until all fish have been harvested at the end of the second growing season.

Alternatively, smaller fingerlings can be used in the *multiple-*

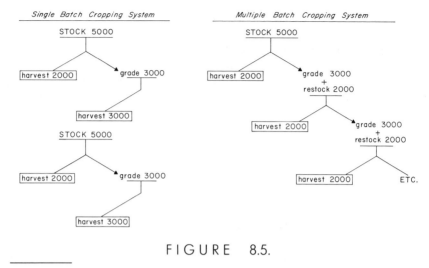

FIGURE 8.5.

Schematic diagram of the two cropping systems used in production of channel catfish ponds.

batch cropping system, in which fingerlings are restocked after each partial harvest. However, even in the multiple-batch system, fingerlings larger than 5 inches are desirable because they compete for feed more aggressively than smaller fingerlings and are less subject to predation by the larger fish that are usually found in multiple-batch ponds.

Cropping Systems

Two cropping systems are used to produce food-sized fish: single-batch (also called "clean harvesting") and multiple-batch (also called "topping," "continuous harvesting," or "understocking"). The two cropping systems are illustrated in Figure 8.5.

Single-batch. In the single-batch production system, the goal is to have only one year-class of fish present in the pond at any one time. Fingerlings are stocked into the pond, grown to harvestable size, and completely harvested before restocking with another batch of fingerlings. This system has two variations. The simpler variation is to harvest the entire crop at one time. This practice is common in watershed ponds that must be drained for harvest; no alternative exists other than to harvest the entire crop. Alternatively, faster-growing fish can be selectively removed by using a seine that allows small fish to grade through the openings. Small fish remain in the pond for further growth.

This grading process may be repeated two or three times until the entire crop is harvested. Then the pond is either drained and restocked or restocked without draining. Draining the pond allows all fish to be removed for absolute adjustment of inventory. If the pond is not drained, water is conserved and less production time is lost between successive crops, but some large fish will remain in the pond because it is generally not possible to remove all fish in a large pond by seining alone. It is important, however, to remove as many fish as possible before restocking to maintain accurate inventory records. If many large fish remain, they also may eat some of the fingerlings subsequently stocked, particularly if the fingerlings are less than 5 inches long.

Multiple-batch. In the multiple-batch system, several different batches of fish of different sizes and ages are present in the pond. Initially a single group of fingerlings is stocked. The faster-growing fish are selectively harvested. An estimate is made of the number harvested and an equal number of fingerlings is then stocked. At this time two groups of fish are present: subharvestable fish from the initial stocking and the newly introduced "understocked" fingerlings. This process of grading and restocking continues indefinitely without draining the pond. Ponds are usually drained only for repair or occasional adjustment of inventory.

Comparison of cropping systems. The multiple-batch production system was developed in the early 1970s when the farm-raised catfish industry was small. The system allowed harvest from small farms throughout the year to meet the year-round needs of processors. The industry is now large enough to assure year-round supplies of fish using single-batch culture with individual ponds being harvested throughout the year, but the multiple-batch system is still the more commonly used strategy.

Maintaining accurate inventory records is difficult for multiple-batch ponds. Several different batches of fish exist after several croppings, and each batch has its own average weight and stocking number. Each time fish are harvested, some large fish escape and some small fish do not grade and are sent to market. Grading is particularly inefficient in cold water because fish are inactive. After 1 or 2 years, it is almost impossible to estimate accurately the overall weight or number of fish in the pond, much less the numbers and weight of particular batches.

Harvested fish are of more uniform size in the single-batch production system because more effort is made to entirely remove one crop before restocking. Although some larger fish evade capture in single-batch ponds (unless the pond is drained), the problem is generally

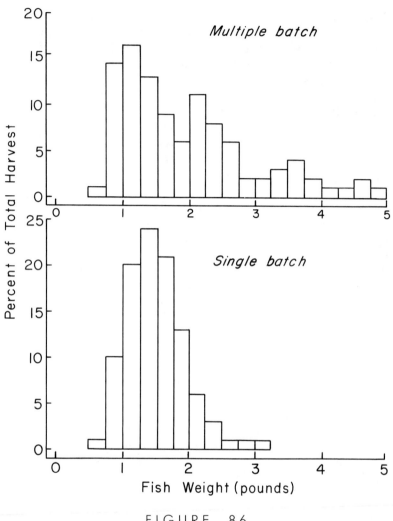

FIGURE 8.6.

Distribution of fish sizes for fish harvested from ponds under two cropping systems. Stocking density in both ponds was 4,500 fish/acre and ponds were in the second year of production without being drained between crops.

worse in multiple-batch ponds (Fig. 8.6). A recent practice introduced by some processors is to price fish according to size. Fillets from fish of 0.75 to 2.5 pounds are in highest demand and fish of this size bring the highest price. Producers are paid less per pound for larger fish. If differential pricing by size becomes a common practice, the large variation in size of fish harvested from multiple-batch ponds will become a serious disadvantage to that cropping system.

The presence of many large fish may directly affect overall production in multiple-batch ponds. The large fish compete aggressively for feed and may prevent the smaller fish from consuming enough for optimal growth. Larger fish also convert feed to flesh less efficiently than small fish, and the overall feed conversion efficiency for fish in a multiple-batch pond may be poor compared to that for fish in ponds with a more homogenous population. Large fish may also eat some of the small understocked fingerlings.

Infectious diseases can be a severe problem regardless of the production system, but the chances of spreading diseases among populations of fish are greater on farms using the multiple-batch production system. Enteric septicemia of catfish (ESC), caused by the bacterium *Edwardsiella ictaluri* (see the section Bacterial Diseases), is particularly easy to pass to or from different populations of fish within a pond. If a group of fish becomes infected with *E. ictaluri*, surviving fish develop a partial immunity to further infection but may become carriers of the bacterium. A severe outbreak of ESC may then occur if these fish are stocked into a pond containing previously unexposed fish with no immunity. Transmission of the disease in this manner is less likely in single-batch ponds.

Despite the disadvantages of the multiple-batch system, it remains the more prevalent pond production system. Theoretically, multiple batches result in greater overall fish production by using the capacity of the pond microbial community to process fish wastes more completely. When fish are initially added to a single-batch pond, amounts of feed added are low because the total weight of fish is low. During this early part of the production cycle, a large volume of water is being used to produce little daily weight gain. In the multiple-batch system, large standing crops of fish are always present, feeding rates are high, and production per acre is, in theory, greater. Of course, the higher average amounts of feed added to multiple-batch ponds mean that problems with poor water quality are more frequent than in single-batch ponds.

On a given farm, more ponds will contain harvest-sized fish when ponds are managed by multiple-batch rather than single-batch production. If fish in some ponds are unmarketable because they are temporarily off-flavor, fish in other ponds can be harvested while waiting for the off-flavor episodes to abate. Use of the multiple-batch system may thus have some advantage on small farms where the number of ponds from which to select fish for harvest is limited.

Presently, it is not known to what extent the theoretical production increases expected in the multiple-batch system are offset by slower growth of the understocked fish and other problems inherent in this system. In practice, successful use of the multiple-batch system depends on timely and effective harvesting of large fish, efficient grading of small fish, maintenance of adequate water quality, and effective disease control. The advantages and disadvantages of the two basic production systems are summarized in Table 8.4.

Stocking Densities

In general, increased fish density in ponds increases total harvest weight but decreases average fish size. Fish grow more slowly (Fig. 8.7) at high densities because they compete for feed and because poor water quality in densely stocked ponds causes stress, which reduces feed intake and efficiency of feed conversion. Optimal stocking density is a compromise among the higher production possible at high fish densities, slower fish growth at high densities, and degree to which water quality deterioration can be tolerated.

Water quality deterioration is a function of the sustained maximum daily feeding rate used in a pond. As a rough guide, little or no supplemental aeration is required if maximum daily feeding rates do not exceed about 35 pounds of feed/acre. Episodes of critically low dissolved oxygen occur with increasing frequency as feed allowances increase above this level. When maximum sustained feeding rates exceed about 100 pounds of feed/acre/day, water quality deterioration is severe and considerable risk is involved in producing fish.

To limit water quality deterioration to some manageable level, some catfish farmers set a maximum allowable daily feeding rate, which determines the stocking density. For example, maximum daily feed allowances must not exceed 35 pounds/acre if aeration devices are not available. For a single-batch pond with all fish harvested at one time, the maximum stocking density can then be determined from the desired average fish weight at harvest and an estimate of the feed consumption rate of fish at their harvest weight:

1. Assume the desired harvest size is 1.5 pounds per fish. Fish of this size consume about 1.3 percent of their weight per day.
2. The total weight possible at harvest without limiting feed intake by fish is equal to the maximum allowable daily feeding rate divided by the daily feed consumption rate of the fish, or 35 pounds feed/acre/day ÷ 0.013 pounds feed/pounds fish per day = 2,692 pounds of fish/acre.
3. The stocking density to achieve this total harvest weight is equal to total harvest weight divided by average desired har-

TABLE 8.4. Advantages and Disadvantages of the Single-Batch and Multiple-Batch Production Systems

Single-Batch		Multiple-Batch	
Advantages	Disadvantages	Advantages	Disadvantages
Easier maintenance of inventory records	Low initial standing crop that does not use maximum pond carrying capacity	Possible greater long-term production	Difficulty of maintaining inventory records
More uniform fish size at harvest	Standing crop very high immediately before harvest unless cropped	More ponds containing harvestable fish	Presence of large fish, which may suppress growth of understocked fingerlings
Lower probability of disease transmission between fish stocks			Poorer overall feed conversion
			More water quality problems
			Increased probability of disease transmission between fish stocks

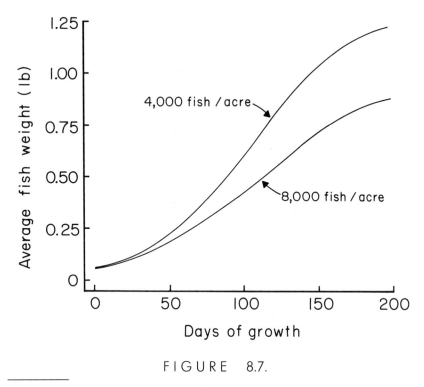

FIGURE 8.7.

Fish growth in ponds managed as single-batch systems at two different stocking densities. Both ponds were stocked with fingerlings weighing 60 pounds/1,000 fish (adapted from Paschal 1984).

vest weight, or 2,692 pounds/acre ÷ 1.5 pounds/fish = 1,794 fish/acre.

So, in a simple single-batch pond with no aeration equipment, stocking density should be no greater than 1,800 fish/acre if the desired fish size at harvest is 1.5 pounds.

Stocking rates for simple single-batch ponds using different combinations of maximum daily feeding allowances and desired harvest sizes are presented in Table 8.5. Commercial catfish farms usually have aeration equipment adequate to handle most problems with low dissolved oxygen, but some managers set 100 pounds/acre/day as a maximum feed allowance to prevent excessive risk of fish loss to other water quality problems or stress-related infectious diseases. If the desired average fish size at harvest is 1.5 pounds and ponds are managed in sin-

TABLE 8.5.

Stocking Densities (Fish/Acre) in Simple Single-Batch Ponds

Desired Fish Weight at Harvest (Pounds/Fish)	Maximum Daily Feed Allowance (Pounds of Feed/Acre/Day)				
	35	50	75	100	150
1.0	2,200	3,100	4,700	6,300	9,400
1.5	1,800	2,500	3,800	5,100	7,700
2.0	1,500	2,100	3,100	4,200	6,300

Note: Based on maximum daily feed allowance and desired average size of fish at harvest. Daily feed consumption rates were assumed to be 1.6, 1.3, and 1.2 percent for 1.0-, 1.5-, and 2.0-pound fish, respectively. Stocking densities are rounded to the nearest 100 fish/acre.

gle-batch systems, stocking densities should be about 5,100 fish/acre at a feed allowance of 100 pounds/acre/day.

Stocking densities can be increased in single-batch ponds that are partially harvested two or three times before all fish are removed because the maximum standing crop of fish is lower than in simple single-batch ponds where all fish are allowed to reach harvestable size before harvest. Stocking densities can be increased by 20 to 30 percent for a given maximum daily feed allowance compared to simple single-batch ponds. Stocking densities can also be increased in single-batch ponds when only small fingerlings are available and ponds are managed on a 2-year production cycle.

In theory, multiple-batch ponds can be stocked at 50 to 75 percent higher densities than simple single-batch ponds for a given maximum daily feed allowance. Larger fish in multiple-batch ponds are continually cropped off so the average weight of individual fish is almost always less than 1 pound. Successful production of fish at the higher stocking densities used in multiple-batch ponds is contingent on frequent and timely harvest to remove larger fish so that a manageable maximum standing crop is maintained. Frequent harvest also allows more rapid growth of understocked fingerlings because the more aggressive, larger fish that consume most of the feed are removed.

Guidelines for stocking densities in different farm situations are presented in Table 8.6. This is only a rough guide. Optimal stocking densities are lower if the growing season is short. New producers initially should use lower densities to reduce risks until gaining some experience. Watershed ponds are stocked less densely because water quality management is usually more difficult than in levee ponds.

TABLE 8.6.
Maximum Feed Allowances and Stocking Densities for Farms and Cropping Systems

	Maximum Feed Allowance (Pounds of Feed/ Acre/Day)	Stocking Density (Fish/Acre)		
		Single-Batch		Multiple-Batch
		One Harvest	Cropped	
Watershed ponds				
No aeration	35	2,000	2,500	3,000
With aeration	75	4,000	5,000	6,000
Levee ponds				
No aeration[a]	50	2,500	3,000	3,500
Limited aeration[b]	75	4,000	5,000	6,000
Sufficient aeration[b]	100	5,000	6,500	7,500

[a]Levee ponds are supplied with pumped water, which affords some aeration capability.
[b]Limited aeration means less than one aerator available per pond; sufficient aeration means at least one aerator per pond with backup aerators also available (see the section Dissolved Oxygen and Aeration).
Note: These guidelines are applicable to ponds stocked with 6- to 8-inch fingerlings in areas with a 150- to 200-day growing season. Actual practices used on farms should also depend on expertise of the producer.

Fish Yields and Production Systems

Net productions greatly exceeding 10,000 pounds/acre/year have been reported in the scientific literature for channel catfish grown in earthen ponds. Such extraordinarily high production is usually achieved only in ponds that are frequently flushed with water and aerated for long periods with relatively large aerators. These pond systems approach raceway culture (see the section Raceway Culture) in terms of water use and energy input and do not reflect conditions common in commercial pond culture.

Reliable estimates of sustained net fish production are difficult to obtain from commercial ponds because market constraints (not being able to sell fish) and off-flavor (processors' rejecting fish even when fish are needed; see the section Off-Flavor) affect production. Realized fish production on commercial farms ranges from less than 3,000 to over 6,000 pounds/acre/year. Sustained annual realized production for well-managed ponds using common production practices probably averages between 4,000 and 5,000 pounds/acre. This level of production can be achieved by using a variety of production schemes, with stocking densities ranging from 4,000 to over 10,000 fish/acre and ponds managed as either single-batch or multiple-batch systems. The most

TABLE 8.7.

Production Data from Experimental 1-Acre
Research Ponds Managed under Four Production
Schemes

	Single-Batch		Multiple-Batch	
	4,500	8,000	4,500	8,000
Net fish production (pounds/acre/year)	5,200	7,100	5,000	5,800
Average harvest weight (pounds/fish)	1.3	1.1	1.4	1.4
Mortality (percentage of fish stocked)	8	12	3	20

Note: Each value is the average for four replicate ponds over the 3-year experimental period.

profitable production scheme is not known; in fact, a single optimal scheme probably does not exist. Choice of production system is usually based on personal bias, farm size, size and availability of fingerlings, production goals, and other factors.

Results of a 3-year experiment conducted at the Delta Branch Experiment Station at Stoneville, Mississippi, can be used to indicate the production possible in ponds managed by production practices common on commercial farms. The ponds in that study were stocked with relatively large fingerlings (averaging about 70 pounds/1,000 fish over the study period) at either 4,500 or 8,000 fish/acre and managed as either single-batch or multiple-batch systems. Fish were harvested regardless of market constraints or presence of off-flavors; thus, the results indicate the potential production under the four combinations of stocking density and cropping system. Pertinent production data from this experiment are presented in Table 8.7.

An economic analysis of the production data obtained in that study is not available at the time of this writing, so conclusive statements cannot be made. However, some general trends can be pointed out:

1. Net production was highest at the highest fish stocking density regardless of cropping system; however, increasing the stocking density by a factor of 1.77 did not result in a commensurate increase in fish production. In the multiple-batch system, production increased by a factor of 1.16 when stocking density was increased from 4,500 to 8,000 fish/acre. In the single-batch system, production increased by a factor of 1.37; however, a significant proportion (10 to 15 percent) of the total weight of

fish harvested from single-batch ponds stocked at 8,000 fish/ acre consisted of fish smaller than 0.75 pound (commonly the minimum weight desired by processors). So net production of fish larger than 0.75 pounds increased only by a factor of about 1.2. Other research (Busch 1984) comparing production in single-batch systems has also shown that a large number of fish do not reach "harvest size" (>0.75 pound) when stocked at densities of 8,000 fish/acre or greater, even when relatively large fingerlings (60 to 80 pounds/1,000 fish) are used.
2. Fish production was somewhat greater in single-batch ponds than in multiple-batch ponds at equal stocking densities. As mentioned, much of the production in single-batch ponds at 8,000 fish/acre consisted of small fish and actual production of fish larger than 0.75 pound was only slightly greater than in multiple-batch ponds at the same stocking density.
3. Although net production increased only by factors of about 1.2 to 1.4 at the highest stocking density, mortality rate increased to a much greater extent. Averaged over all ponds, mortality rate was about 5 percent of fish stocked at 4,500 fish/acre and 15 percent of fish stocked at 8,000 fish/acre. The general trend of increased losses due to disease at higher stocking densities is well known (see Chapter 11).

These data indicate that net catfish production in excess of 5,000 pounds/acre/year is attainable under a variety of production schemes. On the basis of 10 years of catfish production research at the Delta Branch Experiment Station, maximum sustained net production in ponds managed according to commercial practices appears to be in the range of 6,000 to 7,000 pounds/acre/year. We have consistently achieved this level of production in single-batch systems stocked with relatively large fingerlings (60 to 80 pounds/1,000 fish) at 5,000 to 6,000 fish/acre (see, for example, Steeby and Tucker 1988). The profitability of this scheme relative to other possible production schemes is not known.

Growout of Fingerlings to Food Fish

Ponds are stocked after a management strategy is decided upon and fingerlings are available. Ponds can be stocked any time during the year, but fingerlings handle best when water temperatures are below 70°F. Growout ponds should be free of wild fish and weeds, but treatments are not required for insects as they are with fry nursery ponds.

Fish are stocked into ponds by weight because numbers are too large for fish to be counted individually. The following procedure is used to calculate the weight of fish needed to achieve a certain stocking density.

1. Determine the total number of fish needed. This is equal to the desired stocking density (fish/acre) multiplied by the pond water acres.
2. Estimate the average weight of the fingerlings by accurately weighing a random sample of 10 to 15 pounds of fish and then counting the fish in the sample. The accuracy of the estimate is increased if two or three separate samples are weighed and counted.
3. Calculate the weight of fish needed to stock the pond at the desired density.

$$\text{Pounds of fish needed} = \frac{\text{total number needed} \times \text{total weight of samples}}{\text{total number of fish in samples}} \quad (8.1)$$

Average weight of fingerlings is often expressed as pounds of fish per 1,000 fish. The formula becomes

$$\text{Pounds fish needed} = \text{total number needed} \times (\text{pounds}/1{,}000 \text{ fish}) \quad (8.2)$$

For example, assume a 17-acre pond is to be stocked at 5,000 fish/acre and a sample of 13.2 pounds contains 254 fingerlings (the weight per 1,000 fish is equal to 13.2 × 1,000/254 or 52 pounds/1,000 fish). The total number of fish needed is 5,000 fish/acre × 17 acres or 85,000 fish. The weight of fingerlings needed is:

$$\text{Pounds fish needed} = \frac{85{,}000 \times 13.2}{244}$$

$$= 4{,}417 \text{ pounds} \quad (8.3)$$

or, Pounds fish needed = 85,000 × (52 pounds/1,000 fish)
= 4,420 pounds $\quad (8.4)$

Stocking mixed-size groups of fish is advantageous when initially stocking multiple-batch ponds, particularly if several ponds on a farm are stocked at about the same time. If a single size group of fingerlings is stocked, most of the fish approach harvest size at about the same time. This will result in a high weight of fish in the pond and poor water quality, and fish growth may become feed-limited. More importantly, it may be difficult to sell all the fish in several ponds to a processor over a short period. Fish that cannot be sold will continue to grow, possibly to a less desirable size. To spread out harvesting over a longer period, ponds can be stocked with both stocker-sized fish (9 to 12 inches) and fingerlings (5 to 7 inches) at various ratios. After one or two croppings the distribution of fish sizes becomes such that harvest-sized fish will continually be available. Alternatively, large fish and small fish can be stocked into separate ponds to even out harvesting schedules.

When stocking groups of fish of different sizes into one pond, each group must be sampled to estimate average weight. The pounds of each group needed to stock the pond is determined by first calculating the number of each size group needed so that the sum equals the total number of fish required. Then the weight of each size group needed is calculated as shown.

After the required weight of fingerlings has been calculated, they are weighed as accurately as possible and transferred to the growout pond. The fingerlings may not feed actively for several days after stocking, even in warm weather, but some feed should be offered every day so that fingerlings will become reaccustomed to the feeding schedule. Once fish begin active feeding, they are fed once or more daily (see Chapter 10 for more information on feeding).

Environmental conditions are excellent for some time in newly stocked ponds, but water quality, particularly dissolved oxygen concentrations, should be monitored from the day of stocking. As daily feed allowances increase in response to fish growth, problems with poor environmental conditions become more frequent and water quality monitoring programs should be intensified (Chapter 9).

Ponds managed as multiple-batch systems or as single-batch with partial harvest are selectively seined when one-fourth to one-third of the fish reach the desired harvest size. Fish are not restocked in single-batch ponds, but fingerlings should be restocked in multiple-batch ponds as soon after harvest as possible. However, do not restock ponds if either the resident fish or the fingerlings have an infectious disease. Wait until the disease is completely controlled before restocking. Also, do not restock fingerlings if water quality in the pond to be stocked is poor. Fingerlings are stressed during seining and transporting; the added stress of poor water quality may kill the fish or predispose them to disease.

The number of fingerlings to restock in a multiple-batch pond is calculated by first estimating the number of fish harvested. Do this by counting out a random sample of 50 to 100 fish harvested and then weighing the sample. Calculate the number of fish harvested:

$$\text{Number of fish harvested} = \frac{\text{total weight harvested}}{\text{weight of fish sampled}} \times \text{number fish sampled} \quad (8.5)$$

The number of fish harvested is equal to the number of fish to be restocked. The weight of fingerlings to be restocked is then calculated as shown.

RECORD KEEPING

Good records are essential in catfish farming just as in any commercial enterprise. Accounting records are needed for tax purposes and general bookkeeping. Many lending institutions also require good records before they lend money. Thorough, up-to-date records indicate to the lender that the farmer is conscientious and dependable in business. Records also allow the farmer to estimate the growth status and value of the fish crop. Catfish cannot be readily seen or measured during growout. The only way to keep track of production, particularly if many ponds are involved, is through the use of a computer records program.

A records program has been developed for catfish farms (Fouché et al. 1983). The program serves as a record system that will store, summarize, and retrieve data. The program also will calculate future feed requirements and estimate fish growth, fish inventory, and expected harvest dates. This program is available free on request from Computer Applications and Service Department, P. O. Box 5405, Mississippi State, Mississippi, 39762.

INVESTMENTS AND COSTS

It is beyond the intended scope of this book to present a detailed economic analysis of channel catfish farming. Such an analysis also would be of limited general value because costs vary with time and location. The following economic summary, from Keenum and Waldrop (1988), illustrates the major costs of channel catfish production using synthetic farms in northwest Mississippi. Although absolute monetary values will be different in other locations, the relative magnitudes of individual investments and costs will be comparable.

Production System

Three synthetic farm situations were analyzed by Keenum and Waldrop (1988). Results for the smallest and largest farm are presented here (Table 8.8). All ponds were stocked with 6,000 fingerlings weighing 60 pounds/1,000 fish to initiate production. An annual mortality rate of 5 percent was assumed for each pond. Each farm sold 5,000 pounds of fish/acre/year (4,000 fish at 1.25 pounds each). Each pond was restocked with 4,300 fingerlings/acre annually to match the number of fish sold plus fish mortality.

TABLE 8.8.

Physical Characteristics of the Two Synthetic Catfish Farms Used for Economic Analysis

Characteristics	Farm	
	I	II
Total land acres[a]	163	643
Water acres	140	569
Production acres[b]	126	512
Acres per pond	17.5	17.8
Number of ponds	8	32

[a]Includes 3 acres for buildings, parking, etc.
[b]Reflects a 10% reduction in water acres due to annual pond repair.
Source: Waldrop and Keenum (1988).

This establishes an inventory of 1,700 fish/acre continuously maintained. Fish were fed a 32 percent protein floating feed, and a feed conversion ratio of 2.0 was used to estimate feed costs. It was assumed that ponds would be out of production 1 year in 10 for renovation; to account for this lost production, annual production was reduced 10 percent per year.

Investment Requirements

Investment requirements (Table 8.9) were divided into six major groups: land, pond construction, water supply, feeding equipment and facilities, disease and weed control equipment, and miscellaneous equipment (Table 8.10).

Investment costs per land acre decline as farm size increases. Economy of scale is particularly notable in equipment costs: doubling farm size does not require twice the equipment. Some economy of scale also exists for construction costs because the number of common levees between ponds increases as the number of adjacent ponds increases. The fact that production acres of water increase relative to land acres as farm size increases also contributes to economies of scale.

Annual Ownership and Operating Costs

Ownership costs are commonly referred to as *fixed costs*. These costs accrue whether or not the farm operates and consist of depreciation, interest, taxes, and insurance (Table 8.11).

TABLE 8.9.
Estimated Investment Requirements (Dollars) for Catfish Production for Two Farms in Northwest Mississippi, 1988

Item	Farm I	Farm II
Land[a]	130,400	514,400
Pond construction		
Earth moving	100,663	383,317
Drainage structure	9,600	38,400
Gravel	6,131	24,778
Vegetative cover	1,562	5,730
Total pond construction	117,956	452,225
Water supply		
(Well, pump, motor and outlet pipe)	30,720	122,880
Feeding		
Feeder	4,595	10,300
Electronic scales/printer	2,820	5,640
Bulk storage	10,170	20,340
Total feeding	17,585	36,280
Disease, parasite and weed control equipment		
(Boat, motor, trailer)	3,340	3,340
Miscellaneous equipment[b]	188,406	458,670
Total	488,407	1,587,795
Investment per water surface acre	3,479	2,791
Investment per land acre	2,996	2,469

[a]Valued at $800 per acre.
[b]A detailed description of miscellaneous equipment is presented in Table 8.10.
Source: Keenum and Waldrop (1988).

Operating costs, or *variable costs*, accrue if production occurs. Those costs consist of repair and maintenance, fuel, supplies (including fingerlings and feed), labor, harvesting, insurance, and interest (Table 8.12). Annual operating costs are relatively much greater than ownership costs. When such a condition exists, producers tend to react to changes in prices paid by producing when prices are high (revenues exceed operating costs) but reducing production when prices are low in an effort to minimize losses.

Cost Summary

Significant economies of scale exist in channel catfish farming. Unit cost of production consequently decreases with increasing farm size

TABLE 8.10.

Estimated Miscellaneous Equipment Investment Requirements (Dollars) for Catfish Production for Two Farms in Northwest Mississippi, 1988

Miscellaneous Equipment	Farm I	Farm II
Tractors (45–65 hp)	27,104	104,544
(90–100 hp)	32,148	64,296
Trucks (1/2 ton)	11,600	11,600
(3/4 ton)		26,200
(3/4 ton 4 × 4)	14,300	14,300
Service building with office and bath		
(40 × 60 ft)		40,000
(20 × 40 ft)	16,000	
Farm shop equipment	9,000	18,000
Office equipment	800	1,600
Computer with printer	4,000	4,000
Oxygen meter, probe, and accessories	1,490	2,980
PTO-driven aerators	7,550	28,100
Electric floating paddlewheels (10 hp)	25,600	102,400
PTO-driven low lift pump with hose	3,600	3,600
6-ft side mount mower	4,086	4,086
2-way radio communication system	3,132	4,968
Seines (1/2-in mesh)	4,016	4,016
(5/8- or 1-in mesh)	2,037	2,037
(2-in mesh)	1,358	1,358
Hydraulic takeup reel and trailer	3,500	3,500
Live-haul tank (2-compartment)	2,085	2,085
2 1/2-ton used boom truck	15,000	15,000
Total	188,406	458,670

Source: Keenum and Waldrop (1988).

(Table 8.13). Economies of size are mainly attributable to more efficient use of labor and equipment on larger farms. Labor costs decline from 14 percent of total annual costs for the smallest farms to 9 percent for the largest farms. Feed costs account for about 40 percent of the total annual costs; about 10 percent of annual costs are for fingerlings. For every 10 percent change in the price of feed, the total cost/pound of catfish harvested changes by 2.6 cents/pound. Similarly, a 10 percent change in price of fingerlings changes production costs by about 0.7 cent/pound.

The cost estimates produced in this study (Keenum and Waldrop 1988) were assumed to be valid only for the most efficient and produc-

TABLE 8.11.

Estimated Annual Ownership Costs (Dollars) for Catfish Production Facilities and Equipment for Two Farms in Northwest Mississippi, 1988

Item	Farm I	Farm II
Depreciation[a]		
Ponds	11,796	45,223
Water supply (wells, pumps, motors, and outlet pipes)	1,488	5,952
Feeding (feeder with electronic scales and storage)	1,250	2,610
Disease, parasite, and weed control equipment (boat, motor, and trailer)	495	495
Miscellaneous equipment	22,839	60,449
Interest on investment[b]		
Land	14,344	56,584
Pond construction	6,488	24,872
Water supply (wells, pumps, motors, and outlet pipes)	1,514	6,056
Feeding (feeder with electronic scales and storage)	968	1,996
Disease, parasite, and weed control equipment (boat, motor, and trailer)	184	184
Miscellaneous equipment	10,867	25,228
Taxes and insurance	2,867	6,845
Total	74,596	236,494

[a]Computed by the straight line method with zero salvage for depreciable items.
[b]Charged at 11% on the total value of land with all other depreciable items charged at 11% on one-half of the investment.
Source: Keenum and Waldrop (1988).

tive farmers. The industry average fish production in Mississippi is probably closer to 4,000 than 5,000 pounds/acre/year, as assumed in this study. Farmers with below-average production would have significantly higher costs of production.

WATER USE IN PONDS

Pond culture of catfish is a water-intensive endeavor, requiring more water per unit area than most other agricultural crops or animal pro-

TABLE 8.12.

Estimated Annual Operating Costs (Dollars) for Catfish Production for Two Farms in Northwest Mississippi, 1988

Item	I	II
Repairs and maintenance		
Vegetative cover	1,596	5,854
Water supply (wells, pumps, motors, and outlet pipes)	840	3,360
Feeding (feeder with electronic scales and storage)	244	522
Disease, parasite, and weed control equipment (boat, motor, and trailer)	272	272
Miscellaneous equipment	11,819	29,288
Fuel		
Mowing	373	1,369
Feeding	1,979	7,909
Outboard motor	102	408
Electric floating paddlewheels	5,461	21,845
PTO-driven aerators and low lift pump	2,902	11,605
Pumping	8,055	32,659
Transportation	3,400	11,333
Chemicals[a]	15,014	60,847
Telephone expenses	1,800	3,000
Test kits	245	980
Fingerlings	40,751	165,139
Feed (32% protein)	154,791	627,274
Labor		
Operations management	21,000	56,000
Hired labor	38,000	86,000
Harvesting and hauling[b]	25,272	102,412
Liability insurance	2,141	4,545
Interest on operating capital[c]	15,697	57,150
Interest on fish inventory[d]	1,611	6,529
Total	353,365	1,296,300

[a]Medicated feed accounts for approximately 50% of chemical cost for each farm.
[b]Charged at a rate of 3 cents/pound for harvesting and 1 cent/pound for hauling.
[c]Charged at a rate of 10% for 1 month on harvesting with no interest charged on hauling and charged at a rate of 10% for 6 months on all other cost items.
[d]Charged at a rate of 10% annually.
Source: Keenum and Waldrop (1988).

TABLE 8.13.

Summary of Costs (Dollars) of Catfish Production for Two Farms in Northwest Mississippi, 1988

Item	Farm I	Farm II
Total annual cost	427,961	1,532,794
Ownership cost	74,596	236,494
Operating cost	353,365	1,296,300
Total cost per pound[a]	0.68	0.60
Ownership cost per pound	0.12	0.09
Operating cost per pound	0.56	0.51

[a]Based on harvested production of 631,800 pounds for farm I and 2,560,300 pounds for farm II.
Source: Keenum and Waldrop (1988).

duction. In many areas of the United States, the availability of water may limit the feasibility of rearing catfish commercially in ponds. Groundwater is the preferred water source for pond culture. Most aquifers are recharged slowly and water tables are falling in most major aquifers in the United States. Even in Mississippi, which receives over 50 inches of precipitation annually, groundwater levels are falling at an alarming rate and may someday limit production of catfish. Catfish farming is a consumptive use of water because much of the water used evaporates and is lost to further use. Wise use of water in catfish farming is mandatory to conserve water resources.

This section discusses water use and conservation in levee ponds. The water budget of watershed ponds is somewhat more complicated because runoff is difficult to quantify. Sample water budgets for watershed ponds in Alabama are presented by Boyd (1985a) and Boyd and Shelton (1984).

The water budget for a levee pond can include the following terms:

Inflow	Outflow
Precipitation	Evaporation
Runoff	Seepage
Regulated inflow (pumped water)	Overflow
	Regulated discharge (draining)

The general hydrologic equation is

$$\text{inflow} = \text{outflow} \pm \Delta \text{ storage} \tag{8.6}$$

For simplicity assume that the water level is maintained below the drain so that overflow does not occur and that no water is discharged; the only outflows are evaporation (E) and seepage (S). Further assume that no water is pumped into the pond and that runoff from levees is negligible. This is a fair assumption because the watershed contributed by levees is a small fraction of the surface area for large ponds. The only inflow is precipitation (P) falling directly on the pond. The hydrologic equation becomes

$$\tag{8.7}$$

or, rearranging,

$$\Delta \text{ storage} = P - (S + E) \tag{8.8}$$

Change in storage (Δ storage) is the change in water level in the pond over some time period; it is equal to the amount of water (in inches) required to maintain a certain water level. Pote, Wax, and Tucker (1988) developed part of this budget for ponds at Stoneville, Mississippi, using a day-to-day comparison of precipitation and evaporation at that site. Figure 8.8 uses these data and adds a daily seepage rate of 0.05 inches/day to arrive at changes in storage throughout the year. In an average year, precipitation exceeds evaporation plus seepage in the winter and pond water levels rise. Evaporation plus seepage exceeds precipitation in the late spring and summer and water levels decline. Over the year, losses exceed gains and water levels end up about 15 inches below the starting point. This is an estimate of the average amount of water required to maintain the initial water level. Two extreme years are also depicted. In 1979, an extremely wet year, water levels would have stayed above the initial level all year. In 1986, a drought year, water levels declined through the year. If no water was added, a 4-foot-deep pond would have nearly dried up; about 38 inches of water from a well or other source would have been required to maintain water levels at the original height.

By using this approach, the amounts of water required during a 12-month period can be calculated for any site if pond evaporation,* precipitation, and seepage are known. Table 8.14 lists water requirements for different sites in the United States assuming four different

---

*Pond evaporation can be estimated as 0.81 times the evaporation measured with a class A evaporation pan used by the weather service (Boyd 1985b).

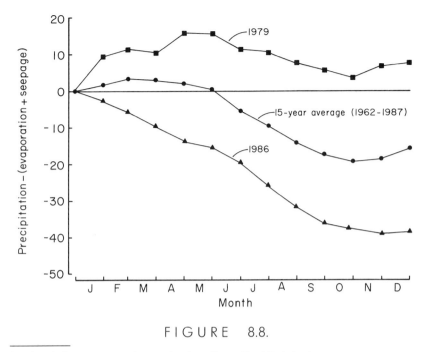

FIGURE 8.8.

Simulated changes in pond water levels at Stoneville, Mississippi, due only to climatic factors and an assumed seepage rate of 0.05 inch/day.

seepage rates. Two points are obvious: seepage is an extremely important factor regulating water requirements and water requirements are dramatically higher in arid regions.

If ponds are drained annually, the water required to refill the ponds must be added to that lost to evaporation and seepage. About 48 inches of water is required for filling most levee ponds.

The requirement for pumped water can be reduced by increasing inflows (precipitation or runoff) or by decreasing outflows (evaporation, seepage, overflow, or discharge). In practice, significant conservation of water can be achieved in only a limited number of ways.

Obviously, precipitation cannot be increased. Runoff is a function of levee morphometry, and increased construction costs prevent increasing the watershed area for levee ponds. However, watershed ponds by definition make use of runoff to reduce the need for pumped water. A properly constructed watershed pond should not require pumped water additions, although the capability to add water is advantageous during prolonged droughts. Conservation of water in levee

TABLE 8.14.

Water Required (Inches/Year) to Maintain Water Levels in Ponds with Different Seepage Rates at Various Locations in the United States

Location	Seepage Rate (Inches/Day)			
	0.05	0.1	0.3	0.5
Tallahassee, Fla.	1	19	92	165
Augusta, Ga.	20	38	111	184
Raleigh, N.C.	20	38	111	184
Columbia, S.C.	17	35	108	181
Knoxville, Tenn.	12	30	103	176
Auburn, Ala.	7	25	98	171
Stoneville, Miss.	15	33	106	179
Little Rock, Ark.	17	35	108	181
Baton Rouge, La.	13	31	104	177
Houston, Texas	25	43	116	189
Wichita Falls, Texas	55	73	146	219
Wichita, Kans.	40	58	131	204
Bakersfield, Calif.	81	99	172	245
Sacramento, Calif.	58	76	149	222

Source: Adapted from Boyd (1986).

ponds thus depends on reducing amounts of water lost. Although evaporation cannot be reduced by any practical means, some control over seepage, overflow, and discharge can be achieved.

Information in Table 8.14 demonstrates the importance of seepage to pond water budgets. Potential losses to seepage are reduced by careful site selection and proper pond construction. Seepage rates should be less than 0.2 inches/day in properly constructed ponds. Seepage in existing ponds can be reduced by sealing the pond bottom with a layer of clay or a waterproof liner, although either procedure is expensive. If the pond bottom soils contain more than 20 percent clay but the pond still seeps excessively, certain chemicals containing sodium can be applied to disperse the clay and reduce permeability. Sodium chloride (common salt) or sodium polyphosphates (tetrasodium pyrophosphate or sodium tripolyphosphate) are most commonly used. Sodium chloride is applied at a rate of 0.2 to 0.3 pounds/square foot. Sodium polyphosphates are used at rates of 0.05 to 0.1 pound/square foot. The chemical is spread evenly and worked into the top few inches of the pond bottom and inside levels. The soils are then thoroughly compacted.

The water requirements estimated in Table 8.14 were developed assuming that all rainfall is captured and no overflow occurs. This is usually not possible because an impractical storage capacity would be required; during extremely wet periods, water levels could overtop the levees unless freeboard was great. Building levees with large freeboard is expensive and excess freeboard is inconvenient. Nevertheless, every effort should be made to capture as much rainfall as possible by maintaining water levels about 3 inches below overflow structures.

Overflow also occurs when water is added to "flush" ponds in efforts to improve water quality. For many years it was assumed that flushing ponds with water two or more times a year was necessary to maintain adequate dissolved oxygen levels and remove toxic metabolites such as ammonia. However, a study conducted by McGee and Boyd (1983) indicates that flushing is not beneficial. In that study, water exchanges of up to four pond volumes were applied between July and September in catfish production ponds. Water quality was not improved and fish production was not increased. It was concluded that biological processes controlled by temperature, feed inputs, and the pond microbial community influenced water quality more rapidly and to a greater degree than the physical process of water exchange.

Overflow may also occur when water from a well is used as a source of oxygen for fish when dissolved oxygen concentrations are low or when fish are crowded in a live car or cutting seine before harvest. Groundwater contains no dissolved oxygen and must be aerated before use, and the amount of oxygen that can be supplied in this manner is relatively low. Mechanical aerators are much more effective at providing oxygen to fish in ponds and also are more efficient because the water they move during the aeration process is moved against a much lower head than water pumped from a well. Nevertheless, the use of groundwater in emergencies is common and is particularly effective when fish are crowded during harvest in hot weather. The cool oxygenated water passing through the fish holding area may prevent loss of fish before they are loaded onto transport trucks. The water used at these times need not be lost as overflow because it is not used to flush the pond but rather to provide a localized area of high-quality water. This water should be captured to offset future evaporative losses.

Discharge is reduced by harvesting fish without draining the pond. This is common practice in food-fish growout ponds, although ponds are occasionally drained to repair levees or to close out fish inventory. Many brood fish ponds and most fry nursery ponds are drained each year. Although large amounts of water are used in these ponds, they represent only about 10 percent of the total acreage of catfish ponds.

FISH-EATING BIRDS

Fish-eating birds are common around catfish farms because ponds provide birds with a concentrated source of food. Since about 1985, populations of fish-eating birds, principally the double-crested cormorant, have increased dramatically on catfish farms in the southern Mississippi River valley. These birds are now one of the major causes of fish loss from ponds in that area.

Cormorants spend summers in the northern United States and Canada and migrate south in September through November to overwinter in the southern United States. Before the mid-1980s, cormorants mainly overwintered on bays and estuaries along the Gulf of Mexico. As the acreage devoted to catfish ponds increased, many birds began to overwinter near catfish farms in Mississippi, Louisiana, and Arkansas. Cormorants roost in wooded areas near rivers and lakes and fly to catfish ponds two or three times a day to feed. Each bird can eat up to 0.5 pound of fish each day. Cormorants swim on the surface and then quickly dive to capture fish; they swim rapidly and can dive to depths over 25 feet and are very efficient predators. Cormorants tend to prey on fish from the more isolated ponds on the edge of the farm. Fingerlings are the preferred food item, but fish up to 0.25 pound are also consumed. Wild fish, such as gizzard shad, that are present in some catfish ponds may be preyed on in preference to catfish. Albino catfish also appear to be eaten preferentially, possibly because they are easier for the birds to see.

At least 30,000 cormorants are estimated to roost each winter near catfish ponds in west-central Mississippi. Good evidence exists that numbers are increasing and that the birds are arriving earlier and leaving later each year. Monetary losses to cormorants in 1988 exceeded $3 million in Mississippi alone and will undoubtedly increase each year. An additional $1.5 million was spent in 1988 in efforts to rid farms of the birds. Birds may also transfer fish disease organisms from pond to pond after eating diseased fish. These indirect losses attributable to fish-eating birds are impossible to estimate.

Great blue herons, great egrets, and other wading birds can also cause considerable loss of fish, but generally wading birds are less of a problem on catfish ponds than cormorants or other diving birds. Wading birds are limited to feeding along pond margins; cormorants feed over the entire pond. However, wading birds can be devastating in the pond culture of bait minnows and other fish that tend to inhabit shallow water around pond margins. Wading birds may transfer fish disease and are generally undesirable on catfish ponds.

Control of fish-eating birds is difficult. The birds cannot be killed indiscriminately because most birds that prey on catfish (including the double-crested cormorant, great blue heron, and great egret) are protected under the Migratory Bird Treaty Act. Special permits for limited killing of these birds can be obtained from the United States Fish and Wildlife Service. To obtain a permit, catfish farmers must demonstrate that the birds have caused economic losses and that other efforts to control them have failed. The permit allows killing of only a few birds per month. Cormorants cannot be eradicated in this manner; limited kills are supposed to be used to make other bird-control programs, such as harassment, more effective.

Harassment involves the use of devices to frighten birds away from a particular area. These devices include gas-operated cannons, fireworks, bird distress calls, gunfire, sirens, flashing lights, balloons, wires with strips of Mylar tape, and human effigies. All of these devices initially work to some degree, but birds soon become accustomed to the noise or visual effects. The devices work best when used in combination. The stimulus also must be varied and moved around on the farm to be effective. Harassment is generally most effective when initiated before the birds begin feeding for the day.

Wading birds usually land on the bank before wading into water. Access can be limited by nets or fencing around the pond perimeter. Nets over the entire pond can prevent cormorants and other diving birds from landing. The net does not necessarily have to be so fine as to exclude the birds. If the birds can see the net, they will be discouraged from landing on the pond. However, the use of nets does not appear practical on large ponds.

POLYCULTURE SYSTEMS

Polyculture is the rearing of two or more species of fish in the same culture unit. Polyculture is widely practiced in other countries as a means of increasing the total yield of fish by allowing species with different feeding habits to use each environmental niche in the culture system. Polyculture can also be used as a management tool. For example, grass carp are commonly polycultured with catfish to control or prevent aquatic weeds.

Commercial catfish ponds offer ideal environmental conditions for polyculture. Channel catfish feed almost exclusively on supplement feeds, but the wastes they generate stimulate the growth of dense phytoplankton and zooplankton blooms. This natural productivity is largely unused by catfish. Research has consistently shown that total yield of fish from catfish ponds can be dramatically increased by adding plankton-feeding fish to catfish culture ponds. Nevertheless, polyculture (except for weed management purposes) is rare in commercial catfish ponds. The major constraints to polyculture in catfish ponds are the lack of established markets for the secondary species and the additional time and labor required to separate the different species for marketing.

Fishes Used in Polyculture

Fish species suitable for polyculture with channel catfish must yield increased profit for the catfish producer. The fish must be either marketable or useful as a management tool. In general, the presence of the secondary fish species should neither decrease the yield of channel catfish nor interfere with normal management of the pond. The fish must tolerate poor environmental conditions at least as well as channel catfish and should be easy to harvest. Also, the fish must be legal to possess, grow, and sell. Check with the appropriate state game and fish agency regarding restrictions or required permits before buying or stocking any fish. A brief description of species potentially useful in catfish polyculture is presented in the following sections. Detailed information on reproduction and culture of these fishes is presented by Bardach, Ryther, and McLarney (1972), Dupree and Huner (1985), and Stickney (1986).

Bigmouth and hybrid buffalo. The bigmouth buffalo is the largest member of the sucker family (Catastomidae) and reaches weights of over 50 pounds. The female bigmouth buffalo can be crossed with the male black buffalo to produce a hybrid that is popular for culture. Buffaloes are native to the Mississippi River system. Bigmouth and hybrid buffaloes are robust fish and resemble the common carp in body configuration. Buffaloes feed throughout the water column and are omnivorous, mainly consuming zooplankton, insects, and detrital plant material. The flesh of bigmouth or hybrid buffalo is firm, white, and mild-flavored, although fillets contain many annoying intermuscular Y-shaped bones.

Bigmouth or hybrid buffalo are perhaps the best candidates for polyculture with channel catfish when the aim is simply to increase total fish production from a pond. Buffaloes are native North American fish, which consume relatively little supplemental feed and thus have little impact on production of channel catfish. Buffaloes are stocked in channel catfish growout ponds at 50 to 200 fingerlings/acre. Hybrid buffalo are preferred because they grow faster than either parent and will not reproduce in the pond. Buffalo will grow from fingerlings to over 3 pounds in a single 150-day growing season. Weight over 4 pounds is the preferred market size, so a second year of growth may be desirable. Large buffalo can be selectively harvested from channel catfish ponds with large-mesh gill nets. Any remaining fish must, however, be hand-sorted after harvesting the catfish. The major constraint to channel catfish–buffalo polyculture is the present low commercial value of buffalo. Fish are sold predominantly through small, local markets.

Common carp. The common carp is native to Europe and Asia and was introduced to North America in 1876. The fish is now widely distributed in the United States and grows to over 20 pounds. In nature, common carp are bottom feeding and omnivorous, subsisting on insect larvae, zooplankton, detritus, and plant material. They learn to feed almost exclusively on supplemental feeds when cultured with channel catfish and reduce yields of catfish. Their habit of uprooting aquatic plants and disturbing bottom muds during feeding can also lead to excessive clay turbidity. Common carp readily reproduce in ponds and will quickly overpopulate unless the pond is drained annually. The market value of common carp in the United States is very low, although common carp are among the most valued and important food fish in the world. The flesh of pond-raised carp is mild-flavored and firm but contains annoying intermuscular Y bones. Common carp are undesirable in channel catfish ponds.

Paddlefish. Paddlefish (Fig. 8.9) are native to the Mississippi River system and are locally known as spoonbill, spoonbill catfish, or shovelnose catfish (they are not, however, catfish). The fish has a long, flattened snout; a large head and mouth; and a sharklike tail. The skeleton is chiefly cartilage. Paddlefish are filter feeders and mainly consume zooplankton and insect larvae. Feeding activity and growth are greatest at moderate water temperatures (70 to 80°F); most growth occurs in spring and fall in the southern United States. Specimens over 100 pounds have been taken from natural waters. The flesh is white and mild-flavored and has a texture reminiscent of that of swordfish or shark. In certain areas of the midwestern United States, paddlefish, particularly smoked paddlefish fillets, are highly regarded as a food item, although markets are limited.

Polyculture of paddlefish with adult channel catfish does not appear to be practical. Relative to channel catfish, paddlefish do not tolerate exposure to low dissolved oxygen concentrations or poor water quality. Water quality in ponds containing paddlefish would have to be managed more intensively than for catfish monoculture. The extra

FIGURE 8.9.

Two paddlefish weighing about 1 pound each. (Photo: Steve Mims)

expenditures for water quality management increase the cost of catfish production and negate the benefit of added paddlefish production. Paddlefish fingerlings less than 6 to 8 inches in length are readily eaten by adult channel catfish. Fingerling paddlefish of at least 12 inches are needed if stocked into ponds with food-sized channel catfish.

Channel catfish nursery ponds offer the best opportunity for paddlefish polyculture. Small paddlefish fingerlings of 4 to 6 inches in length can grow to 1.5 to 2 pounds in less than 9 months. The usual stocking rate is 200 to 500 fingerling paddlefish/acre. Paddlefish are easy to seine and grow faster than fingerling channel catfish, so they can be selectively harvested from the pond by seining with a large-mesh seine. Otherwise, the paddlefish have to be hand-separated from the harvested fingerlings.

Tilapias. Tilapias belong to the family Cichlidae and are widely distributed throughout the tropics and subtropics. They generally resemble sunfish in body configuration (Fig. 8.10). Species of *Tilapia* are perhaps the most widely cultured food fish in the world. The three most commonly cultured species are the blue tilapia, the Nile tilapia, and the Mozambique tilapia. Blue and Nile tilapia grow faster than

Polyculture Systems   **211**

FIGURE 8.10.

A blue tilapia weighing about 0.5 pound.

Mozambique tilapia and are preferred for culture. Most species of tilapia are omnivorous. Mozambique tilapia eat aquatic plants, algae, zooplankton, insects, and detritus. Blue and Nile tilapia feed mainly on zooplankton and phytoplankton, although blue tilapia also eat filamentous algae and some higher aquatic plants. Tilapia do not tolerate cold water; most species die when water temperatures fall below about 50°F. Otherwise, they are extremely hardy fish. They are relatively disease-resistant and tolerate poor water quality and crowding. Tilapia are excellent food fish and may have some future widespread commercial value in the United States. Under the proper conditions, tilapia of 0.5 to 1.0 pound can be produced in a single 150- to 200-day growing season when polycultured with channel catfish (Torrans 1988).

Despite the appeal of tilapia as a candidate for polyculture with channel catfish, production of food-sized tilapia in catfish ponds cannot be generally recommended. Because of their intolerance to cold, most of the tilapia harvest from ponds in temperate regions is at the end of the growing season just before water temperatures fall to lethal levels. Marketing large volumes of a seasonally available fish product has traditionally been difficult. Tilapia become sexually mature at an early age (as little as 5 months old) and produce several broods a year.

This usually results in a huge biomass of small tilapia, which never grow large enough to market. These fish compete with catfish for supplemental feeds during the summer and die during the winter. The large amount of decaying tilapia present in the water can affect the flavor of the catfish in the pond (see the section Off-Flavor). The techniques for controlling reproduction of tilapia (Guerrero 1982) are generally not practical under current commercial conditions. Although tilapia in nature feed low on the food chain, they avidly consume supplemental feeds when cultured with channel catfish. Yields of channel catfish are usually reduced when polycultured with tilapia. Also, tilapia are difficult to harvest by seining. In warm water, they jump over nets or lie flat in the mud and go under the seine. Tilapia are much easier to seine if water temperatures are below 60°F, but complete harvest of tilapias is possible only by draining the pond.

Although cold intolerance and prolific reproduction are disadvantages when tilapia are grown for food, these traits make them an excellent forage for channel catfish brood stock (Torrans 1988). Ten to twenty pair of adult tilapia per acre are stocked in May or June. A large biomass of 1- to 3-inch tilapia is present by fall. The tilapia become lethargic when water temperatures drop and are easily captured by the brood catfish. Tilapia have also been used to prevent environment-derived off-flavors in channel catfish and to control aquatic weeds. However, the effectiveness of tilapia in preventing off-flavors is questionable and grass carp are more effective at weed control and are more easily managed than tilapia.

Grass carp. Grass carp are native to eastern Asia and were introduced into the United States in 1963. The fish is banned in over forty states, but where they are legal, grass carp are commonly used in catfish ponds to control aquatic weeds. This use of grass carp is discussed in the section Aquatic Weed Control. Although the grass carp is not grown for food in the United States, it is an esteemed food fish throughout Asia. The fish is bony but has excellent flavor.

Silver carp. Silver carp are also native to eastern Asia. They were introduced into the United States in 1971. Silver carp are filter-feeding omnivores, mainly consuming phytoplankton and zooplankton. The potential for filter-feeding fish to reduce phytoplankton abundance stimulated interest in silver carp polyculture with channel catfish. Although silver carp thrive in channel catfish ponds, most evidence suggests that their presence does not reduce, and may actually increase, phytoplankton abundance (Smith 1988).

Silver carp are stocked at 100 to 500 fingerlings/acre in catfish ponds and grow to 1 to 3 pounds in a single growing season. Harvesting silver carp by seining is difficult; the fish are strong swimmers and jump over the seine. The fish can grow to well over 10 pounds in catfish ponds, and workers may be injured by the jumping fish. Silver carp are fair to good food fish but have few markets.

Bighead carp. The bighead carp is the third east Asian carp species cultured in the United States. Bighead carp are filter-feeding omnivores much like silver carp, but relative to that of the silver carp, the diet of the bighead consists of a higher proportion of zooplankton. Bighead carp also eat some supplemental feed when polycultured with channel catfish. Bighead carp, like silver carp, have been investigated as a means of controlling the abundance of phytoplankton in catfish ponds. However, by mainly consuming zooplankton (which are important predators on phytoplankton), bighead carp usually cause an increase in phytoplankton biomass (Burke, Bayne, and Rea 1986).

Bighead carp are stocked at 100 to 500 fingerlings/acre when polycultured with channel catfish and grow from fingerlings to over 3 pounds in a single growing season. Fish can grow to over 20 pounds in three years. Bighead carp are good food fish, and a limited market for them exists among Asian emigrants in large cities, such as New York, Chicago, and San Francisco. Fish over 5 pounds are preferred in these markets.

Freshwater prawns. The freshwater prawn *Macrobrachium rosenbergii* (Fig. 8.11) is a shrimplike crustacean native to tropical Indo-Pacific regions. This crustacean spends its adult life in fresh or brackish waters. Larval development, however, requires brackish water of 8 to 22 ppt salinity. The larvae must molt several times before they can be stocked as postlarvae into fresh water. Prawns grow rapidly in warm water and can reach a weight of 2 ounces or more in less than 150 days. Prawns grow most rapidly when water temperatures are greater than 85°F and die when temperatures fall below 60°F. Prawns are a highly prized food item, and markets in the United States could readily be developed if prawns were priced competitively with marine shrimp and supplies were dependable.

Prawn culture in most of the continental United States is constrained in general by the lack of readily available seed stock (postlarvae or juveniles) and the cold intolerance of the animal. Production of postlarvae is relatively difficult, and most prawn farmers purchase them from commercial sources in Central America, Southern California, Hawaii, or the Indo-Pacific region. Postlarvae (plus associated shipping costs) are fairly expensive. Cold intolerance limits the grow-

FIGURE 8.11.

Freshwater prawn *Macrobrachium rosenbergii*. (Photo: Louis D'Abramo)

ing season to only 5 to 7 months in most of the United States, so prawns must be harvested in early fall before they die. The market supply is therefore highly seasonal.

Prawns have been experimentally polycultured with either adult catfish in growout ponds (D'Abramo et al. 1986) or fingerlings in nursery ponds (Heinen et al. 1989). Juvenile prawns (2- to 3-month-old postlarvae) are usually stocked to ensure high survival. Average weights of juveniles should be between 0.007 and 0.035 ounce (0.2 to 1.0 gram). Prawns are stocked at 1,000 to 2,500 juveniles/acre in the spring when water temperatures are expected to remain above 70°F. No special management is afforded the prawns; they subsist on natural pond productivity. Prawns are harvested in the fall before water temperatures fall below 60°F. Yields range between 100 and 250 pounds/acre.

Polyculture of prawns in commercial food-fish growout ponds is probably impractical. Prawns are difficult to harvest with any seine, but the large mesh seines that must be used to harvest and grade food-sized catfish are particularly inefficient because the prawns escape through the mesh. If a smaller mesh seine is used, both small submarket-sized catfish and market-sized catfish will be caught along with prawns. The combined catch must then be hand-sorted. Also, the pond must be seined in fall to harvest the shrimp before they die, regardless of whether the catfish are marketable or not. The catfish may be unmarketable because they are too small, off-flavor, or unwanted by the processor at that time.

Polyculture of prawns with fingerling channel catfish is more promising than culture with food fish. Use of a seine or live car with a 1-inch square mesh allows most fingerlings under 8 to 10 inches in length to escape but retains most of the prawns. The major disadvantage to prawn-fingerling catfish polyculture is the low harvest efficiency of seining. Only 25 to 50 percent of the prawns present in the pond will be captured in one seining. Repeated seining to capture most of the prawns is not practical under most commercial settings and may stress the fingerlings, predisposing them to infectious diseases.

REFERENCES

Anonymous. 1982. *Ponds—Planning, Design, and Construction,* Agriculture Handbook No. 387. Washington D.C.: U.S. Department of Agriculture.

Bardach, J. E., J. H. Ryther, and W. O. McLarney. 1972. *Aquaculture: The Farming and Husbandry of Freshwater and Marine Organisms.* New York: Wiley Interscience.

Boyd, C. E. 1985a. Hydrology and pond construction. In *Channel Catfish Culture,* ed. C. S. Tucker, pp. 107–134. Amsterdam: Elsevier.

Boyd, C. E. 1985b. Pond evaporation. *Transactions of the American Fisheries Society* 114:299–303.

Boyd, C. E. 1986. Influence of evaporation excess in water requirements for fish farming. In *Conference on Climate and Water Management—A Critical Era,* pp. 62–64. Boston: American Meteorological Society.

Boyd, C. E., and J. R. Shelton. 1984. *Observations on the Hydrology and Morphometry of Ponds on the Auburn University Fisheries Research Unit,* Station Bulletin 558. Auburn: Alabama Agriculture Experiment Station.

Burke, J. S., D. R. Bayne, and H. Rea. 1986. Impact of silver and bighead carps on plankton communities of channel catfish ponds. *Aquaculture* 55:59–60.

Busch, R. L. 1984. Production results for three stocking density experiments at Stoneville, Mississippi. Paper read at 6th Annual Catfish Farmers of America Research Workshop, 6–8 January 1984, at New Orleans, Louisiana.

D'Abramo, L. R., H. R. Robinette, J. M. Heinen, Z. Ra'anan, and D. Cohen. 1986. Polyculture of the freshwater prawn *(Macrobrachium rosenbergii)* with a mixed-size population of channel catfish *(Ictalurus punctatus). Aquaculture* 59:71–80.

Dupree, H. K., and J. V. Huner, eds. 1984. *Third Report to Fish Farmers.* Washington D.C.: U.S. Fish and Wildlife Service.

Fouché, L., P. Webb, W. Killcreas, J. Waldrop, K. Chin, N. Kennedy, and R. Paschal. 1983. *A Records System for Catfish Production Management Decision Making,*

AEC Technical Publication No. 43. Mississippi State: Mississippi Agricultural and Forestry Experiment Station.
Guerrero, R. D. 1982. Control of tilapia reproduction. In *The Biology and Culture of Tilapia*, ed. R. S. V. Pullin and R. H. Lowe-McConnell, pp. 309–316. Manila: International Center for Living Aquatic Resources Management.
Heinen, J. M., L. R. D'Abramo, H. R. Robinette, and M. J. Murphy. 1989. Polyculture of two sizes of freshwater prawns (*Macrobrachium rosenbergii*) with fingerling channel catfish (*Ictalurus punctatus*). World Aquaculture 20:72–75.
Keenum, M. E., and J. E. Waldrop. 1988. *Economic Analysis of Farm-Raised Catfish Production in Mississippi*, Technical Bulletin 155. Mississippi State: Mississippi Agricultural and Forestry Experiment Station.
McGee, M. V., and C. E. Boyd. 1983. Evaluation of the influence of water exchange in channel catfish ponds. *Transactions of the American Fisheries Society* 112:557–560.
Paschal, R. G. 1984. Economic analysis of stocking rates and growth functions for farm-raised catfish for food in earthen ponds. Ph.D. Dissertation, Mississippi State University, Mississippi.
Pote, J. N., C. L. Wax, and C. S. Tucker. 1988. *Water in Catfish Production: Sources, Uses, and Conservation*, Special Bulletin 88-3. Mississippi State: Mississippi Agricultural and Forestry Experiment Station.
Smith, D. W. 1988. Phytoplankton and catfish culture: a review. *Aquaculture* 74:167–189.
Steeby, J. A., and C. S. Tucker. 1988. Comparison of nightly and emergency aeration of channel catfish ponds. *Mississippi Agricultural Experiment Station Research Report* Vol. 13, No. 8, Mississippi State, Mississippi.
Stickney, R. R., ed. 1986. *Culture of Nonsalmonid Freshwater Fishes*. Boca Raton, Fl.: CRC Press.
Tisdale, C. B. 1982. Catfish farming in Mississippi—planning, design, and construction. Paper read at the Winter Meeting of Land Improvement Contractors of America, 11 January 1982, at Orlando, Florida.
Torrans, L. 1988. *Blue Tilapia Culture in Arkansas*. Pine Bluff: Arkansas Cooperative Extension Service.

CHAPTER 9

Water Quality Management in Ponds

If a fish pond is built on pesticide-free soil and filled with unpolluted fresh water, the initial water quality conditions are ideal for channel catfish culture. Dissolved oxygen concentrations are near saturation, and the water contains negligible concentrations of carbon dioxide, ammonia, nitrite, or other toxic substances. When fish are stocked and fed, the environment immediately begins to deteriorate and becomes less fit for the fish. The degree to which conditions deteriorate is related to the amount of fish waste reaching the water. This in turn is related to the number and weight of fish in the pond and the amount of feed they are offered.

Deterioration of water quality ultimately limits the production of fish in static-water ponds. The key to successful culture of catfish in ponds is to stock and feed fish at the highest rate possible without degrading the environment to the point where net economic returns decrease as a result of excessive management costs (such as aeration or water pumping), poor fish growth, or loss of fish to environmental or infectious diseases. It is rather extraordinary to report that this "point of diminishing returns" is not known with any certainty. Currently, fish stocking rates in commercial culture ponds range from 3,000 to more than 10,000 fish/acre and maximum daily feeding rates during summer months range from about 50 to over 150 pounds of feed/acre. At the higher stocking densities and feeding rates, water quality is often poor and considerable effort is required to keep fish alive.

Maintenance of good water quality in ponds is technologically simple; however, economic considerations severely limit the practices that can be used under commercial conditions. For example, water temperature is perhaps the most important factor affecting growth, reproduction, and health of channel catfish. The water temperature of a pond can be regulated by using heaters, heat pumps, or other equipment, but this would be extraordinarily expensive and economically unjustified. Certain water quality variables can, however, be economically managed by technological approaches. For instance, mechanical aerators are used to avert dissolved oxygen depletions. Other variables cannot at present be economically managed by means other than limiting the amount of feed added to a level where the incidence and severity of problems are acceptable.

DISSOLVED OXYGEN AND AERATION

Episodes of low dissolved oxygen concentrations are the most common water quality problem in catfish farming. Management of dissolved oxygen is critical because fish may die, grow slowly, or be susceptible to infectious diseases when dissolved oxygen concentrations are low (see Chapter 4).

Dissolved Oxygen Budgets

Concentrations of dissolved oxygen in catfish ponds change constantly and vary from pond to pond. Successful management of dissolved oxygen involves identifying those ponds that may require supplemental aeration to maintain adequate concentrations of dissolved oxygen. Many of the problems with low dissolved oxygen concentrations in catfish ponds can be anticipated if the farmer understands how concentrations are affected by physical and biological processes. These processes constitute the dissolved oxygen budget. The budget appears to be simple: the primary inputs are photosynthesis and diffusion; the primary losses are respiration and diffusion. However, the rates of each process are affected by myriad physical and chemical factors and the dynamics are in fact quite complex.

Diffusion. Water is said to be saturated with dissolved oxygen when it is in equilibrium with oxygen in the atmosphere. Equilibrium

TABLE 9.1.

Solubility of Oxygen in Pure Water at Different Temperatures and a Barometric Pressure of 29.92 Inches of Mercury

°C	°F	ppm	°C	°F	ppm	°C	°F	ppm
0	32.0	14.16	12	53.6	10.43	24	75.2	8.25
1	33.8	13.77	13	55.4	10.20	25	77.0	8.11
2	35.6	13.40	14	57.2	9.98	26	78.8	7.99
3	37.4	13.05	15	59.0	9.76	27	80.6	7.86
4	39.2	12.70	16	60.8	9.56	28	82.4	7.75
5	41.0	12.37	17	62.6	9.37	29	84.2	7.64
6	42.8	12.06	18	64.4	9.18	30	86.0	7.53
7	44.6	11.76	19	66.2	9.01	31	87.8	7.42
8	46.4	11.47	20	68.0	8.84	32	89.6	7.32
9	48.2	11.19	21	69.8	8.68	33	91.4	7.22
10	50.0	10.92	22	71.6	8.53	34	93.8	7.13
11	51.8	10.67	23	73.4	8.38	35	95.0	7.04

exists when the amount of oxygen entering the water from the atmosphere is equal to the amount leaving the water. The amount of dissolved oxygen present at saturation is influenced by water temperature, barometric pressure, and salinity. Water holds more dissolved oxygen as temperature and salinity decrease. For fresh waters the effect of salinity can be ignored. For reference purposes, the solubility of oxygen in water is reported for barometric pressure of 29.92 inches of mercury (Table 9.1). Values in Table 9.1 can be corrected for pressures other than 29.92 inches of mercury by multiplying by the ratio of the actual barometric pressure to 29.92.

When water is below saturation with dissolved oxygen, it is said to be undersaturated; if it is above saturation, it is said to be supersaturated. For convenience the relative amount of dissolved oxygen is sometimes reported as percentage saturation:

$$\text{Percentage saturation} = (C_m/C_s) \times 100 \quad (9.1)$$

where C_m is the measured concentration of dissolved oxygen (in ppm) and C_s is the saturation value found in Table 9.1 (after correction for actual barometric pressure).

Diffusion is the transfer of oxygen across the air-water interface. Oxygen diffuses from the atmosphere into water when the water is undersaturated with dissolved oxygen. Oxygen diffuses from water to the atmosphere when the water is supersaturated. Thus, diffusion can result in either a gain or loss of oxygen, depending on the percentage saturation.

The difference between the saturation dissolved oxygen concentration and the actual dissolved oxygen concentration is called the *saturation deficit* or *saturation surplus*. The rate of diffusion increases as the saturation deficit or surplus increases: highly undersaturated or supersaturated waters gain or lose oxygen faster than those nearly at equilibrium. This is an important concept in aeration because aerators become much less efficient at adding oxygen to water if the dissolved oxygen concentration is near saturation (a low saturation deficit). Some aerators do not mix water effectively and a zone of oxygenated water builds up around the aerator, causing a drastic decrease in efficiency.

The rate of oxygen transfer by diffusion is also related to the ratio of water surface area to water volume. If pond volumes are equal, changes in dissolved oxygen concentrations due to diffusion are greater over a given period of time in shallow ponds than deep ponds because shallow ponds have a greater surface area in contact with air. Wind increases diffusion rates because waves and ripples can increase water surface area severalfold. The relationship between the rate of diffusion and the ratio of surface area to volume is also the basis for the effectiveness of artificial aeration devices. All aerators (except some pure oxygen aeration systems) increase the rate at which oxygen is added to water by increasing the surface area of water in contact with air.

Turbulence, or mixing, is necessary for rapid transfer of oxygen to or from the atmosphere. Under calm conditions, a very thin layer of water at the air-water interface quickly reaches equilibrium with the atmosphere. This effectively hinders further diffusion until the oxygen concentration in the surface film changes by diffusion to or from the bulk water volume. Movement of oxygen between the surface film and bulk water volume is greatly accelerated by turbulent mixing. Wind is the primary source of mixing in fish pond waters. Mixing is also essential for efficient operation of artificial aerators. Mixing brings water with a high saturation deficit to the aerator and this increases aeration efficiency. The aerator then mixes the aerated water with the bulk water volume.

Amounts of oxygen added to ponds by diffusion are difficult to predict because conditions affecting diffusion (saturation deficit or surplus, surface area, and turbulence) are not constant. Oxygen usually is added to the water late at night and in early morning when the water is undersaturated, and lost from the water in late afternoon and early evening when waters are supersaturated. Boyd (1979b) made rough estimates of oxygen transfer between pond water and atmosphere under light wind conditions during 12 hours of darkness in a 3-foot-deep pond. If the initial dissolved oxygen concentration at dusk is 50 per-

Dissolved Oxygen and Aeration **221**

cent of saturation, about 1.7 ppm of oxygen is added to the pond; if the concentration at dusk is 100 percent of saturation, about 0.4 ppm of oxygen is added; if the initial concentration is 200 percent saturation, about 2.4 ppm is lost. The gain of oxygen when the initial concentration is 100 percent of saturation is the result of respiratory losses of oxygen that cause undersaturation later at night. Amounts of oxygen added or lost are lower for deeper ponds and greater during strong winds.

Photosynthesis. Biological production of oxygen by plants (photosynthesis) is the major source of oxygen in catfish ponds. Plants use light energy to produce sugars from carbon dioxide and water with a release of oxygen. Rates of photosynthesis in catfish ponds are controlled primarily by the biomass of plant material and light intensity. In catfish ponds, the predominant plant form is phytoplankton, so at equal light intensities, photosynthetic oxygen production increases with phytoplankton abundance. During periods of high light intensity, dense blooms of phytoplankton may produce so much oxygen that surface waters become supersaturated. However, turbidity caused by the bloom itself restricts light penetration and less oxygen is produced in deeper waters. Under calm conditions, this causes variation in dissolved oxygen concentrations with depth (Fig. 9.1). The amount of variation with depth increases as phytoplankton density increases. Extreme differences in dissolved oxygen concentrations occur when certain species of phytoplankton form floating scums on the pond surface. Most often these phytoplankton are species of blue-green algae (primitive plant forms that are actually plantlike bacteria) that have intracellular gas vesicles to aid in buoyancy regulation. During calm weather, colonies of cells float to the surface and most of the photosynthetic oxygen production occurs in only the upper few inches of water.

In ponds with beds of submerged weeds, dissolved oxygen concentrations may increase with depth. These ponds are usually clear, and light penetrates to the bottom, where oxygen is produced in the weed beds.

Rates of photosynthesis increase rapidly after sunrise as the angle of the sun with the pond surface becomes less acute and less light is reflected from the pond surface. Maximum rates of photosynthesis occur when the sun is high in the sky. Cloudy weather limits photosynthesis by decreasing light intensity.

Respiration. In respiration the energy stored in organic compounds is liberated and oxygen is consumed. Although only plants produce oxygen (and only during daylight), all aerobic organisms in the water (plankton and fish) and mud continuously use oxygen in res-

222 Water Quality Management in Ponds

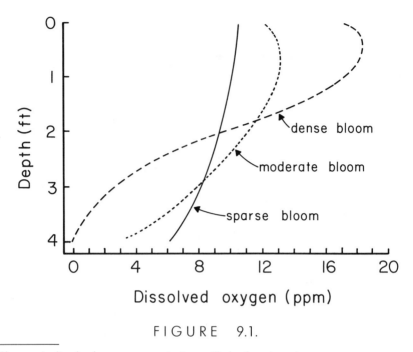

FIGURE 9.1.

Changes in dissolved oxygen concentrations with depth under calm conditions in ponds with different densities of phytoplankton.

piration. The amount of oxygen removed from water depends on the total biomass of respiring organisms and the water temperature. Rates of respiration increase as biomass and water temperature increase.

In a typical catfish pond, plankton are the major consumers of oxygen. Dense blooms may consume over 0.5 ppm of oxygen/hour in warm water. Boyd, Romaire, and Johnston (1978) present equations for estimating hourly rates of oxygen consumption by plankton from estimates of plankton abundance and temperature.

Fish also consume large amounts of oxygen. Oxygen consumption rates by channel catfish can be calculated from fish weight and temperature (see Table 4.2). In a 1-acre pond that is 3 feet deep, 5,000 pounds of 1.0-pound catfish consume about 0.2 ppm of oxygen/hour at 80°F and 0.08 ppm/hour at 55°F. Small fish consume more oxygen than large fish per unit weight. In the pond in the example given, 5,000 pounds of fish weighing 0.1 pound consume about 0.4 ppm of oxygen/hour at 80°F and 0.15 ppm/hour at 55°F.

Oxygen uptake by organisms in the mud is difficult to determine with any accuracy. For a pond with an average depth of 3 feet, a value

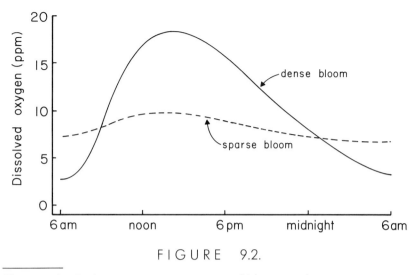

FIGURE 9.2.

Change in dissolved oxygen concentrations over a 24-hour period in surface waters of ponds with different densities of phytoplankton.

of about 0.06 ppm of oxygen/hour for mud respiration was assumed by Boyd, Romaire, and Johnston (1978). Mud respiration rates may be much higher in intensive channel catfish culture ponds in which organic matter has accumulated over several years of operation.

Daily Changes in Dissolved Oxygen

Concentrations of dissolved oxygen in channel catfish ponds usually cycle with a 24-hour period. During daylight, dissolved oxygen concentration rises as photosynthetic oxygen production exceeds losses to respiration. The nighttime decline is caused primarily by respiration. The effect of diffusion depends on whether the water is undersaturated or oversaturated.

Effect of plankton density. Phytoplankton produce oxygen by photosynthesis and are usually the major consumers of oxygen. The amplitude of daily changes in dissolved oxygen is therefore related to phytoplankton density (Fig. 9.2).

Effect of wind. Diffusion of oxygen to and from the atmosphere increases dramatically during windy conditions. Breezy weather throughout a 24-hour period dampens the amplitude of the daily

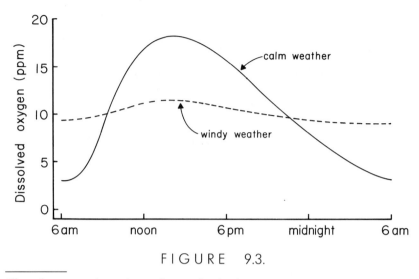

FIGURE 9.3.

Effect of continuously windy weather on dissolved oxygen concentrations in ponds.

change in dissolved oxygen concentration (Fig. 9.3). Wind-induced turbulence also mixes the pond and dissolved oxygen concentrations are similar at all depths in shallow ponds. Consistently breezy conditions are thus beneficial because excessively high or low dissolved oxygen concentrations are less likely to occur than under calm conditions. However, sudden strong winds following hot, calm weather may cause problems. In the southeastern United States, most summer afternoons are hot and calm, and dissolved oxygen concentrations are high in surface waters and low near the pond bottom. Occasionally, severe late afternoon thunderstorms develop. These brief storms are accompanied by strong winds and cold rains that rapidly mix the pond water and dilute oxygen-rich surface water with oxygen-deficient bottom water. These events are called "turnovers." Some of the dissolved oxygen in the surface water is also lost to the atmosphere when winds increase rates of diffusion. The final result is a completely mixed pond containing very little dissolved oxygen (Fig. 9.4). Because these storms usually occur in late afternoon, little oxygen is added by photosynthesis as the weather clears. Conditions are often calm after the storm and diffusion of oxygen from the atmosphere is minimal. Respiration at night then quickly depletes the remaining dissolved oxygen. The consequences of turnovers are usually more severe in deep ponds (such as some watershed-type ponds) because a proportionately larger volume of the pond may be oxygen-deficient than in shallow ponds.

Dissolved Oxygen and Aeration **225**

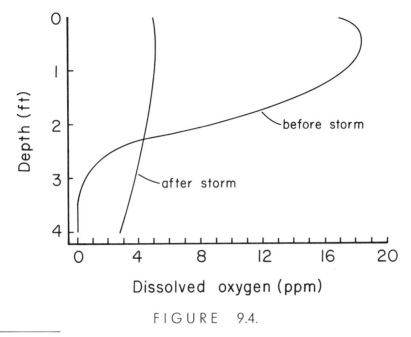

FIGURE 9.4.

Effect of late-afternoon summer storms on dissolved oxygen concentrations in ponds.

Effect of temperature. Rates of photosynthesis and respiration are decreased at low temperatures. Consequently, dissolved oxygen concentrations change little during the day in cold water. Problems with low dissolved oxygen concentrations in channel catfish ponds are extremely rare at water temperatures below about 55°F and are common at water temperatures above 75°F.

Effect of cloudy weather. Clouds reduce light intensity and decrease rates of photosynthesis but do not affect respiration rates (unless the cloudy weather is also accompanied by significantly lower air temperatures). Extended periods of cloudy weather during warm weather can result in depletion of dissolved oxygen (Fig. 9.5).

Effect of phytoplankton die-offs. Populations of phytoplankton continually wax and wane. When the community is composed of many populations of different species, the death of one population is compensated for by the growth of another and overall rates of oxygen production during daylight remain high. At times, one species (usually a blue-green alga) proliferates at the expense of all other species and may

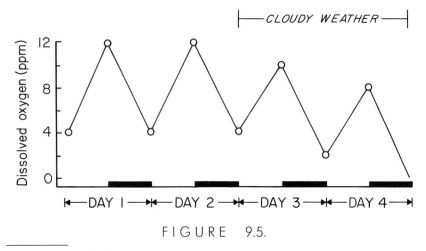

FIGURE 9.5.

Effect of cloudy weather on concentrations of dissolved oxygen at early morning and late afternoon over a 4-day period. Dark bars along the x axis indicate night.

comprise nearly all the phytoplankton biomass in the pond. If this population suddenly dies, community oxygen production nearly ceases and decomposition of the dead algae rapidly lowers dissolved oxygen concentrations. Concentrations of dissolved oxygen remain low until another phytoplankton community develops. The causes of natural phytoplankton die-offs are not known and events are impossible to predict. Die-offs can also be triggered when herbicides are applied to ponds in efforts to control phytoplankton abundance or rid the pond of noxious weeds. Certain chemicals, including copper sulfate, potassium permanganate, and formalin, which are used to control external parasitic or bacterial diseases of channel catfish are toxic to phytoplankton and can also cause algal die-offs and oxygen depletions.

Nighttime Declines in Dissolved Oxygen

The decline of dissolved oxygen concentrations at nighttime is the channel catfish producer's most important water quality management problem. In densely stocked ponds, the combined respiration of plankton, fish, and organisms in the mud often causes dissolved oxygen concentrations to fall to dangerous levels at night during warmer months. Each pond contains different standing crops of fish, different plankton densities, and different densities of mud-dwelling organisms; hence, rates of respiration vary from pond to pond. The key to successful management of dissolved oxygen is early identification of those ponds that may require supplemental aeration to keep fish alive.

Two methods are available to predict the decline in dissolved oxygen concentrations at night. A computer model (Boyd, Romaire, and Johnston 1978) is based on an equation describing the components contributing to nighttime losses of dissolved oxygen:

$$DO_t = DO_{dusk} \pm DO_{df} - DO_f - DO_m - DO_p \qquad (9.2)$$

where DO_t = the dissolved oxygen concentration after t hours of darkness, DO_{dusk} = the dissolved oxygen concentration at dusk, DO_{df} = dissolved oxygen gained or lost by diffusion, and DO_f, DO_m, and DO_p = the dissolved oxygen used by fish, mud, and plankton, respectively. The major losses of dissolved oxygen are plankton and fish respiration, and these can be accurately estimated if the correct information on bloom density, sizes and total weight of fish, and water temperature is available. Although the model provides reliable estimates of dissolved oxygen concentrations throughout the night, it requires too much information for routine use on commercial catfish farms.

A simple estimate of nighttime dissolved oxygen concentrations can be made by assuming that the decline in dissolved concentrations with time is nearly linear. Measurements of dissolved oxygen are made at dusk and then again 2 to 4 hours later. The results can be plotted and a line extended through the points (Fig. 9.6). This method provides a margin of safety as it usually underestimates dissolved oxygen concentrations later at night because water temperatures decrease and diffusion adds increasing amounts of oxygen to water as the water becomes increasingly undersaturated. Few pond managers use this method exactly as described; however, most use the concept to prioritize management decisions. Current practices usually involve measuring dissolved oxygen concentrations in all ponds at intervals throughout the night. These measurements are used to make rough judgments of the relative rates at which concentrations are falling in different ponds. Particular attention is then paid to those ponds that appear to be approaching critically low dissolved oxygen concentrations more rapidly.

Measurements of Dissolved Oxygen

Methods. Two methods of measuring dissolved oxygen are available: chemical test kits and polarographic oxygen meters. Several companies offer chemical test kits based on the classic Winkler procedure. This procedure is subject to serious sampling errors and, more

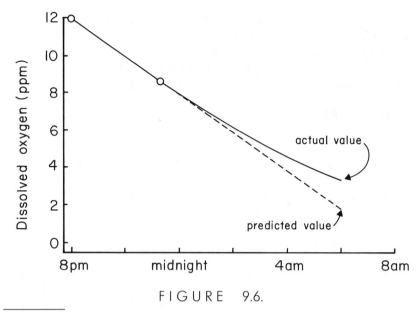

FIGURE 9.6.

Straight-line projection method for predicting dissolved oxygen concentrations at night. The predicted value usually is lower than the actual value.

importantly, is quite time-consuming. Chemical tests can effectively be used only if a few ponds are to be monitored.

Polarographic oxygen meters (Fig. 9.7) offer a fast and reliable method for measuring dissolved oxygen concentrations. The device consists of an electrode that produces an electrical current proportional to the concentration of oxygen in the water and a meter that translates this current into oxygen concentration units that can then be read on a scale.

The electrode, often called the "probe," consists of two electrodes bathed in a salt solution (Fig. 9.8). The electrodes and solution are separated from the environment by a thin plastic membrane that is stretched across the top of the probe. The membrane is essentially impermeable to salts but does allow oxygen (and some other gases) to enter. A potential is applied across the two electrodes by the external voltage source (the batteries), but no current flows unless a certain pair of reactions involving oxygen occurs to complete the circuit. The reaction at the gold cathode consumes oxygen and produces hydroxide ions. Hydroxide ions migrate to the silver anode and react with silver to form silver oxide. This completes the circuit and a current flows. The current is proportional to the overall reaction rate. Because oxygen

Dissolved Oxygen and Aeration   **229**

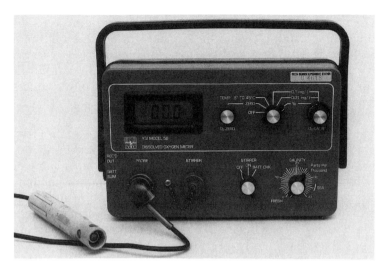

FIGURE 9.7.

A polarographic oxygen meter and probe.

is consumed in the reaction at the anode, the oxygen concentration beneath the membrane is kept low (virtually zero). When the probe is in this condition it is said to be *polarized,* and the only source of oxygen for the reaction is that which diffuses across the membrane. The rate of oxygen diffusion across the membrane is proportional to concentration of oxygen in the bulk solution (the pond water). The current

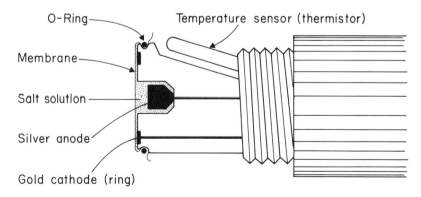

FIGURE 9.8.

Cross section of a dissolved oxygen probe.

is, therefore, proportional to the oxygen concentration outside the membrane. For example, when the oxygen concentration in the pond is low, there is little driving force for diffusion of oxygen across the membrane. The overall reaction rate is low, and little current flows. The current produced at the electrodes is translated into a concentration reading in the meter itself.

The meter "senses" only electrical current, not oxygen. It must be "told" what oxygen concentration corresponds to a certain electrical current. This *calibration* is the most important step in using an oxygen meter. Two methods are used to calibrate an oxygen meter: air calibration, which is easier and more rapid, and a calibration method that uses oxygen-saturated water and takes longer to perform and is subject to substantial error if conducted improperly. The actual calibration procedure varies, depending on the model, and the manufacturer's instructions should be closely followed.

Water must flow across the membrane surface during the dissolved oxygen measurement. Oxygen is consumed at the membrane surface, and unless water constantly moves across it, a zone of oxygen-depleted water develops at the membrane surface and gives a false low reading. Moving the probe back and forth or up and down about 1 foot/ second is usually sufficient. The most reliable readings are obtained in waters containing dissolved oxygen concentrations between about 2 ppm and saturation. Outside this range readings are approximate, particularly in highly supersaturated water. The meter does not give an instant reading. The meter generally takes 10 to 20 seconds to indicate within about 10 percent of the actual dissolved oxygen concentration; in cold weather, it takes even longer. Allow sufficient time for the reading to stabilize for the most accurate measurement. The instrument should be left on during successive measurements to prevent depolarizing the probe. However, if the instrument will not be used for a period of an hour or more, it is best to turn the meter off and recalibrate it when next needed. This extends the life of batteries and probe.

Where to measure. Measurements of dissolved oxygen should be taken at a minimum of two sites on opposite ends of ponds larger than 1 acre because concentrations can vary considerably within a pond. Do not take routine measurements near inflowing water, in scums of algae, or immediately beside the bank. Measurements should also not be taken at the very surface or bottom of the pond. In routine monitoring programs, most managers take measurements about 1 foot beneath the surface.

When to measure. During the warmer months, dissolved oxygen concentrations should be measured at least three times daily: at

Dissolved Oxygen and Aeration **231**

dusk, about 4 hours after dusk, and at dawn. The measurements made at dusk and 4 hours later can be used to assess the danger of low dissolved oxygen concentrations later at night using the method described in Figure 9.6. Additional measurements are then made throughout the night as needed. Record all measurements so they can be checked quickly. Daily records of dissolved oxygen concentrations at dawn and dusk can be kept for each pond to demonstrate trends that will aid in anticipating problems. Such a record will be similar to Figure 9.5.

Dissolved oxygen depletions can also occur during daylight hours if conditions for photosynthesis are poor. Measure dissolved oxygen several times a day during hot, cloudy weather and after treating ponds with herbicides or phytotoxic disease therapeutants. Dissolved oxygen concentrations should also be measured any time fish are in distress because concentrations may be unexpectedly low.

Most producers do not measure dissolved oxygen concentrations during the winter. Remember, however, that dissolved oxygen dynamics is a function of temperature, not calendar month. Dissolved oxygen depletions are not uncommon during unseasonably warm weather in winter. It is good practice to measure dissolved oxygen concentrations at least once a day even when water temperatures are less than 55°F.

Aeration and Feeding Rates

Episodes of critically low dissolved oxygen are more frequent in ponds with dense phytoplankton blooms. Because average phytoplankton bloom density is related to feeding rate, there is a general relationship between average maximum daily feeding rate (pounds of feed/acre/day) and the amount of aeration needed to keep fish alive (Fig. 9.9). If feeding rates are maintained below about 35 pounds/acre/day, dissolved oxygen concentrations rarely fall to critical levels. A maximum feeding rate of 35 pounds/acre/day is sufficient to grow about 2,000 fish/acre to harvestable size in one growing season in a single-batch cropping production system (see the section Food-Fish Production). To increase fish yields and profits, most commercial producers stock fish at higher densities and feed at rates exceeding 50 pounds of feed/acre/day. Consequently, the amount of aeration required to keep fish alive and healthy increases. At some point, the amount of oxygen consumed by plankton, fish, and mud becomes so great that aeration is required for long periods virtually every summer night. Not only is this an economic burden but it also subjects fish to considerable stress. At very high feeding rates, other factors besides oxygen availability (such as ammonia) also begin to limit fish production and the theoretical amount of fish that can be produced at higher stocking and feeding

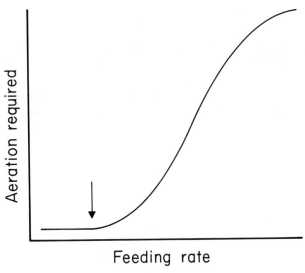

FIGURE 9.9.

Relationship between average maximum daily feeding rates (pounds of feed/acre/day) and amount of aeration (hours/night or hours/year) required to keep fish alive in ponds. This relationship is not well defined; only one point is known with any certainty. If maximum sustained feeding rates exceed about 35 pounds/acre/day (arrow), aeration is needed to keep fish alive.

rates is offset by slower fish growth rates, greater loss of fish to diseases, and high costs of aeration.

Aerator Performance

Because dozens of different aerators are available the decision as to which aerator to purchase often bewilders catfish farmers. Although selection of an aerator is still quite subjective, aerator performance can be quantified through standard engineering tests. The results of these tests can aid in the selection process. Aerator testing methods and interpretation of test results are thoroughly reviewed by Boyd and Watten (1989).

Standard tests produce two values that describe aerator performance. The *standard oxygen transfer rate* (SOTR) is the amount of oxygen an aerator adds to water per hour. The units are pounds of oxygen/hour. *Standard aeration efficiency* (SAE) is the standard oxygen transfer rate divided by the amount of power required. Units are pounds of oxygen/horsepower-hour.

Aerators are tested in a large tank of water that has been deoxygenated. The water is reaerated by running the aerator, and dissolved oxygen concentrations are measured during the trial. The results are then analyzed mathematically to determine SOTR. Because oxygen transfer is affected by the degree of oxygen undersaturation and water temperature, results are projected to a set of standard conditions. The SOTR is the amount of oxygen transferred when clean water has 0 ppm dissolved oxygen and is at 68°F. Aerators usually transfer less oxygen under pond conditions, and SOTR and SAE values are best used to compare aerators as an aid in selecting one for purchase rather than as precise design criteria for pond use. When comparing SAE values, remember that small differences in reported values are not meaningful; results should differ at least 0.5 pounds of oxygen/horsepower-hour to be considered significant. Other factors, such as cost, durability, specific application, and ease of service, must also be considered when selecting an aerator. Standard aeration efficiency and durability are most important when selecting aerators for general day-to-day use. High SOTR values and mobility are important for aerators used to save fish in distress. Boyd and Ahmad (1987) have compiled SOTR and SAE values for a variety of aerators commonly used in channel catfish culture.

Types of Aerators

All aerators currently used in pond culture of channel catfish are air-contact aerators. These aerators add oxygen to water by increasing the rate of oxygen diffusion from the atmosphere into water. Aerators increase the surface area for contact between air and water by agitating surface water, releasing bubbles beneath the surface, or doing both. Aerators also mix the water to maintain high rates of mass transfer of oxygen and circulate water so fish can locate the areas with higher dissolved oxygen concentrations. Pure-oxygen aeration systems are rarely used in ponds, although this equipment is commonly used in some high-density culture systems (Chapter 13).

Diffused air aerators. Diffused air aerators use blowers or compressors to supply air and diffusers or porous pipe to release air bubbles on the pond bottom (Fig. 9.10). Oxygen is transferred as the bubbles are formed and ascend through the water. The amount of oxygen transferred depends on the number and size of bubbles that are formed and the depth at which the bubbles are released. For the same total air

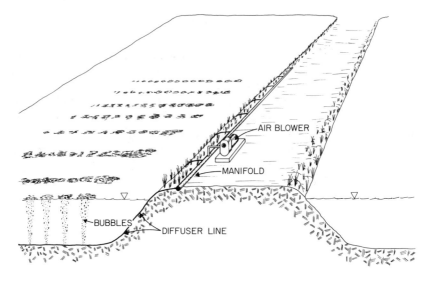

FIGURE 9.10.

Diffused-air aeration system.

volume, oxygen transfer is highest for small bubbles, although in practice more air can be released from coarse-bubble diffusers. Oxygen transfer is also higher when bubbles are released at great depth because there is more time for oxygen transfer as bubbles ascend.

Diffused aeration is not efficient in shallow ponds commonly used in channel catfish production. With diffuser depths of 3 to 4 feet, SAE values are usually less than 2 pounds of oxygen/horsepower-hour. Diffusers also clog easily in fish culture ponds, and the network of supply lines and diffusers interferes with seining.

Vertical pump aerators. Vertical pump aerators consist of a submersible motor with an impeller attached to the output shaft. The motor and impeller are suspended beneath a float, and water is sprayed into the air through an opening in the center of the float (Fig. 9.11).

Vertical pump aerators are fairly efficient; SAE values usually range from 2 to 2.5 pounds of oxygen/horsepower-hour (Boyd and Ahmad 1988). Most vertical pump aerators manufactured for use in fish culture use 1- to 5-horsepower motors and do not produce a large area of oxygenated water. This limits their use to small ponds of 0.1 to 2 acres.

Propeller-aspirator-pump aerators. Propeller-aspirator-pump aerators consist of a motor, shaft, propeller, and flotation (Fig. 9.12).

Dissolved Oxygen and Aeration **235**

FIGURE 9.11.

Vertical pump aerator.

The propeller, which is mounted on the end of a hollow drive shaft, accelerates the water to a velocity high enough to create a partial vacuum at the end of the shaft. Air is pulled down the shaft and dispersed into the water as a stream of fine bubbles. The horizontal flow of bubbles allows more contact time between bubble and water than diffused air aerators in shallow ponds. These aerators are commercially available in sizes from 1.0 to 7.5 horsepower. Propeller-aspirator-pump aerators are among the most commonly used aerators in aquaculture worldwide but oddly are seldom used in catfish culture. The aerators are durable and relatively efficient; SAE values range from 2 to 3 pounds of oxygen/horsepower-hour.

Pump-sprayer aerators. Pump-sprayer aerators consist of a pump that discharges water at high velocity through a pipe or manifold. Pumps may be powered by the power takeoff (PTO) of a tractor or by an electric motor.

Pump-sprayers are simple and usually require less maintenance than other types of aerators. The PTO-driven pump-sprayer known as a *T-pump* or *bankwasher* (Fig. 9.13) is particularly effective during ox-

FIGURE 9.12.

A propeller-aspirator-pump aerator. (Photo: Aeration Industries, Chaska, Minnesota)

ygen depletions. The manifold directs oxygenated water along the shoreline, where distressed fish congregate.

Pump-sprayer aerators range widely in effectiveness and efficiency (Boyd and Ahmad 1988). Those powered by tractor PTOs have SOTR values from 15 to over 160 pounds of oxygen/horsepower. Aerators with higher SOTR values usually require a large tractor (up to 200 horsepower) or require the PTO to be operated at a high speed (1,000 rpm). Electric motor-driven pump-sprayer aerators have SAE values ranging from 1.5 to 3.2 pounds of oxygen/horsepower-hour.

Paddlewheel aerators. Paddlewheel aerators consist of a hub with paddles attached in a staggered arrangement. The aerator is powered by tractor PTO (Fig. 9.14), self-contained diesel or gas engine, or electric motor (Fig. 9.15). Electric paddlewheel aerators are usually mounted on floats and anchored to the pond bank.

Many different designs for PTO-driven paddlewheels exist. Gearboxes and automobile differentials have been used for gear reduction, and paddlewheel diameters (paddle tip to tip) range from 2 to 4 feet.

Dissolved Oxygen and Aeration **237**

FIGURE 9.13.

A T-pump style of pump-sprayer aerator.(Photo: Mastersystems, Greenwood, Mississippi)

Paddles are 2 to 10 inches wide and may be rectangular, triangular, or semicircular (concave) in cross section. Paddles are arranged in various spiraled or staggered arrangements around the hub. Standard oxygen transfer rates for PTO-driven aerators range from 15 to 90 pounds of oxygen/hour (Busch et al. 1984; Boyd and Ahmad 1988). Large-diameter paddlewheels transfer more oxygen than smaller-diameter aerators, and flat paddles are less effective than other cross sections. For a given aerator, oxygen transfer increases with increasing paddle tip submergence and increasing hub rotation speed. Increased diameter, paddle depth, and speed also increase the power required for aerator operation.

    Tractors develop more power than is applied to the PTO and aerator drive shaft, and energy is lost through the drive train, so PTO-driven paddlewheels are not particularly energy-efficient. The current trend is to use electric motors to power paddlewheel aerators. Most electric paddlewheels used on commercial catfish farms are powered

FIGURE 9.14.

A paddlewheel aerator powered by a tractor PTO.

by 10 to 15 horsepower electric motors and hubs are 10 to 15 feet long. Optimization of design (Ahmad and Boyd 1988) has resulted in highly efficient aerators with SAE values of 4 to 5 pounds of oxygen/horsepower-hour. The best design consists of a 3-foot-diameter paddlewheel with paddles triangular (135-degree interior angle) in cross section. Paddles are about 4 to 6 inches wide with four paddles attached per row and spiraled around the hub. Paddle depth is 4 to 6 inches in the water. Paddlewheel speed should be about 90 rpm. On some commercial designs, paddle depth can be varied to optimize the current drawn by the motor. The motor should draw about 90 percent of full load amperage rating to give maximum oxygen transfer and motor life.

Aeration Practices

There are basically four strategies used in dissolved oxygen management of channel catfish ponds: (1) no supplemental aeration, (2) emer-

Dissolved Oxygen and Aeration **239**

FIGURE 9.15.

A floating, electric paddlewheel aerator.

gency aeration, (3) nightly aeration, and (4) continuous aeration. Aeration practices have changed over the last 20 years as farmers have intensified production.

No supplemental aeration. The only way to grow catfish with no supplemental aeration is to use feeding rates that never exceed about 35 pounds of feed/acre/day. This type of low-input fish culture may be desirable under some circumstances, such as when fish production is a secondary farm enterprise. However, very few commercial enterprises currently use this approach to fish culture, although low stocking rates (fewer than 3,000 fish/acre) and low feeding rates (less than 35 pounds feed/acre/day) were fairly common 15 to 20 years ago.

Emergency aeration. Over the years, farmers increased stocking and feeding rates in an effort to grow more fish, and dissolved oxygen concentrations occasionally fell to levels that posed an immediate threat to fish survival. Early efforts at aeration were crude and often ineffective. Running fresh water into the pond from a well or adjacent pond, using a fast motorboat to churn up the water, and applying potassium permanganate were early tactics to avert fish death. As episodes of dissolved oxygen depletion became more frequent, equipment designed for effective aeration was developed. This equipment was made to be powered by tractors, a power source already available to farmers. This also made the equipment mobile and made available a large power potential, thus affording the opportunity for large oxygen transfer rates. The tractor-powered paddlewheel aerator (Fig. 9.14) was one of the first such devices developed and is still a popular aerator.

In emergency aeration, aerators are used only when dissolved oxygen concentrations fall to levels at which fish death is imminent; aerators are not used to "manage" dissolved oxygen. In a typical 17-acre catfish pond, the amount of oxygen consumed at night by plankton, mud, and fish is 100 to 200 pounds of oxygen/hour. Most emergency aerators now in use on commercial ponds transfer 30 to 50 pounds of oxygen/hour under optimal conditions. One aerator obviously cannot meet the oxygen demand of the entire pond. The aerator can, however, meet the oxygen requirements of the fish when they congregate in the area immediately behind it. Even in the area immediately behind the aerator, dissolved oxygen concentrations are usually only slightly higher than in the rest of the pond. Enough oxygen is added to sustain fish life, but not enough to increase dissolved oxygen concentrations in the pond significantly.

Emergency aeration has a serious disadvantage in that fish are routinely exposed to suboptimal concentrations of dissolved oxygen because aeration is initiated only when dissolved oxygen concentrations fall to very low levels. Even then, concentrations of dissolved oxygen near the aerator are maintained just high enough to prevent fish death. Repeated exposure to these conditions undoubtedly results in poorer feed conversion, slower growth, and increased susceptibility to infectious diseases. Despite this serious drawback, emergency aeration remains the most common aeration practice in commercial catfish ponds. It is the only practice that has proved to be profitable.

Although tractor-powered aerators were the first effective emergency aerators used by catfish farmers, a current trend is to replace the mobile, but expensive, tractor-aerator unit with electric motor-driven aerators that are permanently installed in each pond. These aerators have been found much less expensive to operate per the amount of oxygen transferred to the pond water. However, there is a risk of losing aeration capability during power outages.

Nightly aeration and continuous aeration. Nightly aeration and continuous aeration are similar in that the usual objective is to improve or maintain dissolved oxygen concentrations in the entire pond or at least in a large part of it. The two systems differ only in the length of time aeration is made available. In nightly aeration, aeration is initiated at some fixed time at night well before dissolved oxygen levels fall to stressful levels and then continued until photosynthesis begins to add oxygen to the water the next day. In continuous aeration, the aerator runs 24 hours a day during all or part of the year.

The theoretical advantage to these approaches is that dissolved oxygen concentrations are not allowed to fall to very low levels and fish are not stressed. They convert feed to flesh more efficiently and grow faster.

The major disadvantage of these approaches to aeration is the result of the enormous oxygen demand exerted by the plankton bloom relative to the amount of oxygen needed by the fish. Ponds are aerated to supply oxygen to the fish, but fish are responsible for only a fraction of the total oxygen consumed in the pond. Thus, if the fish in the pond consume 25 pounds of oxygen/hour, several times that much must be supplied to the pond to assure meeting their needs. In a practical sense, this means that a much larger aerator is needed than would be required to meet the oxygen demands of the fish alone and that most of the oxygen supplied is being used to meet the requirements of an economically valueless crop (the plankton bloom). Also, continuous aeration is usually wasteful because aerators are inefficient when dissolved oxygen concentrations are near saturation and they actually accelerate the loss of oxygen when waters are supersaturated.

Two studies have compared the effectiveness of emergency and nightly aeration of channel catfish ponds. In one study (Lai-fa and Boyd 1988), nightly aeration increased fish production, improved feed conversion efficiency, and increased profits compared to those of ponds aerated only when fish were in distress. The ponds used in that study were stocked at about 4,000 fish/acre and fish were fed up to 50 pounds of feed/acre per day. Ponds had moderate densities of phytoplankton and were provided with aerators with an SOTR of about 7 pounds of oxygen/hour per acre. Dissolved oxygen concentrations in nightly aerated ponds never dropped below 4 ppm. In another study (Steeby and Tucker 1989), nightly aeration did not improve fish production or feed conversion when compared to those of ponds aerated only when dissolved oxygen concentrations fell below 2 ppm. Ponds used in that study were stocked with 5,000 fish/acre and fish were fed up to 100 pounds of feed/acre/day. Phytoplankton blooms in these ponds were dense and ponds were equipped with aerators with an SOTR of about 5 pounds of oxygen/hour per acre. Dissolved oxygen concentrations were often below 4 ppm in nightly aerated ponds because the relatively small aerators used in this study (although still larger relative to pond size than those used on most commercial catfish farms) were not able to meet the high oxygen demands in the heavily fed ponds.

Aerator Size Requirements

An aerator for emergency aeration must provide sufficient oxygen to exceed the oxygen requirements of the fish. For stocking rates of 4,000 to 8,000 fish/acre this usually requires an aerator with an SOTR of at least 2 to 3 pounds of oxygen/hour/acre. For aerators with average SAE values (2 to 3 pounds of oxygen/horsepower-hour), this amounts to about 1 horsepower/acre. When oxygen demands are high or when fish are otherwise stressed, additional aeration will be required. Current practice is to supply each pond with an efficient (SAE > 3 pounds of oxygen/horsepower-hour) electric aerator at about 1 horsepower/acre and one mobile, tractor PTO-driven aerator for every three or four ponds. The mobile aerators should have an SOTR of at least 2 pounds of oxygen/hour/acre.

In continuous or nightly aeration, the aerator ideally meets the oxygen demand of fish, plankton, and mud. This will require at least three to four times the aeration capabilities needed for emergency aeration.

Placement of Aerators

Catfish become conditioned to moving to the area near an aerator when dissolved oxygen concentrations are low, and permanently installed electric aerators should be located mainly for the convenience of the producer. Aerators should be near a graveled, all-weather road for access during operation and maintenance. They should also be near the power source to decrease the length of power lines.

Mobile aerators should either be placed in the same location each time they are used or placed where fish are located. If fish are in distress, put the aerator as close as possible to the distressed fish. When the dissolved oxygen concentration varies from one end of the pond to another, most fish will be in the end where concentrations are higher;

place the aerator at that end. If an aerator must be replaced while in operation, put the replacement as near the original as possible. Fish may not be able to swim large distances through oxygen-deficient water to locate the replacement aerator.

When to Aerate

A diligent program of dissolved oxygen monitoring is critical when emergency aeration is practiced. On most commercial farms, dissolved oxygen concentrations are measured at frequent intervals throughout the night and aeration is initiated when concentrations fall to some predetermined level. If the farm is equipped with permanent electric aerators in all ponds, aeration is initiated when dissolved oxygen concentrations fall to 2 to 4 ppm. Ponds are checked thereafter and additional mobile aerators are used if necessary. On farms where less than one mobile aerator is available for each pond, it is important to prioritize and move aerators to ponds where dissolved oxygen concentrations are falling most rapidly so they can be aerated first. Aeration should be initiated well before concentrations fall to critical levels.

If continuous or nightly aeration is practiced, aeration is required only during the warmest months, when oxygen demands are highest. Nightly aeration is initiated at dusk or within a few hours after dusk. The use of nightly aeration does not mean the producer can disregard dissolved oxygen monitoring or ignore fish at night. Aerators should be checked frequently throughout the night to assure proper operation. If oxygen demands are high or if the fish are diseased, additional aerators may be required. Even in continuously aerated ponds dissolved oxygen concentrations should be measured two or three times a day.

Alternative Methods of Improving Dissolved Oxygen Concentrations

Channel catfish producers have used a variety of techniques other than aeration in efforts to improve dissolved oxygen concentrations. These techniques are usually used in desperation when oxygen demands exceed aeration capabilities. However, some of these techniques are of limited or no value.

The most common alternative technique for adding oxygen is to pump water from an adjacent pond or from a well. This practice can be beneficial if the water added has a high dissolved oxygen concentration and large amounts are pumped. Water pumped from an adjacent pond can be pumped back to the "donor" pond after the crisis is over.

Most well waters are devoid of dissolved oxygen and should be aerated before use.

Potassium permanganate is sometimes added to ponds with low dissolved oxygen concentrations. Application of the chemical reduces the rate of oxygen consumption by killing plankton microorganisms and oxidizing some organic matter. Potassium permanganate does not add significant amounts of dissolved oxygen to water. Moreover, the effects are short-lived and the chemical is expensive. Potassium permanganate is algicidal and may cause lower dissolved oxygen concentrations after its use than would occur without it (Tucker and Boyd 1977).

Hydrated lime ($Ca(OH)_2$) has also been applied to ponds with low dissolved oxygen concentrations. This practice does not increase dissolved oxygen concentrations, although hydrated lime does remove free carbon dioxide from water. Carbon dioxide aggravates the effect of oxygen deficiency on fish; however, the general practice of using lime during dissolved oxygen depletions cannot be recommended. Application of hydrated lime increases the pH of water. High concentrations of ammonia are sometimes present in channel catfish ponds, and the toxicity of ammonia increases dramatically as pH increases.

TOTAL ALKALINITY AND HARDNESS

Total alkalinity is a measure of the buffering capacity of water. In most waters alkalinity is attributed to the presence of bicarbonate (HCO_3^-), carbonate (CO_3^{2-}), and, to a much lesser degree, hydroxide (OH^-). *Total hardness* is the combined concentrations of calcium (Ca^{2+}) and magnesium (Mg^{2+}). Although alkalinity and hardness are different properties of water, both are expressed as ppm as $CaCO_3$.

The alkalinity and hardness of channel catfish pond water are determined by the quality of the water supply and the nature of the pond bottom soils. Rainfall runoff is dilute, so water in watershed ponds usually has a total alkalinity and hardness of less than 20 ppm as $CaCO_3$, unless ponds are built on limestone-bearing soils. Groundwater usually contains more minerals than surface water, but total alkalinity and hardness of the supply vary greatly depending on the aquifer. The total alkalinity and hardness of pond waters vary significantly with time only when the water is diluted or concentrated: values decrease during periods of excessive rainfall and increase when evaporation exceeds rainfall.

Channel catfish thrive in waters with a wide range of total alkalinity and hardness, but pH fluctuations can be difficult to manage in water with very low total alkalinity and hardness or in water with large imbalances between the two values. The relationships among pH, alkalinity, and hardness are discussed in the next section.

Phytoplankton production is relatively low in water with very low total alkalinity and hardness (Boyd 1979b). Assimilation by phytoplankton of ammonia excreted by fish is an important process preventing dangerous accumulation of ammonia in catfish pond waters (see the section Ammonia and Nitrite). Low phytoplankton productivity in ponds with low alkalinities and hardness sometimes results in chronic episodes of high total ammonia concentrations because assimilation rates are low.

Dissolved metals such as copper are more toxic to fish in waters of low alkalinity and hardness. In waters with ample alkalinity and hardness, metals form nontoxic precipitates and complexes and calcium competes with the metals for uptake by the fish. This process has considerable practical importance because copper sulfate is widely used as an algicide and fish disease therapeutant. The application rate is determined by the total alkalinity of the water (see the section Disease Treatments). Copper sulfate must be used cautiously (if at all) in waters with total alkalinities less than 50 ppm as $CaCO_3$ because fish kills are possible.

Alkalinity and hardness can be increased in ponds by adding agricultural limestone. Boyd (1979b) discusses methods of determining the lime requirement of fish ponds and the effects of liming on water quality. Most ponds filled with soft, low-alkalinity water require between 1 and 2 tons of lime/acre to increase total alkalinity to between 20 and 50 ppm as $CaCO_3$.

CARBON DIOXIDE AND pH

Carbon dioxide (CO_2) is highly soluble in water, but concentrations of carbon dioxide in pure water exposed to air are low (less than 1 ppm) because it is a minor constituent of the atmosphere. Carbon dioxide concentrations in natural waters vary considerably, however, because it is involved in the metabolism of plants and animals. Carbon dioxide is produced by organisms during respiration and consumed during photosynthesis. Thus, concentrations fluctuate on a 24-hour basis in a manner essentially opposite that of dissolved oxygen (Fig. 9.16). In the

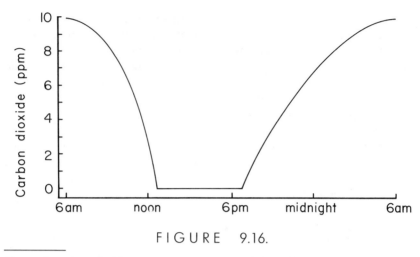

FIGURE 9.16.

Change in carbon dioxide concentrations in a typical channel catfish pond over a 24-hour period.

daytime, carbon dioxide concentrations decrease; at night carbon dioxide concentrations increase.

The amount of carbon dioxide in water depends on the relative rates of respiration and photosynthesis, and, to a lesser extent, the rate of diffusion to and from the atmosphere. The dynamics of carbon dioxide concentrations in channel catfish ponds is, like that of dissolved oxygen, dominated by the activities of plankton, and the amplitude of the daily fluctuation in carbon dioxide concentrations is related to plankton bloom density. Conditions that favor rapid rates of photosynthetic oxygen production favor rapid removal of carbon dioxide. Carbon dioxide is essentially depleted from water on warm, sunny afternoons in ponds with moderate to dense phytoplankton blooms. High respiration rates during summer months in ponds with abundant plankton and high densities of fish result in high rates of oxygen use and carbon dioxide production.

Concentrations of carbon dioxide in channel catfish ponds during the summer usually range from 0 ppm in the afternoon to 5 to 10 ppm at dawn. During extended periods of cloudy weather, carbon dioxide may not be depleted in the afternoon and can exceed 10 ppm at dawn. High concentrations of carbon dioxide also occur after phytoplankton die-offs because rates of photosynthesis are low and great amounts of carbon dioxide are produced as the dead plant material decomposes. Concentrations may exceed 20 ppm for several days after the die-off. Concentrations are consistently higher in the late fall and winter than in the summer. This reflects the less ideal conditions for photosynthesis during the cooler months. Also, concentrations usually vary less during a 24-hour period in the winter because metabolism is lower in cold water. Carbon dioxide concentrations greater than 20 ppm are common at dawn in the winter and may remain high throughout the day, particularly during cold, cloudy weather. Conditions during the winter are different from those during the summer because in winter, dissolved oxygen concentrations are usually near saturation levels (10 to 14 ppm) while carbon dioxide concentrations are concurrently high. In the summer, high concentrations of carbon dioxide usually occur when dissolved oxygen concentrations are low.

Carbon dioxide concentrations are not routinely measured by catfish farmers. Carbon dioxide affects oxygen use by fish, and the two variables usually cycle opposite each other in ponds during the summer, so current management practice is to monitor dissolved oxygen concentrations closely and aerate ponds to prevent dissolved oxygen from falling to very low levels at which the added effect of carbon dioxide will kill fish. Occasionally, however, measurements of carbon dioxide can be useful. Carbon dioxide should be measured whenever apparently healthy fish show signs of dissolved oxygen stress and dissolved oxygen concentrations are above 2 ppm. Healthy channel catfish normally do not come to the surface and gasp or pipe unless dissolved oxygen concentrations are below 1 ppm for an hour or more. Obtain samples for measurement of carbon dioxide from the same depth, location, and time at which dissolved oxygen measurements are taken. Measurements can be made with commercially available chemical test kits. Carbon dioxide concentrations can also be estimated from the pH, temperature, and total alkalinity of the water (Table 9.2). To use Table 9.2, find the factor within the body of the table corresponding to the pH and temperature of the water. Multiply this factor by the measured total alkalinity to estimate the concentration of carbon dioxide. For example, assume that the pH is 7.0, the temperature is 77°F, and the total alkalinity is 200 ppm as $CaCO_3$. The appropriate factor from Table 9.2 is 0.197, so 200 × 0.197 = 39.4, or about 40 ppm carbon dioxide.

Although it is seldom necessary or even desirable to treat for high carbon dioxide levels, hydrated lime (calcium hydroxide or $Ca(OH)_2$) removes carbon dioxide from water:

$$Ca(OH)_2 + 2CO_2 = Ca(HCO_3)_2 \qquad (9.3)$$

On the basis of this reaction, 1 ppm of carbon dioxide reacts with 0.84 ppm of calcium hydroxide. In practice about 1.7 ppm of calcium hydroxide is needed to remove 1 ppm of carbon dioxide because the second step of this two-stage reaction is slow. This treatment should be used cautiously because it will cause the pH of the water to rise. In pond waters of low total alkalinity (<20 ppm as $CaCO_3$), the pH can

TABLE 9.2.

Factors Used to Calculate Approximate Carbon Dioxide Concentrations in Dilute Fresh Waters from pH, Temperature, and Total Alkalinity

	Temperature						
pH	41°F (5°C)	50°F (10°C)	59°F (15°C)	68°F (20°C)	77°F (25°C)	86°F (30°C)	95°F (35°C)
6.0	2.915	2.539	2.315	2.112	1.970	1.882	1.839
6.2	1.839	1.602	1.460	1.333	1.244	1.187	1.160
6.4	1.160	1.010	0.921	0.841	0.784	0.749	0.732
6.6	0.732	0.637	0.582	0.531	0.495	0.473	0.462
6.8	0.462	0.402	0.367	0.335	0.313	0.298	0.291
7.0	0.291	0.254	0.232	0.211	0.197	0.188	0.184
7.2	0.184	0.160	0.146	0.133	0.124	0.119	0.116
7.4	0.116	0.101	0.092	0.084	0.078	0.075	0.073
7.6	0.073	0.064	0.058	0.053	0.050	0.047	0.046
7.8	0.046	0.040	0.037	0.034	0.031	0.030	0.030
8.0	0.029	0.025	0.023	0.021	0.020	0.019	0.018
8.2	0.018	0.016	0.015	0.013	0.012	0.012	0.011
8.4	0.012	0.010	0.009	0.008	0.008	0.008	0.007

Note: Carbon dioxide concentrations (ppm as CO_2) are calculated by multiplying the total alkalinity of the water (expressed as ppm as $CaCO_3$) by the factor in the table corresponding to the measured pH and temperature. For practical purposes, carbon dioxide concentrations are negligible above pH 8.4.

rise to possibly toxic levels if hydrated lime in excess of that needed to remove the carbon dioxide present is added. The rise in pH following treatment with hydrated lime may cause other problems because pond waters with chronically high carbon dioxide concentrations often have concurrent high total ammonia concentrations. At higher pH values a larger fraction of the total ammonia exists in the toxic, un-ionized form and may endanger the fish. Note that treatment with hydrated lime does not address the cause of the high concentrations; it only removes the carbon dioxide present. Concentrations will increase soon after treatment unless the environment changes.

Sodium bicarbonate ($NaHCO_3$, also called bicarbonate of soda) has at times been recommended as a treatment for high carbon dioxide concentrations in ponds. Sodium bicarbonate does not react with carbon dioxide and its use is without basis.

Changes in carbon dioxide concentrations cause the pH of water to fluctuate over a 24-hour period. Net production of carbon dioxide at night causes the pH to decline because carbon dioxide in water forms

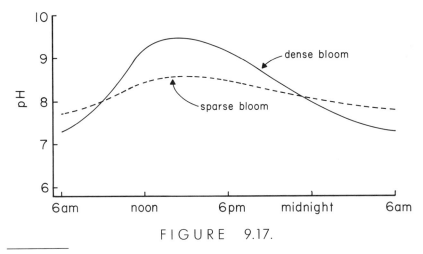

FIGURE 9.17.

Change in pH of catfish pond waters over a 24-hour period as affected by plankton bloom density. These relationships would be typical of ponds with a relatively high total alkalinity and hardness.

an acid. When carbon dioxide is removed from water during photosynthesis, the pH increases. The extent of the daily change in pH depends on the amount of carbon dioxide added and removed, the water temperature, and the buffering capacity (total alkalinity) of the water.

The pH fluctuates to a greater extent in ponds with dense phytoplankton blooms (Fig. 9.17) because large amounts of carbon dioxide are added and removed. Daily changes are usually greater in the summer than in the winter because metabolic rates are higher in warmer waters. Waters of low alkalinity (< 10 ppm as $CaCO_3$) have pH values between 6.5 and 7.5 when at equilibrium with carbon dioxide in the atmosphere. These waters usually have sparse phytoplankton blooms, and pH fluctuates little during the day. The pH may decrease to 6.0 or below at night and rise to 8.0 or more during the day. The pH of waters with moderate total alkalinities (10 to 50 ppm as $CaCO_3$) is between 7.5 and 8.2 at equilibrium with atmospheric carbon dioxide. Dense phytoplankton blooms can develop in these waters, and pH may fluctuate between 6.5 and 10. Waters with higher total alkalinities have an equilibrium pH of about 8.3 to 8.5, but daily fluctuations are moderated because the water has a high pH-buffering capacity. Daily fluctuations are usually between pH 7.5 and 9.5 in ponds with dense summertime phytoplankton blooms.

The pH of most channel catfish pond waters falls in the range of 6 to 9, and toxicity problems with pH are rare. Undesirably low pH values (<6) occur only when ponds are built on acidic soils and filled with poorly buffered water. This condition can usually be corrected by liming the pond. Boyd (1979) details methods of estimating the lime requirement of fish ponds. Very low pH values (<4) occur when ponds are built on certain waterlogged coastal soils developed from marine sediments. These soils, called "cat's clay" or acid-sulfate soils, produce sulfuric acid when drained and exposed to oxygen (Boyd 1979b). Cat's clays are difficult to reclaim for fish ponds and should be avoided as sites for channel catfish ponds.

High pH values (>9) occur temporarily in channel catfish ponds on many summer afternoons when conditions favor rapid photosynthesis. In well-buffered waters, the pH rarely exceeds 10 and the only concern is the effect of pH on the toxicity of ammonia. Some ponds, however, experience long episodes of afternoon pH values above 10. The water in these ponds typically has a moderate to high total alkalinity (50 to 200 ppm as $CaCO_3$) and a low total hardness (less than 20 ppm as $CaCO_3$). The principal ions in these waters are sodium and bicarbonate. This type of water is common in ponds supplied by wells drawing from deep (>500 feet) aquifers along the Atlantic and Gulf coastal plains of the United States. These waters have been subject to "natural softening" while in contact with sodium-rich clays and minerals of marine origin. Calcium and magnesium originally in the water were exchanged for sodium, thereby reducing total hardness. The reason this type of water often has undesirably high pH values is not known with certainty and the problem is difficult to treat. Addition of calcium (as gypsum, $CaSO_4$, or calcium chloride, $CaCl_2$) in amounts to equalize the total alkalinity and hardness sometimes moderates the high afternoon pH (Mandal and Boyd 1980), but this treatment is not always effective. Addition of organic matter, such as hay, to ponds results in increased carbon dioxide production and may lower afternoon pH values. This treatment is not practicable on a large scale. The decomposing organic matter may also cause dissolved oxygen depletion.

Routine measurements of pH are usually not made on most commercial catfish ponds. However, measurements should be made whenever samples for total ammonia analysis are collected. Water temperature and pH are needed to calculate toxic, un-ionized ammonia concentrations. Measurement of pH can be made by inexpensive color comparison methods or by use of a portable pH meter. A good pH meter is much more reliable than colorimetric methods and some pH meters are quite reasonably priced.

AMMONIA AND NITRITE

The use of supplemental aeration allows maximum sustained daily feeding rates to exceed 50 pounds/acre without fish kills. Nevertheless, aeration does not allow unlimited fish stocking densities and feeding rates because other factors begin to affect fish health and limit growth. After meeting the oxygen requirements of fish, accumulation of nitrogenous waste products (principally ammonia) is the second factor that limits intensification of pond culture of channel catfish. The point at which buildup of nitrogenous wastes begins to limit production is not known with certainty. In experimental ponds supplied with abundant aeration, net channel catfish production does not increase proportionately with fish stocking densities when they exceed about 7,000 fish/acre and sustained maximum daily feeding rates exceed about 100 pounds/acre/day (Cole and Boyd 1986). This is probably about the level of intensification at which nitrogenous waste products begin to affect production in commercial culture ponds severely.

Forms and Cycling of Nitrogen

Nitrogen is present in pond waters as dissolved gases, dissolved inorganic combined nitrogen compounds, and various dissolved and particulate organic compounds. The relationships among the various forms are qualitatively described in the nitrogen cycle (Fig. 9.18).

Molecular nitrogen gas (N_2) is largely derived from the dissolution of atmospheric nitrogen in water. Some nitrogen gas is produced under anaerobic conditions in the process of denitrification, described later. Nitrogen gas is relatively insoluble in water, but equilibrium concentrations in water are higher than for oxygen because nitrogen is the principal gas in air. Water contains between 10 and 20 ppm nitrogen gas at equilibrium with the atmosphere at temperatures encountered in fish ponds. Nitrogen gas is very unreactive and normally poses no health problem to fish. However, waters can at times become supersaturated with nitrogen and other atmospheric gases and cause a condition known as gas bubble trauma. This condition is discussed in Chapters 4 and 7, but nitrogen gas supersaturation is rarely, if ever, a problem in ponds.

Nitrogen fixation is the process by which certain microorganisms reduce molecular nitrogen to ammonia, which is then used to synthesize proteins. Certain blue-green algae and bacteria can fix considerable nitrogen in some aquatic ecosystems, but this process is probably of little significance in channel catfish ponds.

Inorganic combined nitrogen compounds include un-ionized ammonia (NH_3), ionized ammonia (NH_4^+), nitrite (NO_2^-), and nitrate (NO_3^-). Un-ionized ammonia and ionized ammonia exist in a pH- and temperature-dependent equilibrium. The sum of the two forms is called *total ammonia* or simply *ammonia*. The amount of ammonia existing in water in the un-ionized form increases as pH and tempera-

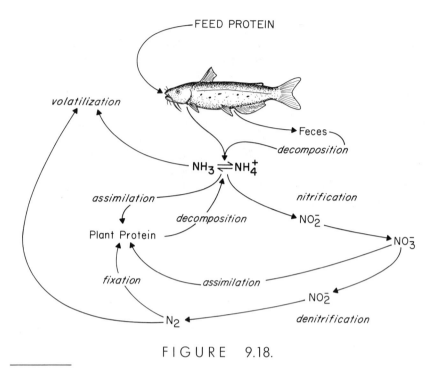

FIGURE 9.18.

Schematic illustration of the nitrogen cycle in a catfish pond.

ture increase. Un-ionized ammonia gas can be lost to the atmosphere. Un-ionized ammonia is much more toxic to fish than ionized ammonia (see the section Ammonia).

Ammonia is a source of nitrogen for plant growth and is rapidly assimilated by phytoplankton or other aquatic plants and used to synthesize plant proteins. At death and decay the proteins are broken down and ammonia is released to the water in the process called *mineralization*. Ammonia is also oxidized to nitrate by bacteria in the two-step process of nitrification. Nitrate is also used by plants as a source of nitrogen. Ammonia is first oxidized to nitrite primarily by species of *Nitrosomonas*, and nitrite is then oxidized to nitrate by species of *Nitrobacter*. These bacteria use ammonia or nitrite as energy sources and use carbon dioxide as the source of carbon. Nitrification requires oxygen and occurs most rapidly when dissolved oxygen concentrations are near saturation. Nitrification proceeds most rapidly at pH values of 7 to 8 and at temperatures of 80 to 95°F. *Nitrobacter* is less tolerant of low temperatures and high pH than *Nitrosomonas*, and these conditions can result in accumulation of nitrite. Nitrite can be toxic to channel catfish at low concentrations; nitrate is considered to be nontoxic.

Under anaerobic conditions, many species of bacteria can use nitrate instead of oxygen in respiration. This process, *denitrification*, occurs principally in anaerobic bottom muds. The end products are gaseous and are ultimately lost to the atmosphere.

Ammonia

Aside from dissolved molecular nitrogen of atmospheric origin, virtually all the nitrogen in catfish ponds is derived from the feed added to the pond. Fish consume the feed and use the amino acids from feed protein to synthesize new protein. Body proteins are continually turned over; some of the protein is degraded and the waste nitrogen is excreted. The major end product of protein degradation in fish is ammonia, which is excreted from the gills as un-ionized ammonia. Smaller amounts of ammonia reach the water when uneaten feed or fish fecal material is decomposed. The amount of ammonia produced is roughly proportional to the amount of feed consumed. Doubling the number of fish and the amount they are fed doubles the amount of ammonia entering the pond water.

Approximately 0.03 pound of nitrogen in the form of ammonia is produced for each pound of feed consumed. This amounts to a considerable quantity in ponds receiving large amounts of feed. For example, in a 1-acre, 3-foot-deep pond receiving 100 pounds of feed/acre per day, the total ammonia-nitrogen* concentration should increase by about 0.4 ppm/day. At this rate it would require only a few days before total ammonia concentrations would increase to levels at which the resulting concentration of un-ionized ammonia would kill the fish.

Clearly, then, transformations and losses serve to reduce concentrations of ammonia in catfish ponds and allow fish culture. Ammonia is continuously lost to the atmosphere when un-ionized ammonia volatilizes (Murphy and Brownlee 1981). Rates of ammonia volatilization are highest during periods of high pH and strong wind-induced turbulence. However, in the short term concentrations of ammonia in static-water fish culture ponds are largely controlled by the activities of the phytoplankton bloom (Tucker, Lloyd, and Busch 1984; Neori et al. 1989). At maximum sustained daily feeding rates up to about 75 to 100

---

*Concentrations of combined inorganic nitrogen compounds are expressed in terms of the nitrogen they contain. This manner of expression is common in water chemistry. To convert concentrations on a nitrogen basis to concentrations of the specific component, use these factors: 1 ppm NH_3—N = 1.21 ppm NH_3; 1 ppm NH_4^+ − N = 1.29 ppm NH_4^+; 1 ppm NO_2^- − N = 3.29 ppm NO_2^-.

pounds/acre, ammonia accumulates only during periods when assimilation by phytoplankton is reduced.

Phytoplankton actively assimilate ammonia during daylight and early evening (Neori et al. 1989). In ponds with a healthy phytoplankton community, concentrations of total ammonia increase slightly during the late night and then decrease during the day. These small daily fluctuations are probably of little practical significance. However, when phytoplankton metabolism is decreased for longer periods, total ammonia concentrations increase dramatically. Reduced ammonia assimilation by phytoplankton occurs when they die naturally or are killed, when some nutrient other than nitrogen limits their growth, or when cold water temperatures decrease metabolic activity.

Concentrations of total ammonia-nitrogen in catfish ponds usually vary between 0 and 4 ppm during spring, summer, and early fall. Un-ionized ammonia-nitrogen concentrations range from 0 to over 1 ppm and are highest in late afternoon when pH is highest. Concentrations of total ammonia increase dramatically after natural phytoplankton die-offs or after phytoplankton or plants are killed with an herbicide. This is the result of mineralization of plant protein nitrogen and reduced assimilation of ammonia excreted by fish. Intense net production of carbon dioxide after phytoplankton die-offs causes the pH to decrease, and this reduces the proportion of total ammonia in the un-ionized form. Sometimes it causes the concentration of un-ionized ammonia actually to decrease after die-offs. However, concentrations of total ammonia-nitrogen may become so high after some massive phytoplankton die-offs that un-ionized ammonia levels are high enough to stress or kill fish.

Highest sustained concentrations of total ammonia-nitrogen in catfish culture ponds usually occur in late fall and winter. Low water temperatures limit assimilation of ammonia by phytoplankton, but ammonia excretion by fish continues, although at a reduced rate. Total ammonia-nitrogen concentrations can exceed 5 ppm for extended periods during cold weather, but the lower pH that is typical of winter conditions somewhat moderates the concentrations of un-ionized ammonia. Nevertheless, extended periods of un-ionized ammonia-nitrogen concentrations greater than 0.2 ppm are common in catfish ponds during the winter (Tucker, Lloyd, and Busch 1984).

In ponds with maximum sustained daily feeding rates above about 100 pounds/acre, the amount of ammonia excreted by fish exceeds the assimilation capacity of phytoplankton because algal growth is limited by factors (such as light or carbon availability) other than supply of nitrogen, phosphorus, or other mineral nutrients. Average concentrations of total ammonia-nitrogen increase rapidly as feeding rates exceed about 100 pounds/acre/day (Cole and Boyd 1986).

Ammonia and Nitrite   **255**

Total ammonia concentrations can be high for long periods in a muddy pond, even if the amounts of feed added to the pond are relatively low. The turbidity caused by the suspended clay particles blocks light penetration and thereby limits phytoplankton growth and ammonia assimilation.

Ammonia concentrations should be measured regularly in commercial culture ponds. Sample each pond at least once a week throughout the year. One sample per pond is sufficient for practical purposes, and samples should be obtained from about 1 foot beneath the surface and several feet from the bank. Measurements are best made in late afternoon because pH and temperature (and therefore un-ionized ammonia concentrations) are highest at that time. Ammonia should also be measured after phytoplankton die-offs and after treatment of ponds with herbicides or fish disease therapeutic chemicals. An ammonia analysis should also be conducted as part of every fish disease investigation to aid in the diagnostic process.

Inexpensive, portable test kits can be used to make ammonia measurements. These kits measure total ammonia concentrations; water temperature and pH must be measured at the same time to calculate un-ionized ammonia concentrations. To calculate un-ionized ammonia concentrations, find the appropriate factor in Table 9.3 from the pH and temperature of the water. Multiply this factor by the total ammonia-nitrogen concentration. For example, assume that the total ammonia-nitrogen concentration is 2.0 ppm, the pH is 8.8, and the water temperature is 75°F. The factor corresponding to pH 8.8 and 75°F is 0.250. The un-ionized ammonia-nitrogen concentration is 2.0 × (0.250) = 0.50 ppm.

Management of ammonia is difficult in large commercial channel catfish ponds. In small ponds (less than 2 acres) used to culture fish in other countries, concentrations of total ammonia are maintained at low levels by frequent flushing. Flushing with "green water" from an adjacent pond with a healthy phytoplankton bloom and low total ammonia concentrations is preferred to flushing with clean well water (Neori et al. 1989). Flushing with clean water from a well reduces ammonia by dilution but can remove microorganisms that assimilate or transform ammonia. Fish continue to excrete ammonia, and concentrations quickly increase after flushing is stopped. Flushing with "green water" dilutes ammonia and also seeds the pond with healthy phytoplankton that continue to assimilate ammonia after flushing. In larger ponds (>10 acres) typically used in commercial channel catfish culture, potential rates of pumping are too low relative to pond volume to have a significant effect on ammonia concentrations by dilution. Adding "green water" to seed a pond with healthy phytoplankton may be beneficial after phytoplankton die-offs, but this has not been verified. Phy-

TABLE 9.3.

Factors Used to Calculate Un-Ionized Ammonia-Nitrogen
Concentrations in Dilute Fresh Waters from pH,
Temperature, and Total Ammonia-Nitrogen Concentrations

pH	Temperature							
	39°F (4°C)	46°F (8°C)	54°F (12°C)	61°F (16°C)	68°F (20°C)	75°F (24°C)	82°F (28°C)	9((3
6.0						0.001	0.001	0.
6.2					0.001	0.001	0.001	0.
6.4			0.001	0.001	0.001	0.001	0.002	0.
6.6		0.001	0.001	0.001	0.002	0.002	0.003	0.
6.8	0.001	0.001	0.001	0.002	0.002	0.003	0.004	0.
7.0	0.001	0.002	0.002	0.003	0.004	0.005	0.007	0.
7.2	0.002	0.003	0.003	0.005	0.006	0.008	0.011	0.
7.4	0.003	0.004	0.005	0.007	0.010	0.013	0.017	0.
7.6	0.005	0.006	0.009	0.012	0.017	0.021	0.027	0.
7.8	0.007	0.010	0.014	0.018	0.024	0.032	0.042	0.
8.0	0.012	0.016	0.021	0.029	0.038	0.050	0.065	0.
8.2	0.018	0.025	0.033	0.045	0.059	0.077	0.100	0.
8.4	0.028	0.038	0.052	0.069	0.090	0.117	0.150	0.
8.6	0.044	0.059	0.079	0.105	0.136	0.174	0.218	0.
8.8	0.068	0.091	0.120	0.157	0.200	0.250	0.306	0.
9.0	0.103	0.137	0.178	0.227	0.284	0.346	0.412	0.
9.2	0.155	0.201	0.256	0.318	0.386	0.456	0.526	0.
9.4	0.224	0.285	0.353	0.425	0.499	0.570	0.637	0.
9.6	0.314	0.387	0.463	0.539	0.612	0.678	0.736	0.
9.8	0.420	0.500	0.578	0.650	0.714	0.769	0.815	0.
10.0	0.534	0.613	0.684	0.746	0.798	0.841	0.875	0.
10.2	0.645	0.715	0.775	0.823	0.863	0.893	0.917	0.

Note: Un-ionized ammonia nitrogen concentrations (ppm as N) are calculated by multiplying the total amm
nitrogen concentration (ppm as N) by the factor found in the table corresponding to the measured pH and
temperature.

toplankton communities usually recover rapidly after die-offs without seeding and ammonia concentrations then decrease. Nevertheless, running high-quality water into a pond may be useful if fish are suffering from acute un-ionized ammonia toxicosis. The area of high-quality water around the outfall provides a temporary haven for the fish. The concentration of un-ionized ammonia can be decreased by adding an acid or acid-forming substance to decrease the pH of the water, but this is only a temporary solution and is usually impractical even on a small scale.

Concentrated suspensions of nitrifying bacteria have been applied to ponds in attempts to reduce ammonia concentrations. The efficacy of this practice has not been verified (Boyd et al. 1984; Tucker and Lloyd 1985). Other practices such as the use of ion-exchange or zeolite filters to remove ammonia physically or of large "biological" filters to nitrify ammonia are not economically feasible in pond culture.

Chronic high levels of total ammonia nitrogen (>5 ppm) during the summer growing season indicate that fish densities and feeding rates are too high for static pond culture. Temporarily reducing daily rations or not feeding the fish at all for several days reduces ammonia production, and total ammonia concentrations may decrease. However, this reduces fish growth and ammonia concentrations increase when rations are subsequently increased. A more logical approach is to decrease stocking rates and maximum daily feeding rates to levels where total ammonia-nitrogen concentrations remain below 4 ppm except in winter and after occasional phytoplankton die-offs.

Nitrite

Nitrite-nitrogen concentrations rarely exceed 1 ppm in channel catfish ponds during summer months because ammonia concentrations are relatively low as a result of assimilation by phytoplankton and this somewhat limits the amount of substrate for nitrification. Any ammonia nitrified is quickly converted to nitrate because both steps in the nitrification process operate at the same rate. Nitrite may briefly accumulate after sudden increases in ammonia concentrations following phytoplankton die-offs or herbicide treatments. The increased substrate (ammonia) stimulates growth of *Nitrosomonas* bacteria and nitrite is produced. Growth and metabolism of the *Nitrobacter* group lags behind until sufficient substrate (nitrite) accumulates. Nitrite concentrations soon decrease as the nitrite is converted to nitrate.

In the southeastern United States, episodes of elevated levels of nitrite are common in the late fall, winter, and early spring. Concentrations of ammonia tend to be high at these times because assimilation of ammonia is reduced when water temperatures are cool. In fall, winter, and spring, water temperatures drop rapidly as cold fronts pass and then slowly warm before passage of the next front. Water temperatures may change more than 20°F in just a few days. The *Nitrosomonas* group of bacteria are more tolerant of cold temperatures than the *Nitrobacter* group, and the combination of high ammonia concentrations and unequal rates of substrate oxidation when water temperatures are changing rapidly can result in large buildups of nitrite. Nitrite-nitrogen concentrations exceeding 4 ppm are not uncommon. During prolonged periods of cold water temperatures (<45°F), both steps of nitrification proceed very slowly and accumulations of nitrite are less frequent. Large seasonal differences in the concentration of nitrite in ponds are less common in regions like south Florida where winters are milder and water temperatures do not change as rapidly or to such a large extent as in areas farther north.

The frequency and severity of episodes of high nitrite levels increase as fish stocking densities and feeding rates increase. This is the result of larger inputs of ammonia, which can be converted to nitrite, in densely stocked ponds.

Nitrite concentrations can be reduced in small ponds by flushing if an adequate water supply is available. This practice is ineffective in large ponds. Fortunately, an inexpensive and effective treatment that protects channel catfish from nitrite toxicosis exists. Chloride, from common salt (NaCl), inhibits uptake of nitrite by channel catfish, thereby decreasing the amount of methemoglobin ("brown blood") formed in fish during nitrite exposure (see the section Nitrite). Maintaining a ratio of at least 20 ppm chloride for every 1 ppm of nitrite-nitrogen[*] adequately protects fish from nitrite toxicosis. To calculate treatment rates, use the following procedure:

Calculate the chloride needed to maintain a 20:1 ratio of ppm chloride to ppm nitrite-nitrogen. The amount of chloride already in the pond water must be taken into account:

$$\text{Parts per million chloride needed} = (20 \times \text{ppm nitrite-nitrogen}) - \text{ppm ambient chloride} \qquad (9.4)$$

If the answer is a positive number, salt must be added. If the answer is zero or negative, sufficient chloride is already present to protect fish.

Calculate how much salt (NaCl) is needed to give the required ppm of chloride:

$$\text{Pounds of NaCl} = \text{ppm chloride needed} \times 4.5 \times \text{pond acreage} \times \text{average depth in feet} \qquad (9.5)$$

The number 4.5 is the pounds of salt required to give 1 ppm chloride in 1 acre-foot of water.

As an example, assume the measured nitrite-nitrogen concentration is 1 ppm and the chloride concentration in the pond is 15 ppm. The pond surface area is 15 acres and the depth averages 4 feet.

---

[*]Some confusion is unavoidable here because nitrite concentrations can be expressed as either nitrite-nitrogen or as nitrite. If concentrations are expressed as nitrite, the desired ratio is 6 ppm chloride to 1 ppm nitrite. The calculations that follow are then similar except that in equation (9.4), 20 is replaced by 6 and nitrite concentrations are used rather than nitrite-nitrogen concentrations.

Parts per million chloride needed = (20 × 1) − 15 = 5 ppm
chloride (9.6)

Pounds NaCl = 5 × 4.5 × 15 × 4 = 1,350 pounds of salt
required (9.7)

The salt should be evenly distributed over the pond for quickest effect. However, if a boat is not available, salt can be dumped in front of an aerator, where it will dissolve and be distributed by the currents. Salt treatment is long-lasting because chloride is lost from the pond only by dilution with rainfall, runoff, or addition of well water. Do not attempt to flush nitrite from large ponds after adding salt to treat nitrite toxicosis. Flushing will remove the chloride that protects fish, but nitrite concentrations may not decrease because ammonia will continue to be excreted by fish and oxidized to nitrite.

The existing chloride concentration in the pond offers some protection from nitrite exposure to fish and determines how frequently nitrite concentrations should be measured. Ponds with low chloride concentrations (<20 ppm) should be sampled at least weekly and two or three times a week in fall, winter, and spring. Frequent measurements are necessary because nitrite-nitrogen concentrations have only to rise above 1 ppm to exceed the 20:1 ratio of chloride to nitrite-nitrogen recommended. Ponds with moderate chloride concentrations (up to 100 ppm) should be sampled at least weekly in spring and fall and biweekly at other times. Nitrite-nitrogen concentrations greater than 5 ppm are extremely rare in channel catfish ponds, and managers of ponds with chloride concentrations over 100 ppm seldom monitor nitrite levels. In areas where the water source contains low concentrations of chloride, adding salt to all ponds as a preventative measure to achieve a final chloride concentration of 20 to 30 ppm provides a margin of safety if nitrite concentrations suddenly rise. This safety measure also decreases the frequency with which nitrite measurements need to be made.

The frequency of nitrite measurements should be increased if nitrite concentrations begin to rise and also during episodes of high levels of nitrite. Measurements of total ammonia concentrations can be used to anticipate rises in nitrite levels because increased concentrations of nitrite may occur any time total ammonia concentrations rise to unusual levels. Do not stop measuring nitrite levels after adding salt to ponds; nitrite concentrations may continue to rise, and additional salt may be required.

Chloride concentrations usually change slowly in ponds and monthly chloride measurements are normally sufficient. However, chloride concentration should be checked more frequently during rainy periods because heavy rains can significantly reduce chloride levels by dilution. Chloride levels should also be checked in any pond where nitrite concentrations begin to rise.

Nitrite and chloride concentrations can be determined reliably with simple, inexpensive portable test kits. Nitrite and chloride vary little with location within a pond and a single sample from each pond is sufficient.

OFF-FLAVOR

Off-flavors are objectionable or undesirable flavors in fish. Off-flavor can be caused by fat rancidity resulting from prolonged or improper storage of processed fish. Fish may also acquire off-flavors from certain feed ingredients. For instance, feeds high in marine fish oil may impart an excessively "fishy" flavor to the otherwise mild-tasting channel catfish. Off-flavors can also result when fish absorb odorous chemicals from water. These chemicals may enter the water as pollution or may be produced by microorganisms growing in the water. The most common cause of off-flavor in farm-raised channel catfish are odorous natural chemical products of aquatic bacteria or algae. Off-flavors related to pollution are rare in pond-raised fish, but some culture facilities, such as open water net cages, are susceptible to unforeseen pollution that can cause off-flavors.

Off-flavors are not harmful to fish, yet they are considered among the most serious problems in commercial channel catfish culture. Off-flavors are common in fish raised in densely stocked ponds, but because of market demands for high-quality products, fish that have objectionable flavors are not accepted for processing.

Most processing plants use preharvest flavor testing to screen fish for off-flavors. Before harvest, the producer submits a sample of one or two fish to quality control personnel at the processing plant for evaluation of flavor and appearance. If they are deemed acceptable, and the plant processing schedule permits, fish are tentatively approved for purchase. One or two additional samples for flavor evaluation may be required, usually on the day before or the day of harvest. A final sample is usually taken from the transport truck immediately before the live fish are unloaded at the plant. If the flavor of fish sampled at any time is unacceptable, fish are rejected for processing; they are either not harvested or returned to the pond from the transport truck. This imposes a marketing constraint on producers and is a considerable economic burden.

Types and Causes of Off-Flavor

Dozens of off-flavors have been described in channel catfish. The following list is adapted from flavor descriptors compiled by Lovell (1983a) and Johnsen, Civille, and Vercelotti (1987). Most research has focused on the organisms and chemicals responsible for earthy-musty flavors; little is known of the causes of other off-flavors. A more detailed review of off-flavors in fish is presented by Tucker and Martin (1990).

Earthy-musty off-flavors. Earthy-musty flavors are the most common off-flavors in pond-raised channel catfish. The flavor varies and is similar to that associated with old books, beets, peat, freshly turned soil, or damp cellars. Two somewhat distinct subclasses can be recognized by experienced tasters: a flavor suggestive of old books and often related to the presence of the chemical geosmin, and a muddy flavor often related to the chemical 2-methylisoborneol.

Geosmin and 2-methylisoborneol are fat-soluble alcohols produced by actinomycetes (a group of common, filamentous bacteria) and blue-green algae. The chemicals are rapidly absorbed by fish from the water and deposited in fat-rich tissues. Geosmin and 2-methylisoborneol are then lost from tissues when fish are moved to clean water. The off-flavor disappears within 2 to 7 days in clean water, depending on the intensity of off-flavor and water temperature. Purging is slower in cooler waters.

Musty-smelling dehydration by-products of 2-methylisoborneol have also been isolated from off-flavored channel catfish (Martin, Bennet, and Graham 1988) although the source of the chemicals is unknown. The dehydration products are hydrocarbons and are strongly held in fats and are not quickly purged from fish. The incidence of off-flavors related to 2-methylisoborneol dehydration products is not known.

Petroleum off-flavors. Petroleum flavors are reminiscent of diesel oil, gasoline, or kerosene. These off-flavors undoubtedly result from accidental spills of petroleum products used in tractors or engines to run aerators, wells, or other equipment. Care should be taken to prevent even small spills of petroleum products in fish ponds because some hydrocarbon components are highly concentrated from water into fats and minute levels impart off-flavors to fish.

Sewage off-flavors. Sewage flavors resemble the odor of a sewage lagoon. Specific chemical compounds associated with these off-flavors have not been identified but most likely are products released during bacterial decomposition of organic materials in the pond. Noxious sewagelike odors are present around ponds (particularly near aerators) after phytoplankton die-offs or during other periods of acute oxygen depletion. After these events, flavors described as "sewage," "rotten," or "fishy" are sometimes present in fish. Sewage or septic flavors may then be a natural consequence of the eutrophic conditions in culture ponds. The presence of these flavors in fish after phytoplankton die-offs may have implications for certain practices used to alleviate other off-flavors. Some producers apply algicides to ponds containing fish with earthy-musty flavors in attempts to kill the blue-green algae suspected of producing geosmin or 2-methylisoborneol and allow the fish to purge the off-flavor. Even if the treatment accomplishes this goal, fish may develop other off-flavors resulting from decomposition of the dead phytoplankton.

Grainy-vegetative off-flavors. Grainy-vegetative flavors are also described as grassy or plantlike. This is judged as a favorable characteristic at low levels when the dominant character is more grainlike. Most likely, a wide variety of compounds are responsible for these flavors, some of which could be related to feed ingredients. Weedy or grassy odors in water have been associated with the presence of several blue-green and other algal genera, but no relationship has been verified for fish.

Stale, rancid, or putrid. Stale, rancid, or putrid flavors are dissimilar, but all may reflect changes in flavor occurring after death of the fish rather than flavor compounds derived from the environment. Stale is characterized as lacking freshness, rancid is the flavor or odor associated with oxidized fats or oils, and putrid is the flavor associated with decaying protein. Channel catfish are frequently rejected for processing because fish samples submitted before harvest have these off-flavors. The flavor may be present because the sample was handled improperly and partially spoiled before submission to quality control personnel at the processing plant. In this instance, thousands of pounds of fish are not harvested because of one poor sample. Samples caught from the pond must be washed, quickly put on ice, and transported to the plant without delay so that the flavor of the sample is representative of the pond population.

Incidence and Cause of Off-Flavors

Sensory scores for channel catfish sampled from four experimental ponds at Stoneville, Mississippi, illustrate the dynamic nature of the incidence and severity of off-flavor (Fig. 9.19). On the initial sampling

Off-Flavor **263**

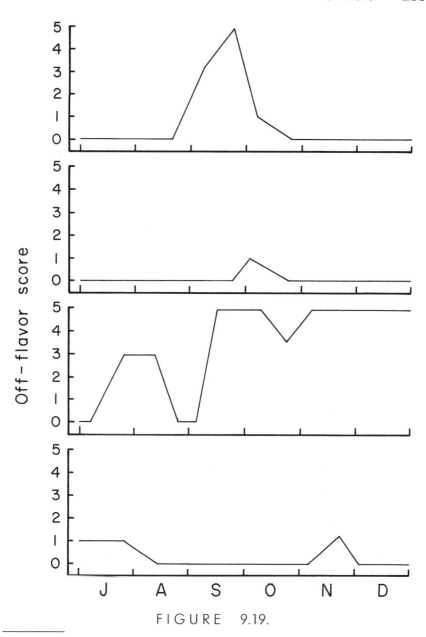

FIGURE 9.19.

Changes in flavor score of fish sampled from four experimental ponds over a 6-month period. A score of 0 = no off-flavors (fish acceptable for processing); 5 = intense off-flavor.

date in July, all four ponds were judged to contain acceptably flavored fish; in early October, three of the four samples were judged off-flavor to varying degrees. Duration of off-flavor episodes varied from 2 weeks to over 3 months. These ponds were part of a larger study involving fourteen ponds that were sampled from June through February. The highest incidence of off-flavor occurred in September and October, with samples from ten of the fourteen ponds judged off-flavor. The most intense off-flavors also occurred at this time; the flavor described as "musty" (confirmed as being caused by 2-methylisoborneol) was most common and most intense. The lowest incidence of off-flavor occurred in December, with samples from five ponds judged off-flavor.

Studies of catfish culture ponds in Alabama (Lovell et al. 1986) showed that fish from 50 to 75 percent of the ponds sampled in late summer or early fall were off-flavor and unacceptable for harvesting. The predominant off-flavor was "earthy-musty." Geosmin was confirmed in 80 percent of the fish with earthy-musty off-flavors; 2-methylisoborneol was not detected in any fish sampled. The total incidence of ponds with off-flavor fish and the incidence of earthy-musty flavors increased during the summer growing season as phytoplankton biomass increased in response to higher water temperatures and amounts of feed added to ponds.

The incidence of off-flavor in Mississippi farm-raised channel catfish is similar to that found in Alabama, and earthy-musty off-flavors are also the most common. However, 2-methylisoborneol, rather than geosmin, is usually the cause of these off-flavors in Mississippi. The reason for the different chemical causes of earthy-musty off-flavors in Alabama and Mississippi is not known, although differences in the density and composition of channel catfish pond phytoplankton communities in the two states have been reported (Tucker and Boyd 1985).

Brown and Boyd (1982) found that ponds in Alabama with low fish densities and low maximum daily feed allotments generally yielded the best-tasting fish. No strong relationship between either phytoplankton density or type of phytoplankton and degree of off-flavor was found, but in several ponds with intense off-flavor, the blue-green alga *Lyngbya* species was abundant. The general findings that incidence of off-flavor is seasonal and that it is related to fish stocking and feeding rate are not surprising. Blue-green algae appear to cause most flavor problems in pond-raised channel catfish, and this algal group is favored in warm, nutrient-rich environments.

Cost of Off-Flavor to the Producer

The cost effects of off-flavor are difficult to evaluate because the incidence is episodic and dynamic. Keenum and Waldrop (1988) attempted to estimate the effects of off-flavor on production costs of Mississippi pond-raised channel catfish. They used three simple scenarios with the only variable being the time of year when off-flavor occurred. In each scenario it was assumed that 25 percent of the production would be held in inventory for 3 months as a result of off-flavor.

When off-flavor occurred during the winter (assumed to be the nongrowing season) the only additional cost charged was an opportunity cost for delayed income. For this scenario, an additional cost of 1.75 cents/pound of catfish harvested was calculated. When off-flavor occurred during the summer growing season, interest was charged on the quantity of fish that could not be sold. Additionally, net income from foregone sales must be forfeited by the producer because restocking and growth of the restocked fingerlings cannot be accomplished. Because off-flavor fish held during the growing season continue to grow, additional revenue offsets part of the interest and foregone net income. For this scenario, an additional cost of 4.10 cents/pound of catfish harvested was added to production costs. The cost of off-flavor occurring during part of both the nongrowing and the growing seasons was estimated as the sum of the costs for the first two scenarios, or 7.95 cents/pound of catfish harvested. Using these three estimates and an annual production of 300 million pounds of channel catfish in 1988, the added production cost of off-flavor for that year was between $5 million and $11 million.

Other potential costs accruing from off-flavor are difficult to estimate. Fish may be lost to infectious diseases or poor water quality while they are in inventory waiting for off-flavor to abate. This represents loss of investment in fish that should have been marketed. As fish grow, they convert feed less efficiently and compete with smaller fish for feed allotments. This characteristic lowers the overall feed conversion rate and growth rate of the pond population. Additional labor may also be required for sampling and management programs. Market constraints also impact the effect of off-flavor on actual production costs. Market constraints (not being able to sell fish when they reach market size, also called *quotas*) increase production costs in a manner similar to off-flavor. This cost may be added to or subtracted from the cost of off-flavor because fish may not have been marketable regardless of flavor.

Management and Treatment of Off-Flavors

Three approaches have been used to deal with off-flavor in pond-raised channel catfish: manage around off-flavor episodes, prevent off-flavor episodes, or remove the flavor from fish once they have developed it.

Removing the off-flavor from fish can be accomplished by treating the pond with a biocide to kill the organism producing the odorous substance or by moving the fish to a "clean" environment.

Managing around off-flavor episodes. Off-flavor episodes in ponds are dynamic and when several ponds are considered, not all are likely to contain off-flavor fish at the same time. Ponds that contain a sufficient quantity of market-sized fish should be checked frequently for off-flavor and sold promptly if the fish are suitable. Market constraints may impact the feasibility of this procedure, but every effort should be made to harvest and market fish free of off-flavor as soon as possible.

Preventing off-flavor episodes. A significant correlation exists between amounts of feed added to ponds and flavor scores of channel catfish (Brown and Boyd 1982). This suggests that the incidence and severity of off-flavor can be decreased by lowering fish stocking rates and adding less feed to ponds. However, for pond-raised channel catfish, it appears that stocking and feeding rates would have to be lower than those currently considered profitable for this procedure to have a significant effect. Costs of channel catfish production are very sensitive to harvested yield, and although off-flavor reduces long-term yield, savings in reduction of the incidence of off-flavor probably would not offset decreased revenue from lower yields.

Earthy-musty flavors related to geosmin or 2-methylisoborneol are the most common off-flavors in pond-raised fish. Because blue-green algae are frequently implicated as the major source of these compounds in eutrophic waters, most research in preventing off-flavor has concerned the use of algicides to reduce blue-green algal abundance. Only one algicide (Solricin 135®) is marketed in the United States with specific claims that its use will prevent or eliminate off-flavor problems associated with blooms of blue-green algae. The active ingredient, potassium ricinoleate, is similar in structure to naturally occurring compounds isolated from certain aquatic plants that inhibit the growth of blue-green algae. However, when tested in ponds, Solricin 135® did not reduce the abundance or percentage of blue-green algae in phytoplankton communities and did not prevent severe episodes of off-flavors in channel catfish (Tucker and Lloyd 1987).

Copper sulfate, a relatively nonselective algicide, has been used in attempts to prevent episodes of off-flavor by applying the chemical frequently (weekly or so) at low concentrations (less than 0.2 ppm as copper). Although producers report some success using this technique, there is no sound evidence that this procedure is successful in reducing the incidence or severity of off-flavor. Frequent treatment of channel catfish ponds with nonselective algicides also cannot be recommended because chronic deterioration in water quality causes decreased fish yields (Tucker and Boyd 1978).

Torrens and Lowell (1987) reported a decrease in the incidence of off-flavor in channel catfish polycultured with the planktonivorous fish blue tilapia. Apparently, the feeding habits of tilapia altered the environment in a manner that did not favor the growth of odor-producing microorganisms. However, tests in Mississippi failed to demonstrate a similar benefit of tilapia polyculture on off-flavor in channel catfish. Furthermore, with the onset of cold water temperatures, the tilapia died and catfish developed a severe "rotten fish" or "putrid" flavor from feeding on the decaying tilapia. Nevertheless, other planktonivorous fish, such as silver carp, may hold some promise as biological control agents and should be further investigated.

Eliminating off-flavors from fish. Some compounds responsible for off-flavor, such as geosmin and 2-methylisoborneol, are eliminated fairly rapidly from fish once uptake of the compound from water ceases. Other compounds are more slowly eliminated, and several weeks may be required for off-flavors to dissipate. Obviously, the nature of the compound responsible for off-flavor influences the feasibility of procedures undertaken to purge flavors.

Algicides are sometimes applied to ponds with off-flavor fish in an attempt to kill the organism responsible for producing the odorous compound. Treatment is often unsuccessful, probably because no effort is made to identify the chemical or biological cause of the problem. Limited field observations suggest that treatment of ponds with a quick-acting algicide such as copper sulfate has some merit when off-flavors are due to geosmin or 2-methylisoborneol and there is good reason to suspect a specific blue-green alga as the causative organism. This procedure often fails when compounds other than geosmin and 2-methylisoborneol are involved, possibly because other compounds either are eliminated from fish too slowly or originate from sources other than algae. This treatment procedure should be contrasted to the practice of using frequent low doses of an algicide to "thin" phytoplankton communities and prevent off-flavor. Frequent treatments affect water quality for extended periods and reduce fish growth; a single treatment immediately before harvest has little effect on fish growth. However, treatment of fish culture ponds with algicides may cause oxygen depletion.

Removal of fish from the "off-flavor pond" to a clean environment is a proven procedure for dissipating certain off-flavors. Pond-cultured channel catfish are sometimes simply moved to an adjacent pond, but there is a risk that the new pond will also contain or develop a commu-

TABLE 9.4.

Percentage Weight Loss of Channel Catfish Held, without Feeding, in Clean, Flowing Water at Three Temperatures

Temperature °F	Percentage Weight Loss				
	0 Day	3 Days	6 Days	10 Days	15 Days
60	0	5	7	8	9
72	0	10	10	13	15
79	0	13	15	17	18

Source: Lovell (1983b).

nity of odor-producing microorganisms. Alternatively, special facilities, such as tanks or small ponds, may be used to hold fish while the off-flavor dissipates. Water exchange or algicides are used to prevent phytoplankton blooms from developing during the short period the fish are held. These treatments add to the cost of production because of the labor involved in harvesting and moving the fish.

When fish are moved to a pond or tank to purge off-flavors, great care must be taken to prevent stressing of the fish. Ideally fish should not be harvested and moved during hot weather, but most flavor problems occur during the summer. Seine and move fish as quickly as possible; do not hold them in live cars for more than 1 or 2 hours and do not transport them at excessively high densities. If the water temperature of the pond from which fish are harvested differs by more than 5°F from the water into which the fish are to be transferred, temper the fish by changing the water in the transport truck by no more than 1°F per minute until it matches the temperature of the receiving water.

Fish moved to tanks or ponds and held at high densities often are not offered feed and may not accept feed even if offered. Weight loss by fish in these facilities is proportional to the length of time fish are held and to water temperature (Table 9.4). Because fish may lose considerable weight, this procedure is useful only for those off-flavors that are quickly dissipated.

AQUATIC WEED CONTROL

Channel catfish ponds are ideal habitats for aquatic plants and some plant life will always be present. Any type of plant can cause problems

Aquatic Weed Control 269

during culture of channel catfish, and in some instances measures must be used to eliminate the weed or control its abundance.

Types of Weeds

The plants that grow in catfish ponds can be categorized into two groups, depending on their taxonomic classification. The algae are primitive plants that have no true roots, stems, or leaves and do not produce flowers or seeds. Algae can be categorized as phytoplankton and filamentous algae. The higher aquatic plants are more advanced; they usually have roots, stems, and leaves and produce flowers and seeds. Higher aquatic plants may be either submersed or emergent.

Phytoplankton. Phytoplankton are algae that are microscopic simple plants suspended in the water or form floating scums of near-microscopic colonies on the surface of the ponds. Communities of phytoplankton are referred to as "bloom." There are hundreds of species of phytoplankton; identification of the different species is difficult and requires a microscope.

Phytoplankton are the most common type of plant found in catfish ponds. Moderate densities of phytoplankton are desirable in catfish ponds if for no other reason than to keep the more troublesome types of plants from becoming established. Phytoplankton become a weed problem when they become excessively abundant or when certain undesirable species become dominant in the community. Excessive phytoplankton abundance causes serious water quality problems such as frequent periods of dangerously low concentrations of dissolved oxygen. Also, some blue-green algae cause off-flavor in fish and are highly undesirable. For those reasons, methods for control of phytoplankton abundance or selective control of certain types of phytoplankton would be desirable. Regrettably, it is very difficult to manage phytoplankton communities without some negative consequence to water quality.

Filamentous algae. Most filamentous algae begin growing on the bottom of the pond and rise to the surface when gas bubbles become entrapped in the plant mass. They form "clouds" or mats of finely divided cottony or slimy plant material. These filamentous algae are also known as "pond scum" or, more commonly, "moss." One type of filamentous alga, *Chara*, resembles submersed higher plants in growth habit. It is weakly anchored in the mud and grows up through the water. Positive identification of the different types of filamentous algae usually requires a microscope. Control methods are similar for all filamentous algae except *Pithophora*, which is very resistant to copper-based algicides and requires special treatment. It is thus important to identify the species of filamentous algae present in the pond to ensure that proper treatment is selected. The following are the most common filamentous algae in channel catfish ponds:

Hydrodictyon (waternet): each cell is attached repeatedly to two others, forming a repeating network of five- or six-sided meshes that looks like a nylon stocking.

Spirogyra: usually a dark green slimy mass that can be pulled apart and drawn out into fine filaments. This alga usually is easy to identify microscopically because the chloroplast is spiraled in a characteristic "corkscrew" along the inside of the cell wall.

Pithophora: probably the most noxious and difficult filamentous alga to control. *Pithophora* is irregularly branched, not slimy, and somewhat coarser than masses of *Spirogyra*. A mass of *Pithophora* feels like wet cotton. The distinguishing microscopic character is the presence of barrel-shaped "spores" along the filament.

Chara: a more advanced group of algae that resemble submersed higher plants in growth habit. This plant is commonly called "muskgrass" because of the garlic or skunklike odor released when the plants are crushed. Masses of *Chara* feel rough or crusty when crushed in the hand (Fig. 9.20).

Filamentous algae are undesirable because they interfere with fish harvest. Seines may ride up over the mass of plants, allowing fish to escape, and the weight of plant material caught in the seine may stress equipment or prevent seining altogether. Even if the amount of plant material is small enough to allow seining, fish may become entangled in the mass of plants seined up and will be stressed as workers pick through the plants to recover them. This is particularly a problem with fingerlings and fry.

Submersed plants. Submersed plants spend their entire lifetime beneath the surface of the water, although the flowering parts of the plants may extend above the surface. Usually the plants are rooted in the mud, although masses of plants may tear loose and float free. These plants are objectionable because they interfere with fish harvest. The most common submersed higher aquatic plants in channel catfish ponds in the southeastern United States are the following:

Najas (bushy pondweed): rooted, submersed plants with slender branching stems and narrow, ribbonlike leaves. Bushy pond-

Aquatic Weed Control

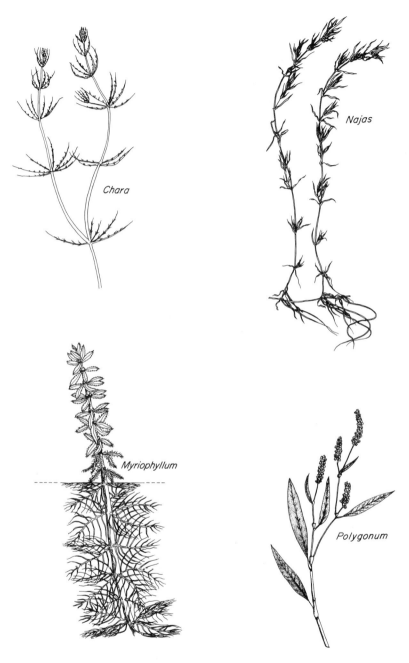

FIGURE 9.20.

Four common weeds in catfish ponds in the southeastern United States.

weed is the most common submersed weed problem in catfish ponds (Fig. 9.20).

Ceratophyllum (coontail): these plants have long, thin stems that are not rooted. The leaves are in whorls and are forked.

Cabomba (fanwort): these plants also have long, thin stems but are usually rooted in the mud. The leaves are opposite or whorled and are repeatedly divided to form a fanlike shape.

Myriophyllum (parrotfeather): rooted plants with long, thin stems. The leaves are whorled or opposite and finely divided into threadlike segments. In some species the end of the stem is rigid and protrudes above the water surface (Fig. 9.20).

Emergent plants. Emergent aquatic plants are rooted in the bottom and grow above the water. Many can also grow under strictly terrestrial conditions. The plants are rigid and not dependent on the water for support. In catfish ponds these plants usually infest only the margins of ponds and do not interfere with normal pond management. If stands of emergent plants become too dense or widespread, they may interfere with the seining or feeding of fish and should be removed or killed. Nuisance stands of emergent weeds are more common in older ponds in which erosion of the bank has created large areas of shallow water. They may also occur in ponds with a limited supply of water relative to pond size. When these ponds are dry, emergent weeds, especially highly competitive plants such as *Polygonum* (smartweed), colonize the pond bottom. Because the ponds fill so slowly, plant growth keeps pace with the rising water.

The most common emergent weeds in channel catfish ponds are the following:

Polygonum (smartweed): leaves are alternate and elliptical. The stem is erect and jointed, with each swollen node covered by a thin sheath. Flowers are white, pink, or greenish (Fig. 9.20).

Typha (cattails): cattails are familiar plants with stout, erect stems up to 8 feet tall. Leaves are flat and ribbonlike; flowers are brown and cigar-shaped.

Salix (willows): shrubs or trees with simple, elliptical leaves in alternate arrangement.

Occurrence of Weed Problems

Some plant life will always be present in catfish ponds, but the type of aquatic plant community that becomes established in a pond depends on the relative abilities of particular plants to compete for resources.

The growth of phytoplankton is favored in waters with high concentrations of nitrogen, phosphorus, and other plant nutrients dissolved in the water. Phytoplankton are efficient at using dissolved nutrients and reproduce rapidly. Once established, the phytoplankton community competes effectively for nutrients and also restricts the penetration of light so that plants germinating on the bottom do not receive enough light to continue growing.

The growth of rooted submersed plants is favored in ponds with low supplies of nutrients in the water. These ponds often are clear with light penetrating to the bottom, and rooted plants can use the nutrients in the bottom muds for growth. Established stands of submersed weeds compete for nutrients and light and prevent phytoplankton from becoming established. Some submersed plants also produce chemicals that inhibit growth of algae.

Emergent plants usually colonize only the margins of ponds where the water is less than 2 feet deep. If the levees or banks of the pond are eroded and have large areas of shallow water, expansive growths of emergent plants may be present. Emergent plants are rooted and can use nutrients in the mud. Thus, their establishment is also favored by low nutrient levels in the water.

In an informal survey of catfish culture ponds in west-central Mississippi, only about 2 percent of all ponds were found to contain submersed or emergent weed problems. Most of the weed problems were in fry nursery ponds. Nursery ponds are particularly susceptible to weed problems because they are managed differently than food-fish growout ponds. Growout ponds contain large numbers of fish and receive large amounts of feed. This results in high concentrations of plant nutrients in the water, which favor the growth of phytoplankton. Food-fish growout ponds are rarely drained, and once the phytoplankton community is established, it remains the dominant plant form in the pond year after year. Most fry nursery ponds are drained every year and refilled with water in late winter or early spring so they will be ready for stocking with fry in late spring or summer. Nursery ponds receive no feed for several weeks before stocking. This results in clear water with low nutrient levels, and this favors the growth of rooted submersed weeds such as *Najas* or filamentous algae such as *Chara*.

Prevention of Aquatic Weeds

Several factors should be considered when deciding on a course of action regarding aquatic weed control in fish ponds. The presence of any plant can be tolerated, depending on its abundance, and ponds should not be treated simply because a plant is present. Only experience will tell whether a small infestation of a potentially noxious weed will come to dominate a pond and cause problems at a later date.

Certain management procedures can be used to minimize the chances of infestations of submersed and emergent plants and filamentous algae. Such procedures should become part of common pond management and may obviate the need for chemical control measures.

Pond construction. Noxious weed growth usually starts in the shallow areas of ponds. Rooted plants have less chance to become established if the area of the pond where light can penetrate to the bottom is decreased. Every effort should be made to minimize the shallow areas of the ponds by constructing ponds with as much slope on the inside levee as is practical. Ponds should also have an adequate supply of water so they can be quickly filled.

Refilling an empty pond. An empty pond should not be refilled unless it is to be restocked immediately. A pond full of clear water with no nutrient input is quickly colonized by nuisance weeds. It is easier and less expensive to disc under terrestrial plants that may start growing in the empty pond bottom than it is to rid a pond of submersed weeds. When the pond is to be filled, it should be flooded as quickly as possible. It is better to fill ponds one at a time rather than slowly fill several ponds at once from a common water supply. Plants that grow up from the pond bottom have less chance to become established when water depth is increased quickly. If the supply of water is limited, ponds should be filled in winter if possible. Many higher aquatic plants are dormant or grow slowly in cold water, allowing for the development of a phytoplankton community that will shade the water and prevent weed growth when the water warms.

Fertilization. The use of a proper fertilization program is perhaps the best method of preventing the growth of troublesome weeds in fry nursery ponds. Weed problems can be prevented by establishing a phytoplankton bloom as quickly as possible after filling the ponds. The best way to do this is to add inorganic fertilizers to the pond.

The key ingredient in fish pond fertilizers is phosphorus. The most common phosphorus source in bagged, granular fertilizers is triple superphosphate; however, this is not an efficient form of phosphorus for fish pond fertilizers. When granular phosphorus fertilizers are broadcast over the ponds, the granules settle to the bottom because triple superphosphate is very insoluble. Most of the phosphorus reacts with bottom muds and never reaches the water. Any phosphorus that dissolves while the granules settle through the water quickly reacts with calcium in the water and is changed into unavailable calcium phosphates.

Liquid fertilizers are more effective than granular fertilizers at stimulating a phytoplankton bloom, especially in hard, alkaline waters. The phosphorus in liquid fertilizers is already in solution and immediately available for uptake by the phytoplankton. Although the phosphorus from liquid fertilizers also will eventually become unavailable because of reactions with calcium in the pond water, it remains in solution long enough to be taken up in adequate quantities by the phytoplankton.

The most common, and best, grade for liquid fertilizers runs from about 10-34-0 to 13-38-0. This proportion of about three times as much phosphorus (expressed as P_2O_5) as nitrogen (expressed as N) has been found to have an excellent balance. The rate used successfully by commercial catfish producers in Mississippi is 2 to 4 pounds/acre applied every other day for 8 to 14 days or until a noticeable phytoplankton bloom develops.

Liquid fertilizers should not be poured directly into the water because they are denser than water and will sink to the bottom. They can be sprayed from the bank or applied from a boat outfitted for chemical applications.

Manual harvesting. Removing potentially noxious emergent weeds by hand is another management practice that may reduce the possibility of having to use chemicals. As small areas of the pond margin become infested, plants are removed manually. Manual harvesting of weeds is only suited to controlling emergent vegetation in relatively small ponds. Care should be taken to remove as much of the rootstock or rhizome as possible to minimize regrowth.

Biological Control

Biological weed control in catfish ponds involves the use of fish to consume unwanted aquatic vegetation. Several fish species have been evaluated as biological control agents in warm-water ponds, including the grass carp, common carp, and various tilapias. These fish are most often used to control submersed plants or filamentous algae. With the exception of the grass carp, all these fish have undesirable characteristics that make their use impractical in commercial catfish ponds.

The grass carp or "white amur" was introduced into the United States from southeast Asia in 1963 and is now widespread, especially in the southeastern states. The fish is banned in more than thirty states, but provides a valuable tool for control of nuisance aquatic weeds where it is legal and available.

The controversy over the distribution and use of grass carp is based on the potential effect of this fish on native fish and wildlife. Considerable discretion should be used when considering stocking these fish into catfish ponds. Because grass carp need running water to spawn, every effort should be made by managers of ponds to prevent their escape into natural waters. To diminish further the likelihood that grass carp will reproduce and thrive in natural waters, it is recommended that only sterile, triploid carp be used in channel catfish ponds.

The grass carp has several traits that make it a good species to polyculture with channel catfish. Small grass carp (less than 1 to 2 pounds) are almost completely herbivorous and do not compete with catfish for feed to a significant degree. Grass carp tolerate a wide range of environmental conditions: they can survive at water temperatures of 32° to 105°F and are nearly as tolerant as catfish of low dissolved oxygen concentrations. The fish grows rapidly, as much as 5 to 10 pounds a year, yet it is inefficient at digesting plants. It must consume large quantities of plant material to grow and may eat two or three times its weight in plant material per day.

Grass carp prefer to eat succulent submersed plants such as *Najas* and *Chara*. Fibrous plants such as grasses and smartweed are less preferred, and grass carp will not eat them if preferred plants are available. Food consumption by grass carp is greatest at water temperatures of 80 to 85°F, and the fish stops feeding when the water temperature falls below about 55°F.

Grass carp can be stocked into nursery ponds before stocking catfish fry to prevent weed growth rather than waiting until weeds develop to treat the problem. Fry nursery ponds usually are drained each year and grass carp must be restocked each year. Grass carp also are used by some catfish producers to control existing weeds in food-fish ponds. However, they require considerable time to reduce weed infestations, particularly if coverage is extensive. Results may take a year to be realized. Food-fish ponds are usually not drained each year and grass carp become permanent inhabitants of the pond. Larger grass carp learn to feed on pelleted feeds and often do little to control weeds in the second or third year they are present. Usually weed problems have been controlled by this time and a phytoplankton community that prevents further weed problems has developed. As the catfish are harvested, any grass carp also captured in the seine can be removed. Grass carp have little or no commercial value in the United States.

The appropriate stocking density for grass carp depends on the severity of the weed problem. When they are used to prevent the establishment of submersed weeds, 5 to 10 small (3- to 6-inch) carp/acre should be stocked. The same stocking rate is also adequate if the pond is lightly infested with weeds. For more severe weed problems 10 to 15 fish/acre should be stocked. For heavily weed infested ponds, stocking rates can be increased to 15 to 25 fish/acre.

Chemical Control of Aquatic Plants

Chemical control is the most common means of eradicating weeds in catfish ponds. Many pond managers do not use any weed prevention measures, except possibly fertilization, and if weeds become established, it is usually quicker and more effective to use herbicides than any other method.

Many different herbicides have been used in and around waters. Use of most of these chemicals is illegal in waters that constitute a "fishery," and even fewer can be used in waters used to raise food fish. In the United States, registration of chemicals for fishery use is granted by the Environmental Protection Agency or the Food and Drug Administration under the Federal Environmental Pesticide Control Act (FEPCA) of 1972. The lack of registration does not necessarily mean that the chemical is harmful to the environment or that it is extremely toxic. Aquaculture is considered a minor use by most chemical companies, and they are simply not willing to spend the large amount of money needed to compile the data necessary for registration review. However, some unregistered herbicides are toxic to fish or their use may result in chemical residues in the edible portion of the fish that are harmful to humans. For these reasons, only herbicides labeled for use in food-fish ponds should be used by catfish farmers, and label instructions should be carefully followed.

The following herbicides or herbicide groups are labeled for use in food-fish ponds. Table 9.5 summarizes herbicide use for common weeds in channel catfish ponds.

Copper sulfate (various trade names). Copper sulfate ($CuSO_4 \cdot 5H_2O$) is available in various particle sizes from fine powder to large crystals. The fine powder is more effective because it dissolves more rapidly. Copper sulfate should only be used to control algae, because the rates necessary to kill other plants may also be toxic to fish. The filamentous alga *Pithophora* is also resistant to copper sulfate.

Because copper is extremely toxic to fish in soft waters of low alkalinity, it is recommended that copper sulfate not be used in waters with a total alkalinity of less than 50 ppm as $CaCO_3$. Copper sulfate becomes less effective as an algicide in hard, alkaline waters because

TABLE 9.5.
Summary of Herbicides Used to Control Some Common Weeds in Catfish Ponds

Weed	Herbicide	Response[a]
Algae		
Phytoplankton	Copper sulfate	G
	Chelated copper	G
	Potassium ricinoleate	P
	Simazine	G
Filamentous algae	Copper sulfate	G
(except *Pithophora*)	Chelated copper	G
	Diquat	F
	Endothall (alkylamine)	F
	Simazine	G
Pithophora	Diquat	F
	Endothall (alkylamine)	F
	Simazine	F
Chara (muskgrass)	Copper sulfate	G
	Chelated copper	G
	Diquat	F
	Endothall (alkylamine)	F
	Simazine	G
Submersed plants		
Cabomba (fanwort)	Endothall (potassium)	G
	Fluridone	G
	Simazine	F
Ceratophyllum (coontail)	Diquat	G
	Endothall (potassium)	G
	Fluridone	G
	Simazine	G
	2,4-D (granular)	F
Myriophyllum (parrotfeather)	Diquat	G
	Endothall (potassium)	G
	2,4-D (granular)	G
Najas (bushy pondweed)	Diquat	G
	Endothall (potassium)	G
	Fluridone	G
	Simazine	G
	2,4-D (granular)	F
Emergent weeds		
Polygonum (smartweed)	Fluridone	F
	Glyphosate	G
	2,4-D (liquid)	G
Salix (willows)	Diquat	F
	Glyphosate	G
	2,4-D (liquid)	G
Typha (cattails)	Diquat	F
	Fluridone	F
	Glyphosate	G
	2,4-D (liquid)	G

[a] G = good control; F = fair control; P = poor control.

copper rapidly precipitates out of solution; hence, the treatment rate increases with total alkalinity. The formula used to calculate the treatment rate is

Parts per million $CuSO_4 \cdot 5H_2O$ = (total alkalinity) ÷ 100 (9.8)

In waters with a total alkalinity greater than about 300 ppm as $CaCO_3$, copper from copper sulfate precipitates out of solution so rapidly that it is difficult to achieve an effective treatment.

Chelated copper (Algimycin®, Copper Control®, Cutrine®, K-Tea®). Various brands of chelated copper herbicides are available in both liquid and granular form; the liquid form is more commonly used. The copper in these herbicides is bound in organic complexes so that it will not precipitate out of solution as rapidly in hard, alkaline waters. Although these herbicides usually are more effective than copper sulfate, they are considerably more expensive to use. Chelated copper herbicides kill phytoplankton and filamentous algae, with the exception of *Pithophora*. They should not be used in waters of less than 50 ppm $CaCO_3$ total alkalinity because of their toxicity to fish.

Potassium ricinoleate (Solricin 135®). The herbicide potassium ricinoleate is available as a liquid and is prepared commercially by causing castor oil to react with potassium hydroxide to form a soap. The chemical structure of potassium ricinoleate is similar to that of certain natural chemicals produced by higher aquatic plants that inhibit blue-green algae growth. Field trials in Mississippi have not been successful (Tucker and Lloyd 1987), and use of the product is limited at present.

Simazine (Aquazine®). Simazine is sold as a wettable powder and is mixed into a slurry with water for application. It is a broad-spectrum herbicide that kills most algae and submersed plants. The herbicide is slow-acting and very persistent in the pond. Because simazine is persistent and unselective in herbicidal activity, water quality may be poor in catfish ponds for weeks after its use (Tucker, Busch, and Lloyd 1983). This trait has limited its usefulness in fish culture ponds.

Diquat (Aqua-clear®, Aquaquat®, Ortho diquat herbicide®). Diquat is sold as a liquid containing 2 pounds active ingredient/gallon. It is a broad-spectrum herbicide that controls most filamentous algae, including *Pithophora* and *Chara*, and submersed weeds, such as *Najas* and coontail; and can be mixed with a surfactant and sprayed to control emergent weeds such as cattail. Diquat should not be used in muddy water and mud should not be stirred up during application because diquat will bind tightly with clay particles suspended in the water, rendering the herbicide ineffective at controlling plants growing beneath the surface.

Endothall, dipotassium salt (Aquathol® and Aquathol K®). The dipotassium salt of endothall is available in either liquid (Aquathol K®) or granular form *(Aquathol®).* The dipotassium salt formulations of endothall do not kill algae, but they control a wide spectrum of submersed higher plants, including *Najas,* coontail, and fanwort. The granular formulation is relatively expensive, but it is a particularly effective treatment for *Najas.*

Endothall, alkylamine salt (Hydrothol 191®). The alkylamine salt of endothall is most commonly used in the liquid formulation and is a more potent herbicide than the dipotassium salt. It controls most filamentous algae, including *Pithophora* and *Chara,* as well as a broad spectrum of submersed higher plants. But the alkylamine salt formulation also is quite toxic to fish and should not be routinely used if there are alternative control measures available. If the herbicide is to be used, only a portion of the pond should be treated at one time. Fish will avoid the treated area and may not be killed.

Fluridone (Sonar®). Fluridone is available as an aqueous suspension or as pellets. It does not kill phytoplankton or filamentous algae but controls a very broad spectrum of submersed higher plants. This herbicide is very slow-acting and results may take a month or more to be noticeable. Fluridone is an effective herbicide, but it is quite expensive.

2,4-D (Aquacide®, Aqua-Kleen®, Weed-Rhap®, Weedtrine II®). The herbicide 2,4-D is formulated for aquatic use as the dimethylamine salt or isooctyl ester. It is available in liquid or granular form. The granular form is effective at controlling submersed higher plants such as *Najas* and coontail. The liquid formulation can be sprayed on smartweed or other emergent plants.

Glyphosate (Rodeo®). Glyphosate is sold as a liquid and is for use only on emergent and shoreline plants. The herbicide is mixed with a surfactant and sprayed on the vegetation. Glyphosate is a broad-spectrum herbicide that is useful for the control of cattails, grasses, smartweed, and willows around pond margins.

Consequences of Herbicide Use

Herbicides are seldom directly toxic to fish when used according to the manufacturer's specifications. However, the addition of any herbicide to a plant-infested body of water alters water quality. Oxygen production by photosynthesis decreases and decomposition of the dead plant material increases oxygen consumption. The result is a noticeable decrease in dissolved oxygen concentrations compared to those of pretreatment levels. The extent to which dissolved oxygen levels are reduced depends on the amount of plant material killed, the amount of plant material unaffected by the herbicide, the rate at which death occurs, the water temperature, and other factors. Decomposition of the dead plants also raises carbon dioxide and total ammonia concentrations. The increase in total ammonia concentrations following sudden plant death may be considerable, but the increased carbon dioxide concentrations tend to decrease the pH, shifting the equilibrium to favor the nontoxic, ionized form of ammonia. Phosphorus, potassium, and other minerals are also released on plant decomposition, and concentrations of all essential plant nutrients are usually higher after herbicide treatment. At some time after treatment, the concentration of herbicide decreases to a nontoxic level and these nutrients are available for new plant growth.

The deterioration of water quality following herbicide use can have serious consequences in catfish ponds. Obviously, if dissolved oxygen concentrations fall to very low levels, fish will be killed. This is particularly a problem in catfish fry ponds because fry may not be able to find the area of oxygenated water behind emergency aerators. Even if dissolved oxygen concentrations are maintained above lethal levels, the fish may be severely stressed and be more susceptible to fish diseases.

Stressed fish also feed poorly and decreased fish growth can be expected, particularly if water quality is affected for an extended length of time. Simazine, a particularly persistent herbicide, can affect water quality in catfish ponds for 1 month after treatment. Catfish production has been decreased by 20 percent in simazine-treated ponds, presumably the result of water quality deterioration after treatment (Tucker and Boyd 1978).

Control of Phytoplankton Abundance

Low concentrations of dissolved oxygen and development of off-flavor are the two most important water quality problems in channel catfish culture. Both problems are the result of uncontrolled phytoplankton growth in heavily fed ponds, and numerous efforts have been made to manage phytoplankton communities in fish ponds. Most methods are ineffective and many actually further degrade water quality (Smith 1988).

A variety of algicides have been used to reduce phytoplankton density, but the ultimate results are always undesirable. When sufficient algicide is added to a pond with a dense bloom, the sudden die-off usually causes severe oxygen depletion and high levels of carbon dioxide and ammonia. Phytoplankton repopulate the pond as soon as algicide levels decrease because nutrient levels remain high. Episodes of poor water quality resulting from this cycle of death and regrowth stress fish and cause reduced growth or increased susceptibility to infectious diseases (Tucker and Boyd 1978). Similar problems occur when algicides are used in attempts to eliminate specific noxious phytoplankton species. All of the effective algicides registered for use in food-fish ponds are broad-spectrum and cannot be used selectively to eliminate one species or one type of phytoplankton.

Biological control of phytoplankton growth is an attractive alternative to the use of herbicides. Most efforts have involved the use of plankton-feeding fish such as silver carp, bighead carp, and tilapias. In theory the plankton-feeding fish continually harvests the bloom, improves water quality, and provides additional fish production. However, most attempts at biological control of phytoplankton growth have failed (Smith 1988). Quite often phytoplankton abundance increases when plankton-feeding fish are present because these fish effectively remove large phytoplankton and zooplankton that compete with or consume small phytoplankton. The presence of plankton-feeding fish may thus change the structure of the plankton community, but not decrease its density (Smith 1988).

Decreasing nutrient levels by limitations on daily feed allotments is the only reliable method available for reducing, on average, the incidence of phytoplankton-related water quality problems. Such problems are rare if maximum daily feeding rates are less than about 50 pounds/acre, but this feeding rate is uneconomical in most commercial enterprises. Although water quality problems are common at higher feeding rates, the use of chemical or biological measures to control phytoplankton density cannot be recommended.

TURBIDITY

Turbidity in most channel catfish ponds is caused by plankton. Moderate plankton turbidity is desirable because it restricts light penetration and prevents growth of undesirable plant forms such as filamentous algae and submersed higher aquatic plants. Some ponds are turbid from suspended clay particles. Clay turbidity seldom directly affects channel catfish in ponds; however, suspended clay particles reduce light penetration and limit phytoplankton growth. This causes low rates of photosynthesis and long periods of low dissolved oxygen concentrations. Aeration is necessary during these periods, and water currents produced by aerators may keep the clay particles in suspension and perpetuate the problem. Ammonia and nitrite concentrations may also be chronically high in turbid ponds receiving large feed inputs. Ammonia accumulates because its assimilation by phytoplankton is reduced. High nitrite levels are then more likely because substrate concentrations for nitrification are increased. Interestingly, clay turbidity has a beneficial result in that problems with off-flavors in fish are less common in turbid ponds.

Turbidity in watershed ponds is usually caused by erosion of the watershed during periods of heavy rainfall. Vegetative cover over the entire watershed greatly decreases turbid runoff. Poorly constructed roads often are the major source of turbid runoff into watershed ponds. Roads should be graveled if possible, and ditches can sometimes be constructed to direct the runoff around the pond.

Runoff from roads or levees can also cause turbidity in levee ponds. Vegetative cover should be established over as much of the levee as possible to reduce turbid runoff. Levee ponds often become turbid in the winter when winds are high and large waves erode the bank. Clay turbidity usually clears in spring when winds are lighter and phytoplankton blooms become established.

Turbidity caused by aerators is more common in old, shallow ponds partially filled in with sediments or in new ponds that were not properly compacted before use. The loose bottom muds in these ponds are easily disturbed by aerator currents. If possible, locate aerators in water at least 3 feet deep to minimize erosion of pond bottoms and resuspension of sediments.

Certain fish, particularly common carp, disturb bottom sediments during feeding and can cause chronic turbidity problems. The effect of these fish is worse in ponds with soft, muddy bottoms. In extremely shallow ponds with soft bottoms, even the activities of channel catfish can cause excessive clay turbidity. The problem can become self-perpetuating because aeration of shallow ponds exacerbates the condition.

Improper pond construction and poor pond maintenance account for most turbidity problems and can be corrected during pond renovation. Otherwise, turbidity problems are usually self-limiting and as the clay settles out, a phytoplankton bloom becomes established. The presence of phytoplankton accelerates settling of clay particles through the formation of algae-clay aggregates (Avnimelech, Troeger, and Reed 1982), and an established phytoplankton bloom hinders the development of further clay turbidity.

TABLE 9.6.
Treatments Used to Remove Clay Turbidity from Ponds

Material	Application Rate	Comments
Dry hay	150–500 pounds/acre	Spread over pond; apply at 10-day intervals until clear; may cause oxygen depletion
Cottonseed meal	75–100 pounds/acre	Spread over pond with 20 pounds/acre superphosphate fertilizer; apply at 10-day intervals until clear; may cause oxygen depletion
Barnyard manure	1,000 pounds/acre	Spread over pond; may cause oxygen depletion
Gypsum	200–1,000 pounds/acre	Spread over pond on calm day; repeat at 7-day intervals until clear
Filter alum	25–35 ppm	Spread over pond on calm day; may reduce pH to dangerous levels

Several techniques are available for the removal of clay turbidity (Table 9.6). All these methods work to varying degrees, but the source of turbidity must be removed to achieve a permanent solution to clay turbidity problems. Other factors also limit the practicability of these treatments. Organic matter must decompose to be effective in removing clay turbidity. This is a slow, unpredictable process, only effective in warm water. Decomposition of the large amounts of organic matter necessary for effective treatment may also induce oxygen depletions. Addition of gypsum (calcium sulfate, $CaSO_4$) works well in soft water, but in hard waters (total hardness greater than 100 ppm as $CaCO_3$) calcium concentrations are already high and addition of gypsum will not flocculate the clay. Treatment with filter alum (aluminum sulfate, $Al_2(SO_4)_3$) is the most effective means of removing clay turbidity (Boyd 1979a). Alum produces sulfuric acid in water, so it destroys alkalinity and reduces pH, possibly to the point where fish will be killed. To prevent alum treatment from reducing pH to harmful levels, sufficient residual alkalinity should exist after treatment. Each ppm of alum destroys 0.5 ppm of total alkalinity as $CaCO_3$, so alum treatment rates should not exceed about 1.5 times total alkalinity. In waters of low alkalinity, calcium hydroxide ($Ca(OH)_2$) can be applied simultaneously at a rate of 0.4 ppm/1.0 ppm of alum to prevent large changes in alkalinity and pH.

HYDROGEN SULFIDE

Hydrogen sulfide (H_2S) is a toxic gas produced under anaerobic conditions during the bacterial decomposition of organic matter. It accumulates in sediments of channel catfish ponds and can be released when the sediments are disturbed. The gas is easily detected by a noxious, rotten-egg odor when trapped bubbles of carbon dioxide, methane, and hydrogen sulfide rise to the surface from disturbed muds. Hydrogen sulfide is extremely toxic to fish but is rapidly oxidized to nontoxic sulfate (SO_4^{2-}) in aerobic waters. Hydrogen sulfide has not been identified as a problem during routine pond growout of channel catfish.

Dangerous levels of hydrogen sulfide could possibly be released from sediments under fish confined in seines or live cars after seining. The activities of the fish in the small area stir up the mud and release the trapped gases. Aerators are usually placed near confined fish to supply oxygen-rich water. Currents produced by the aerator also serve to dilute and remove hydrogen sulfide from the confinement area.

TOXIC ALGAE

Some marine and freshwater algae produce chemicals that are toxic to vertebrates. Fish kills caused by algal toxins have been well documented for certain marine phytoplankton, called *dinoflagellates*, that cause "red tides" along coastal regions (White 1984). Freshwater blue-green algae in the genera *Microcystis*, *Anabaena*, and *Aphanizomenon* are common in catfish ponds and are known to produce compounds toxic to birds and mammals. However, fish kills caused by blue-green algal toxins appear to be extremely rare. Waterborne toxins produced by blue-green algae apparently cannot cross fish gill membranes and enter the circulatory system (Phillips et al. 1985); toxicosis can be induced only when the toxins are injected into fish or, possibly, when fish eat the toxin-containing algal cells.

Only one fish kill that possibly involved blue-green algal toxins has been described (Schwedler, Tucker, and Beleau 1985). Fingerling channel catfish were observed in distress after injesting large quantities of *Microcystis* while feeding on floating feed in a dense surface scum

TABLE 9.7.
Common Agricultural Chemicals Grouped According to Relative Toxicity to Channel Catfish

Common Name	Trade Name	Common Name	Trade Name
Group A			
2,4-D	Weedar 64	MCPA	
2,4-DB	Butoxone	MSMA	
Acephate	Orthene	Malathion	Cythion
Acifluorfen	Blazer	Methamidophos	Monitor
Alachlor	Lasso	Methazole	Probe
Aldicarb	Temik	Methyl parathion	
Bentazon	Basagran	Metribuzin	Lexone
Bifenox	Modown	Molinate	Ordram
Bromoxynil	Buctril	Monocrotophos	Azodrin
Carbaryl	Sevin	Naptalam	Alanap
Chlordimeform	Galecron/Fundal	Norflurazon	Zorial
Chlorimuron	Classic	Oxamyl	Vydate
Chlorothalonil	Bravo	Oxydemeton-methyl	Metasystox-R
Clomazone	Command	Oxyflourfen	Goal
Cyanazine	Bladex 4L	Paraquat	Gramoxone
DSMA		Phosmet	Imidan
Dicamba	Banvel	Phosphamidon	Swat
Dicrotophos	Bidrin	Prometryne	Caparol
Diflubenzuron	Dimilin	Propanil	Stam
Dimethoate	Cygon	Propiconazole	Tilt
Diuron	Karmex	Sethoxydim	Poast
Ethalfluralin	Sonolan	Sulprofos	Bolstar
Fenoxaprop-ethyl	Whip	Thiabendazole	Mertect
Flauzifop-butyl	Fusilade 4E	Thidiazuron	Dropp
Fomesafen	Reflex	Thiobencarb	Bolero
Glyphosate	Roundup	Thiodicarb	Larvin
Imazaquin	Scepter	Thiophanate-M	Topsin-M
Iprodione	Rovral	Triadimefon	Bayleton
Lactofen	Cobra	Triclopyr	Rely
Linuron	Lorox	Trifluralin	Treflan
Group B			
Azinophos-methyl	Guthion	Flucythrinate	Payoff
Benomyl	Benlate	Fluometuron	Cotoran
Bifenthrin	Capture	Fluvalinate	Mavrik
Carbofuran	Furadan	Mancozeb	Dithane M-45
Chlorpyrifos	Lorsban	Methomyl	Lannate
Cyfluthrin	Baythroid	Pendimethalin	Prowl
Cypermethrin	Ammo-Cymbush	Permethrin	Ambush/Pounce
Diazinon	Spectracide	Profenofos	Curacron
Dicofol	Kelthane	Propargite	Comite
Endosulfan	Thiodan	Quizalofop ethyl	Assure
Esfenvalerate	Asana	Tralomethrin	Scout
Fenvalerate	Pydrin		

Note: Group A consists of pesticides least toxic or least likely to accumulate as residues in fish. Pesticides in group B are highly toxic or likely to accumulate as residues.

Pesticides **287**

of this alga. The fish exhibited tetany and convulsions, symptoms often seen after exposure to a neurotoxin. Microcystis is known to produce neurotoxins as well as hepatotoxins (substances that affect liver function). Relatively few fish died during this incident.

The potential for fish kills related to algal toxins certainly exists because known toxin-producing species of blue-green algae are common in channel catfish ponds; nevertheless, incidents such as the one described are extremely uncommon. To be safe, however, do not apply feed in areas with dense scums of algae. Surface scums usually accumulate along the lee shore of ponds, so offer feed from the upwind shore. This is consistent with common feeding practices anyway, because feeds are usually offered from this shore so that the floating pellets will not be carried quickly to shore by wind-generated surface currents.

Inexperienced or unqualified "experts" ascribe many fish kills in catfish ponds to the presence of "toxic algae"; in fact, "toxic algae" has become a catchall for any fish kill whose cause cannot easily be identified. For instance, until very recently, proliferative gill disease of channel catfish (see the section Diseases of Uncertain Origin) was thought by some to be caused by algal toxins. It now appears that the disease is caused by a protozoan parasite. As more research information has become available over the years, fewer and fewer fish kills have been "caused" by toxic algae.

PESTICIDES

Some pesticides used on agricultural crops are extremely toxic to channel catfish. Facilities used to rear channel catfish should, if possible, be located to minimize the chance of inadvertent contamination. However, most of the ponds used to rear channel catfish in the southeastern United States are adjacent to areas devoted to cotton, soybeans, rice, or wheat farming, and the potential exists for accidental contamination. Ponds can become contaminated by drift of chemicals sprayed on fields or by runoff from fields if ponds are filled by watershed runoff.

The toxicity of pesticides is usually expressed as 96-hour LC50 values. The *96-hour LC50* is the concentration of chemical that kills 50 percent of fish tested after 96 hours of exposure. Tests are conducted under a standard set of conditions in aquaria, and LC50 values may not reflect concentrations that kill fish under field conditions. To simplify interpretation of laboratory toxicity data, the Mississippi Department of Agriculture and Commerce places pesticides into two broad categories based on relative toxicity to channel catfish (Table 9.7). Pesticide use guidelines are then based on category. Generally, 96-hour LC50 values are above 1 ppm for pesticides in group A and less than 1 ppm for those in group B. Schwedler, Tucker, and Beleau (1985) list 96-hour LC50 values for most of these chemicals.

The following recommendations have been developed to protect catfish from pesticide toxicosis and reduce levels of potentially harmful residues.

1. If toxicity data for a particular chemical are not available, it should not be used around ponds.
2. Read and follow all label instructions before using any chemical.
3. Adequate precautions to prevent contamination of ponds with any pesticides should always be taken, even to the extent of delaying or omitting an application.
4. Group A pesticides are the pesticides of choice for use within 0.25 mile of ponds.
5. All applications of pesticides should be made when the wind is blowing away from ponds.
6. Aerial applications of pesticides should be made so that no pull-ups, approaches, turns, or ferrying will be over ponds.
7. Pesticide application equipment (aerial or ground) should be calibrated, adjusted, and operated in a manner to prevent drift.
8. Pesticide drift to nontarget areas is prohibited by most pesticide labels and may constitute "use inconsistent with labeling," which is prohibited by state and federal pesticide laws. Violators are subject to penalties and/or loss of license or certification if found guilty.
9. Fish kills or residues from pesticide drift may result in civil liability for the applicator.

REFERENCES

Ahmad, T., and C. E. Boyd. 1988. Design and performance of paddle wheel aerators. *Aquacultural Engineering* 7:39–62.

Avnimelech, Y., B. W. Troeger, and L. W. Reed. 1982. Mutual flocculation of algae and clay: evidence and implications. *Science* 216:63–65.

Boyd, C. E. 1979a. Aluminum sulfate (alum) for precipitating clay turbidity from fish ponds. *Transactions of the American Fisheries Society* 108:307–313.

Boyd, C. E. 1979b. *Water Quality in Warmwater Fish Ponds*. Auburn: Alabama Agricultural Experiment Station.

Boyd, C. E., and T. Ahmad. 1987. *Evaluation of Aerators for Channel Catfish Farming*. Auburn: Alabama Agricultural Experiment Station.

References 289

Boyd, C. E., R. P. Romaire, and E. Johnston. 1978. Predicting early morning dissolved oxygen concentrations in channel catfish ponds. *Transactions of the American Fisheries Society* 107:484–492.

Boyd, C. E., and B. J. Watten. 1989. Aeration systems in aquaculture. *Reviews in Aquatic Sciences* 1:425–472.

Boyd, C. E., W. D. Hollerman, J. A. Plumb, and M. Saeed. 1984. Effect of treatment with a commercial bacterial suspension on water quality in channel catfish ponds. *Progressive Fish-Culturist* 46:36–40.

Brown, S. W., and C. E. Boyd. 1982. Off-flavor in channel catfish from commercial ponds. *Transactions of the American Fisheries Society* 111:379–383.

Busch, R. L., C. S. Tucker, J. A. Steeby, and J. E. Reames. 1984. An evaluation of three paddlewheel aerators used for emergency aeration of channel catfish ponds. *Aquacultural Engineering* 3:59–69.

Cole, C. A., and C. E. Boyd. 1986. Feeding rate, water quality, and channel catfish production in ponds. *Progressive Fish-Culturist* 48:25–29.

Johnsen, P. B., G. V. Civille, and J. R. Vercellotti. 1987. A lexicon of pond-raised catfish flavor descriptors. *Journal of Sensory Studies* 2:85–91.

Keenum, M. E., and J. E. Waldrop. 1988. *Economic Analysis of Farm-Raised Catfish Production in Mississippi*, Technical Bulletin 155. Mississippi State: Mississippi Agricultural and Forestry Experiment Station.

Lai-fa, Z., and C. E. Boyd. 1988. Nightly aeration to increase the efficiency of channel catfish production. *Progressive Fish-Culturist* 50:237–242.

Lovell, R. T. 1983a. New off-flavors in pond-cultured channel catfish. *Aquaculture* 30:329–334.

Lovell, R. T. 1983b. Off-flavors in pond-cultured channel catfish. *Water Science and Technology* 15:67–73.

Lovell, R. T., I. Y. Lelana, C. E. Boyd, and M. S. Armstrong. 1986. Geosmin and musty-muddy flavors in pond-raised channel catfish. *Transactions of the American Fisheries Society* 115:485–489.

Mandal, B. K., and C. E. Boyd. 1980. Reduction of pH in waters with high total alkalinity and low total hardness. *Progressive Fish-Culturist* 42:183–185.

Martin, J. F., L. W. Bennett, and W. H. Graham. 1988. Off-flavor in channel catfish (*Ictalurus punctatus*) due to 2-methylisoborneol and its dehydration products. *Water Science and Technology* 20:99–105.

Murphy, T. P., and G. G. Brownlee. 1981. Ammonia volatilization in a hypertropic prairie lake. *Canadian Journal of Fisheries and Aquatic Sciences* 38:1035–1039.

Neori, A., M. D. Krom, I. Cohen, and G. Gordin. 1989. Water quality conditions and particulate chlorophyll a of new intensive seawater fish ponds in Eilat, Israel: daily and diel variations. *Aquaculture* 30:63–78.

Phillips, M. J., R. J. Roberts, J. A. Stewart, and G. A. Codd. 1985. The toxicity of the cyanobacterium *Microcystis acruginosa* to rainbow trout, *Salmo gairdneri* Richardson. *Journal of Fish Diseases* 8:339–344.

Schwedler, T. E., C. S. Tucker, and M. H. Beleau. 1985. Non-infectious diseases. In *Channel Catfish Culture*, ed. C. S. Tucker, pp. 497–541. Amsterdam: Elsevier.

Smith, D. W. 1988. Phytoplankton and catfish culture: a review. *Aquaculture* 74:167–189.

Steeby, J. A., and C. S. Tucker. 1988. Comparison of nightly and emergency aeration of channel catfish ponds. *Mississippi Agricultural Experiment Station Research Report* Vol. 13, No. 8, Mississippi State, Mississippi.

Torrans, L., and F. Lowell. 1987. Effects of blue tilapia/channel catfish polyculture on production, feed conversion, water quality, and channel catfish off-flavor. *Proceedings of the Arkansas Academy of Science* 41:82–86.

Tucker, C. S., and C. E. Boyd. 1977. Relationships between potassium permanganate treatment and water quality. *Transactions of the American Fisheries Society* 106:481–488.

Tucker, C. S., and C. E. Boyd. 1978. Consequences of periodic applications of copper sulfate and simazine for phytoplankton control in catfish ponds. *Transactions of the American Fisheries Society* 107:316–320.

Tucker, C. S., and C. E. Boyd. 1985. Water quality. In *Channel Catfish Culture*, ed. C. S. Tucker, pp. 135–227. Amsterdam: Elsevier.

Tucker, C. S., and S. W. Lloyd. 1985. Evaluation of a commercial bacterial amendment for improving water quality in channel catfish ponds. Mississippi Agricultural and Forestry Experiment Station Research Report Vol. 10, No. 9, Mississippi State, Mississippi.

Tucker, C. S., and S. W. Lloyd. 1987. Evaluation of potassium ricinoleate as a selective blue-green algicide in channel catfish ponds. *Aquaculture* 65:141–148.

Tucker, C. S., and J. F. Martin. 1990. Environment-related off-flavors in fish. In *Aquaculture and Water Quality*, eds. D. E. Brune and J. R. Tomasso, in press. Baton Rouge, Louisiana: World Aquaculture Society.

Tucker, C. S., R. L. Busch, and S. W. Lloyd. 1983. Effects of simazine treatment on channel catfish production and water quality in ponds. *Journal of Aquatic Plant Management* 21:7–11.

Tucker, C. S., S. W. Lloyd, and R. L. Busch. 1984. Relationships between phytoplankton periodicity and the concentrations of total and un-ionized ammonia in channel catfish ponds. *Hydrobiologia* 111:75–79.

White, A. W. 1984. Paralytic shellfish toxins and finfish. In *Seafood Toxins*, ed. E. P. Raegalis, pp. 171–180. Washington D.C.: American Chemical Society.

CHAPTER 10
Feeds and Feeding Practices

Although natural food organisms may provide certain micronutrients, the contribution of pond organisms to the nutrition of intensively cultured catfish is considered minuscule. The nutritional requirements of cultured catfish are met by using a feed that is formulated to provide all required nutrients (a complete feed) in the proper proportions necessary for rapid weight gain, high feed efficiency, and desirable composition of gain. Feed cost represents about half of variable production costs in catfish culture; thus careful consideration should be given to feed selection and use. High-quality feeds are essential for culture of catfish, but if consumption is poor, rapid growth will not be achieved. In addition, uneaten feed increases production costs and contributes to deterioration of water quality. Proper feeding practices are as important to the catfish producer as are good feeds. The prudent catfish producer should select the appropriate feed for a specific culture system and use it in a manner to ensure efficient conversion of feed to fish flesh. The following sections on feeds and feeding practices, although based on the best available information, are intended only as guidelines since feeding catfish is as much an art as it is a science. If additional information is desired, the reader is referred to the following: National Re-

TABLE 10.1.

Protein Efficiency Ratio (PER) Values for Selected Protein Sources Determined with Catfish

Protein Source	International Feed Number	PER^a (1)	(2)
Fish meal, menhaden	5-02-009	2.48	2.41
Fish meal, anchovy	5-01-985	2.44	
Catfish waste, dry		2.08	2.01
Soy-catfish scrap^b			2.47
Soy-liquid fish^c			2.40
Soybean meal, 44%	5-04-604	1.70	2.04
Soybean meal, 48%	5-04-612	1.80	
Meat meal with bone	5-00-388	1.64	
Poultry feather meal	5-03-795	0.97	

^{a}PER = gain/protein fed.
^{b}Mixture of 42% defatted soyflakes, 28% dehulled full-fat soybeans, and 30% catfish scrap (offal).
^{c}Mixture of 43% defatted soyflakes, 30% dehulled full-fat soybeans, and 27% liquid fish (catfish offal hydrolyzed by enzymatic process).
Source: (1) Lovell (1980); (2) Robinson et al. (1985).

search Council (1983), Robinette (1984), Dupree (1984), Robinson and Wilson (1985), Lovell (1989), and Robinson (1989).

FEEDSTUFFS

Relatively few different feedstuffs are used in catfish feeds, primarily because there is a limited number of feedstuffs that can provide the high level of nutrition required by catfish. No single feed ingredient can supply all the essential nutrients; thus a mixture of feedstuffs of animal and plant origin is used to meet nutritional requirements of catfish. Feedstuffs of animal origin are generally of higher quality for catfish than those of plant origin, as is indicated by protein efficiency ratios (Table 10.1) and by adequacy of animal proteins in meeting the indispensable amino acid requirements of catfish. For example, fish meal provides higher levels of lysine and sulfur amino acids than soybean meal or cottonseed meal (Table 10.2).

Criteria used to determine the suitability of a feedstuff for catfish include availability, palatability, digestibility, nutritional value, mill-

TABLE 10.2.

Available Essential Amino Acid Content of Soybean, Cottonseed, and Menhaden Fish Meal Proteins and Percentage of Dietary Requirement for Channel Catfish

Amino Acids	Soybean		Cottonseed		Fish Meal	
	Percentage of the Protein	Percentage of the Requirement for Catfish	Percentage of the Protein	Percentage of the Requirement for Catfish	Percentage of the Protein	Percentage of the Requirement for Catfish
Arginine	7.25	168	9.17	213	5.59	130
Histidine	2.18	142	1.87	121	2.01	130
Isoleucine	4.01	155	2.03	78	4.11	159
Leucine	6.35	181	3.90	111	6.53	186
Lysine	5.82	113	2.92	57	6.69	130
Methionine-cystine	2.52	108	1.86	80	3.15	135
Phenylalanine-tyrosine	7.08	141	6.01	120	6.30	126
Theonine	3.25	144	2.13	95	3.58	159
Tryptophan	1.18	219	0.86	160	0.91	168
Valine	4.09	138	2.94	99	4.59	155

Note: Underlined values are less than the requirement.
Source: Adapted from Lovell (1989).

ing properties, and cost. Feedstuffs may be classed into several categories based on their composition and their primary nutrients. The discussion presented here will be confined to ingredients that are classified as either protein or energy supplements.

Protein Supplements

Feedstuffs that contain 20 percent or more protein are classified as protein supplements. Generally, feed ingredients that contain high levels of protein are more expensive than those high in fats or carbohydrates. The following feedstuffs are most commonly used as protein supplements for catfish: soybean meal, cottonseed meal, peanut meal, fish meal, meat and bone meal, and blood meal. Catfish offal meal and full-fat soybean meal have been used sparingly in catfish feeds.

Soybean meal. Soybean meal is prepared by solvent extraction of ground full-fat soybeans to give a protein content of 44 percent or by dehulling and extracting to yield a meal containing 48 percent protein. Soybean meal is one of the highest-quality plant proteins for use in catfish feeds, because it contains relatively high levels of lysine and other essential amino acids (Table 10.2) and is palatable and digestible. It is the major source of protein used in commercial catfish feeds, making up from 40 to 60 percent of the feed. Soybean meal contains antinutritional factors, but these factors are usually destroyed by heat applied during processing of the meal. However, there are indications that additional heat may be beneficial in improving the nutritional value of soybean meal for young catfish.

Cottonseed meal. Cottonseed meal that is currently used in commercial catfish feeds is a solvent extracted meal containing 41 percent protein. Cottonseed meal is highly palatable to catfish; however, its use in feeds has been limited by low lysine content, high fiber content, and presence of free gossypol (a chemical found in pigment glands of cottonseed that can be toxic to animals). Catfish can tolerate up to 900 ppm free gossypol without detrimental effects. Higher levels depress growth and decrease feed efficiency. Relatively high levels of cottonseed meal can be used in catfish feeds without problems with gossypol toxicity. A more serious concern is the low level of available lysine found in cottonseed meal; only about 66 percent of the total lysine is biologically available to catfish. Low availability of lysine is due in part to the binding of lysine to free gossypol during processing of cottonseed meal. Thus if levels of cottonseed meal above about 20 percent are used in catfish feeds, available lysine becomes limiting. There is evidence that levels as high as 50 percent can be used, if the diet is supplemented with synthetic lysine. Presently, it is recommended that no more than 15 to 20 percent cottonseed meal be used in commercial catfish feeds.

Peanut meal. Peanut meal contains approximately 45 percent protein, up to 7 percent fat, and 10 to 13 percent fiber. Peanut meal is palatable to catfish, but it is used sparingly in feeds because of its low lysine content. About 15 percent peanut meal can be used in a typical commercial catfish feed without detrimental effects. Perhaps higher levels can be used if peanut meal is supplemented with lysine.

Fish meal. Fish meals are prepared by cooking whole fish or fish processing waste, removing the water and oil by pressing, and then drying. Fish meals prepared from whole fish are high-quality protein sources for catfish. They contain 60 to 80 percent protein, are high in digestible energy and certain minerals, and are highly palatable and digestible. Fish meals are high in lysine and methionine; thus they are excellent supplements for diets based primarily on plant proteins, which are generally deficient in lysine and methionine. Fish meals prepared from marine species are good sources of long-chain omega-3 fatty acids, which are essential for many fish and may be essential for catfish. Fish meals prepared from fish offal are also good nutrient sources but are of lower nutrient value than those prepared from whole fish. The quality of fish meal is dependent on processing. Overheating during processing can reduce nutrient availability. Commercial catfish feeds used for growout of advanced fingerlings to a harvestable size contain from 4 to 10 percent fish meal. In a properly balanced catfish growout feed, 4 percent fish meal appears to be sufficient. Catfish fry feeds contain about 50 percent fish meal.

Meat and bone meal. Meat and bone meal is an abundant animal protein source in the United States. It is defined as the dry rendered product derived from mammalian tissue, exclusive of hair, hoof, horn, manure, and stomach contents. It contains about 50 percent protein, which is of relatively good quality. Meat and bone meal contains considerably less lysine than fish meal, but it has a better balance of amino acids than most plant proteins, except perhaps soybean meal. It is a good source of calcium, phosphorus, and digestible energy for catfish. It may be variable in quality because of differing amounts of bone and tendon present. Also, some meat and bone products are blended with feather meal to increase the protein content of the meat and bone meal. The protein quality of feather meal is rather poor, and it is not digested by catfish unless it is hydrolyzed during processing. Meat and bone meal is high in ash, which limits its use in catfish feeds somewhat because of the possibility of mineral imbalances. Meat and bone meal is used in some catfish feeds at a rate of 4 percent to replace part of the fish meal. Levels as high as 15 percent have been used successfully in catfish feeds. A combination of meat and bone meal with blood meal is also used in catfish feeds to replace part of the fish meal.

Blood meal. Flash- or spray-dried blood meal contains 80 to 86 percent protein and is rich in lysine but deficient in methionine. It is used in catfish feed primarily to provide lysine. Blood meal is unpalatable to certain fish. When catfish are changed from a feed containing fish meal to one containing blood meal at moderate levels (8 percent or so), they are slow to consume the blood meal feed. However, after a day or so they adapt to the feed and readily consume the blood meal diet. Low levels (1 to 2 percent) of blood meal are used in some commercial catfish feeds.

Catfish offal meal. Catfish offal meal is prepared by cooking offal, pressing to remove water and oil, and then drying. The meal is approximately 50 percent protein and contains 8 to 15 percent fat. It is rich in calcium and phosphorus. Catfish offal meal is of better nutritional quality for catfish than meat and bone meal, but not as good as menhaden fish meal. Its nutritional quality may vary, depending on the composition of the raw products used. It has been used sparingly in commercial catfish feeds, primarily because of the lack of a consistent supply.

Full-fat soybeans. Full-fat soybeans are not used to any extent in commercial catfish feeds, primarily because of the high level of fat (18 percent) inherent in soybeans. Too much fat can cause an imbalance in the energy-to-protein ratio, which may reduce the intake of essential nutrients as well as increase fattiness in the fish. A limited amount of full-fat soybeans can be used in commercial catfish feeds as long as the total fat level in the diet does not exceed 5 to 6 percent.

Energy Feeds

Energy feedstuffs are defined as feedstuffs that are high in energy and contain less that 18 percent fiber and usually less than 20 percent protein. Common energy feedstuffs used in catfish feeds include grains and grain milling by-products, and fats and oils.

Grains and milling by-products. Grains and milling by-products are relatively low in protein (8 to 15 percent) and fat (4 to 12 percent) but high in carbohydrate, primarily starch (60 to 70 percent). Starch is relatively well digested by catfish (60 to 70 percent); cooking increases starch digestibility 10 to 15 percent. Starch is also important in feed manufacturing because it is useful as a binder and is essential for good expansion of extruded catfish feeds. The most common grain used in commercial catfish feeds is yellow corn or corn screenings. Corn is used in catfish feeds at levels of 25 to 40 percent. Corn gluten meal contains 40 to 60 percent protein and is rich in methionine, but it contains 200 to 350 ppm xanthophylls. If the level of xanthophylls in catfish feeds exceeds about 11 ppm of diet, an undesirable yellow color is imparted to the flesh. Thus the use of corn gluten meal in catfish feeds is restricted.

A small amount (2 percent or so) of wheat grain is sometimes used in extruded catfish feeds to improve pellet quality. Wheat is a good energy source for catfish and higher levels can be used, but corn is usually more economical. Wheat middlings are used at levels up to 20 percent in some catfish feeds. Middlings contain about twice the protein of corn but also contain more fiber and are generally more expensive. Rice bran is used at a level usually not exceeding 7 to 8 percent because of its relatively high content of fat and fiber.

Fats and oils. Fats and oils are used in catfish feeds primarily because they are a highly digestible source of energy. They also provide a source of essential fatty acids and are sprayed on the finished feed pellet to reduce feed dust ("fines"). Vegetable, animal, or a mixture of these fats or oils can be used in catfish feeds. Vegetable oils are generally more expensive than fat of animal origin. Menhaden fish oil, which is high in omega-three fatty acids, and catfish oil, which is similar to beef and poultry fat in fatty acid composition, can also be used in catfish feeds. Catfish oil (1 to 2 percent) is sprayed onto the finished catfish feed pellet throughout most of the growing season in Mississippi. During times of shortages of catfish oil, menhaden oil is mixed with the catfish oil to extend the supply. It is not clear whether the menhaden oil is beneficial in providing essential fatty acids or whether it improves feed palatability, as has been suggested.

FEED FORMULATION

To formulate cost-effective feeds the nutritionist must consider nutritional and nonnutritional factors. Nonnutritional factors such as economics, cultural practices, and feed manufacture are important considerations when formulating catfish feeds because all of these factors impact the selection of feed ingredients used to provide necessary nutrients. Traditionally, commercial catfish feeds have been based on fixed formulas without significant alteration of the formula irrespective of the price of ingredients. Other animal industries use a least-cost approach to feed formulation; that is, ingredient changes are based on cost. Regardless of the method used to formulate catfish feeds the nutritionist must consider nutrient requirements, nutrient composition and availability of various feed ingredients, and milling constraints.

Recommended Nutrient Levels

Nutrient levels recommended for production-type catfish feeds are summarized in Table 10.3. Feeds formulated on the basis of the recommendations provided are considered to be adequate for growth of 4- to 6-inch fingerlings to a harvestable size. As discussed in Chapter 5, several factors influence the desired nutrient density of catfish feeds. Theoretically, nutrient levels should be adjusted to compensate for the impact of these factors. For example, it would be desirable to alter dietary protein levels as the fish grow or to provide specialized feeds to counter various environmental stresses. However, presently the logistics involved with catfish feed manufacture prevent the manufacture and distribution of a number of different feeds. Also, catfish producers are not equipped to store a variety of feeds. In addition, little information is available concerning the interrelationship between nutrition and environmental factors. As the catfish industry becomes more sophisticated and more nutritional information becomes available, it may be possible to formulate feeds to meet specific nutritional requirements of fish reared under various culture conditions.

Several of the nutrient levels suggested for catfish feeds (Table 10.3) exceed the actual nutrient requirement. Excesses of labile nutrients are necessary to ensure adequate levels in processed feeds. Vitamins are particularly sensitive to the heat and moisture encountered during catfish feed manufacture (Table 10.4). About 60 percent of vitamin C added to extrusion-processed catfish feeds is destroyed. Losses of vitamin C are also fairly high during storage. For example, loss of vitamin C in extruded catfish feeds stored in polyethylene bags at 77°F is as high as 70 percent after 14 days. Stable forms of vitamin C are currently available and others are being developed. The use of the stable vitamin C products in commercial catfish feeds depends on their biological activity and on economics.

Least-Cost Feeds

Least-cost computer formulation of feeds for livestock has been in general use for several years. To utilize computer programs to formulate cost-effective feeds effectively, information on the cost of feed ingredients, nutrient content in feedstuffs, nutrient requirements, nutrient availability from feedstuffs, and restrictions on the levels of various ingredients is needed. Computer feed formulation for catfish was not feasible until recently because of the lack of the necessary nutritional information. Although such information is available, least-cost feed formulation of catfish feeds is not a widespread practice, because few feedstuffs can be substituted and there is a lack of knowledge concerning nutrient levels that result in maximum profit as opposed to those that maximize weight gain. In addition, catfish feed mills generally do not have the storage capacity to store the number of ingredients needed for least-cost formulation. Also, the logistics of obtaining a wide assortment of feedstuffs on a timely basis is often difficult. During peak feeding periods, the turnover time for major ingredients in large catfish feed mills may be as short as 2 days.

Limitations to least-cost catfish feed formulation exist, but a modified method of formulating catfish feeds on cost basis is used. For example, cottonseed meal, milo, and meat and bone meal are used to replace a part of soybean meal, corn, and fish meal, respectively, depending on cost.

Mathematical (linear) programming is the method most often used for computer formulation of least-cost animal feeds. This method requires that restrictions be set and that the least-cost combinations of ingredients are selected to meet the restrictions. Some restrictions are nutritional; others concern processing or milling, inherent problems with feedstuffs, and miscellaneous factors.

Minimum and maximum levels are set for several nutrients, including protein and certain amino acids, available phosphorus, and digestible energy (Table 10.5). It is not necessary to set restrictions on protein level (although a minimum is usually set), if amino acid requirements are met. For catfish, it is not necessary to set restrictions on amino acids except for lysine and the sulfur-containing amino acids, methionine and cystine. If the requirements for those amino acids, which are usually limiting in plant feed ingredients, are met, then other indispensable amino acid requirements are satisfied. Phosphorus restrictions are expressed on an available basis, because from one-third to one-half of the phosphorus in feed ingredients is not available to catfish. Digestible energy is also restricted to maintain a balance between dietary energy and protein.

Some restrictions are concessions to feed processing. Adequate amounts of starch must be provided to manufacture a floating feed; thus at least 25 percent corn or other grains or grain by-products should be included. High levels of fat and fiber interfere with the manufacturing process, so they are limited. Also, sinking feeds generally require addition of a binder to improve the water stability of the feed pellet.

TABLE 10.3.
Recommended Nutrient Levels for Catfish Feeds Used for Growout of Advanced Fingerlings to a Market Size

Nutrient	Recommended Level	Units	Comments
Protein	32	% of feed	May vary, depending on numerous factors. See discussion in protein section. Primary sources are soybean meal and menhaden fish meal.
Indispensable amino acids:			
Lysine	5.1	% of protein	Meet requirements presented in Table 10.2. If lysine and sulfur amino acid requirements are met, other amino acids will be adequate using feedstuffs commonly used in catfish feeds.
Methionine + cystine	2.3	% of protein	
Energy, digestible	8–10	kcal/g protein	Use carbohydrate and lipid as energy sources to spare protein for growth.
Lipid	≤6	% of feed	Optimal level is not well defined. Need enough to supply EFA. Consider effects on product quality and constraints of feed manufacture. Mixture of animal and vegetable fats, catfish oil, or other fish oil may be used. High levels of marine fish oil may impart a "fishy" flavor to the flesh.
Carbohydrate	25–35	% of feed	No requirement per se. Supply in form of grains. Floating feeds require approximately 25% grain for binding. Crude fiber levels should be <6–8%.
Vitamins[a]			
Thiamin	11	ppm	Thiamin mononitrate may be used as supplement.
Riboflavin	13.2	ppm	
Pyridoxine	11	ppm	Pyridoxine HCl is generally used as supplement.
Pantothenic acid	35	ppm	Calcium d-pantothenate used because it is stable.
Nicotinic acid	88	ppm	Either nicotinic acid or nicotinamide may be used; both are stable and have equivalent biological activity.
Folic acid	2.2	ppm	Actual requirement not known.

Nutrient	Amount	Unit	Comments
B_{12}	0.01	ppm	Synthesized in catfish intestine in presence of cobalt.
Choline	275	ppm	Abundant in natural feed ingredients but biological availability not known.
Ascorbic acid	375.6	ppm	Particularly sensitive to destruction during feed manufacture; use stable form to reduce loss during manufacture. Higher levels may be beneficial under certain conditions.
A	2,000 (4,400)	IU/lb (IU/kg)	Acetate ester is generally used because it is more stable during processing. A gelatin-sugar-starch coating also improves stability.
D_3	1,000 (2,200)	IU/lb (IU/kg)	D-activated animal sterol used as source of D_3. Gelatin-sugar-starch beadlet used to improve stability.
E	30 (66)	IU/lb (IU/kg)	DL-α-tocopherol acetate is form generally used because esterification improves stability.
K	4.4	ppm	Menadione sodium bisulfite is form used.
Minerals			
Phosphorus, available	0.5	%	Plant phosphorus approximately 30% available; animal phosphorus approximately 40% available. Dicalcium phosphate is used as supplemental source of phosphorus and is about 80% available.
Magnesium	0.05	%	No supplements added; abundant in natural feedstuffs.
Zinc	200	ppm	Phytic acid in feed reduces zinc availability. ZnO is used as a supplement.
Selenium	0.1	ppm	Until recently this was maximum allowable by FDA. Presently, up to 0.3 mg/kg may be used.
Manganese	25	ppm	Phytic acid can reduce manganese availability. Manganese oxide is used as supplement.
Iodine	2.4	ppm	Requirement not established. Calcium iodate used as supplement.
Iron	30	ppm	Ferrous sulfate and ferrous carbonate used as iron supplements.
Copper	5	ppm	Copper sulfate used as supplement.
Cobalt	0.05	ppm	Cobalt carbonate used as supplement.

^aRecommended amounts of vitamins to be added in extrusion processed feeds.
Source: Adapted from Robinson (1989).

TABLE 10.4.

Stability of Vitamins Added in Extrusion Processed Catfish Feed

Vitamin	Percentage Recovery in Extruded Feed
Vitamin A acetate (in gelatin-starch beadlet)	65
Thiamin mononitrate	64
Pyridoxine hydrochloride	67
Folic acid	91
DL-α-tocopherol acetate	100
Ascorbic acid (ethylcellulose coated)	43 (37)[a]

[a]Average of assays of feed from three mills located in Mississippi Delta. Assays conducted at Mississippi State University, Delta Branch Experiment Station.
Source: Lovell (1989).

TABLE 10.5.

Restrictions for Least-Cost Formulation for Production Feed for Catfish

Qualifier	Restriction	Amount	Unit
Crude protein	Minimum	32.0	%
Crude fiber	Maximum	7.0	%
Lipid	Maximum	6.0	%
Available phosphorus	Minimum	0.5	%
Available phosphorus	Maximum	0.7	%
Digestible energy	Minimum	2.8	kcal/g
Digestible energy	Maximum	3.0	kcal/g
Available lysine	Minimum	1.63	%
Available methionine	Minimum	0.30	%
Available methionine and cystine	Minimum	0.74	%
Grain or grain by-products	Minimum	25.0	%
Cottonseed meal	Maximum	15.0	%
Whole fish meal	Minimum	4.0	%
Nonfish animal protein	Maximum	4.0	%
Xanthophylls	Maximum	11.0	ppm
Vitamin premix[a]	Include		
Trace mineral premix[a]	Include		

[a]Should provide levels given in Table 10.3.
Note: Percentages are expressed as a percentage of diet.
Source: Adapted from Robinette (1984); Lovell (1989); Robinson (1989).

TABLE 10.6.

Model Feed Formulation Containing 48 Percent Protein for Feeding Catfish Fry

Ingredient	International Feed Number	Percentage
Fish, menhaden meal	5-02-009	60.0
Meat meal with bone	5-00-388	10.0
Blood, spray dried	5-00-381	5.0
Wheat, middlings	4-05-205	19.6
Fat		5.0
Vitamin premix[a]		Include
Trace mineral premix[b]		Include

[a]Should contain excess vitamins (three times the recommendations given in Table 10.3) because of solubilization of vitamins in water.
[b]Should meet levels recommended in Table 10.3; actual amount to be added depends on premix used.

Certain feedstuffs contain substances that inhibit growth or can be toxic if sufficient amounts are included in the diet. For example, cottonseed meal is limited somewhat by the direct or indirect effects of free gossypol.

Miscellaneous restrictions may be imposed by the feed mill or by its clientele. For example, some catfish producers insist on a certain level of fish meal or other ingredients.

Quadratic programming has been used to formulate poultry feeds and may offer certain advantages over linear programming, since it considers diminishing productivity with increasing nutrient input and changes in value of product. The use of quadratic programming requires knowledge of the biological response from feeding trials, that is, the regression of weight gain on dietary nutrient concentration. Insufficient information is available to use the quadratic programming approach for the formulation of catfish feeds correctly.

Model catfish feeds are given in Tables 10.6, 10.7, and 10.8. The feeds depicted are complete feeds that provide all known nutrient requirements for the catfish; they are representative of those used for commercial catfish culture.

Physical Characteristics

Catfish feeds must be water-stable; as a result, feedstuffs that have binding properties must be selected or pellet binders must be added.

TABLE 10.7.

Model Feed Formulation Containing 36 Percent Protein for Feeding Small Catfish Fingerlings

Ingredient	International Feed Number	Percentage
Fish, menhaden meal[a]	5-02-009	12.0
Soybean meal (48%)[b]	5-04-612	54.5
Corn, grain[c]	4-02-935	30.8
Fat[d]		1.5
Dicalcium phosphate		1.0
Vitamin premix[e]		Include
Trace mineral premix[e]		Include

[a]Other types of fish meal can be used.
[b]15% Cottonseed meal can be substituted for soybean meal on a nitrogen basis.
[c]Other grains or grain by-products can be used.
[d]Catfish oil or other animal and plant fats can be used.
[e]Should meet levels recommended in Table 10.3; actual amount depends on premix used.

TABLE 10.8.

Three Model Feed Formulations Containing 32 Percent Protein for Growout of Advanced Fingerlings to Market Size

Ingredient	International Feed Number	Percentage (1)	Percentage (2)	Percentage (3)
Fish, menhaden meal	5-02-009	8.0	4.0	—
Meat meal with bone	5-00-388	—	4.0	15.0
Soybean meal (48%)	5-04-612	50.0	37.0	—
Soybean meal (44%)	5-04-604	—	—	47.5
Cottonseed meal (41%)	5-01-621	—	15.0	—
Corn, grain[a]	4-02-935	34.2	33.3	33.4
Rice, bran	4-03-928	5.0	—	3.25
Wheat, middlings	4-05-205	—	4.0	—
Dicalcium phosphate	6-01-080	1.1	1.0	1.0
Fat (sprayed on finished feed)		1.5	1.5	—
Vitamin premix[b]		Include	Include	Include
Trace mineral premix[b]		Include	Include	Include

[a]Other grains or grain by-products may be used.
[b]Should meet recommendations in Table 10.3; actual amount to be added depends on premix used.
Note: Formulations for extruded feeds (floating). If pelleted feed (sinking) is desired, a pellet binder must be added.

Extruded (floating feeds) must contain an appreciable amount of starch for suitable binding and good floatability. It is usually not necessary to add a binder to floating feeds; however, some manufacturers use a nutritive binder to reduce feed dust (fines). The manufacture of a sinking feed (hard pelleted feed) requires the use of a binder other than those naturally present in feedstuffs. About 2 to 3 percent of lignin sulfonate, hemicelluloses, or other binders are used. Sinking catfish feeds remain stable in water for 15 to 20 minutes, which is generally enough time for the fish to consume the feed before the pellets disintegrate. Floating catfish feeds may remain stable in water for several hours. A "slow-sink" extruded feed is generally used for winter feeding of catfish.

FEED PROCESSING

Feed processing involves the physical and chemical processes that are essential for manufacture of a feed: grinding, mixing, agglomeration, and forming of ingredients into homogenous pellets that are suitable for catfish. Catfish feeds used for growout of fingerlings to harvestable-size fish are either steam-pelleted into sinking pellets or extrusion-processed into floating pellets.

Steam Pelleting

Steam-pelleted feeds are manufactured by using moisture, heat, and pressure to form ground feed ingredients into larger homogenous feed particles. Steam is added to the ground feed ingredients ("mash"), to increase the moisture level to 15 to 18 percent and temperature to 160 to 185°F. The addition of steam to the feed "mash" gelatinizes the starches, which bind the feed particles together. The hot "mash" is pressed through the pellet die. Feed pellets are then dried to a moisture level of about 10 to 12 percent.

Binding agents such as lignosulfonates, bentonites, and certain cellulose derivatives are nonnutritive binders that are often used to improve the water stability of catfish feeds and decrease the amount of feed dust. Binders that have nutritive value (e.g., specially processed milo) are also used. Steam-pelleted catfish feeds generally are stable in water for only 15 to 20 minutes, depending on the amount of binder used. Feed mixtures containing high levels of fat or fiber are more difficult to pellet. If needed, fat should be added after pelleting and highly

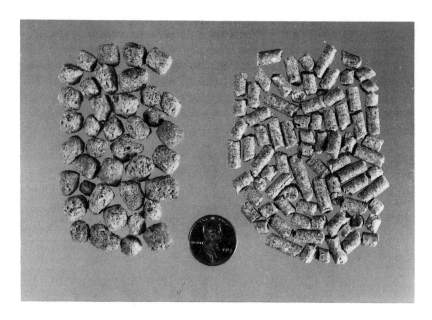

FIGURE 10.1.

An extruded floating feed (left) and a pelleted sinking feed (right).

fibrous feedstuffs should not be used in large quantities. Manufacture of steam-pelleted feeds requires less energy than manufacture of extruded feeds, they are less expensive, and less nutrient destruction occurs during feed processing. A typical pelleted catfish feed is shown in Figure 10.1.

Extrusion

Extrusion requires higher levels of moisture, heat, and pressure than steam pelleting for proper gelatinization of starch. Feed ingredients are finely ground, mixed, and treated with steam to a moisture of about 25 percent to form a "mash." Depending on the type of extruder used, the mash may be cooked before extrusion. The mash is heated to about 200 to 300°F under pressure in the extruder barrel. The superheated material is then forced through the die openings at the end of the barrel; an immediate reduction in pressure results in vaporization of part of the water in the material and expansion occurs. Extruded feeds contain more moisture than steam-pelleted feeds and thus require more drying. Extruded feeds are stable in water for several hours, are more expen-

Feed Processing **307**

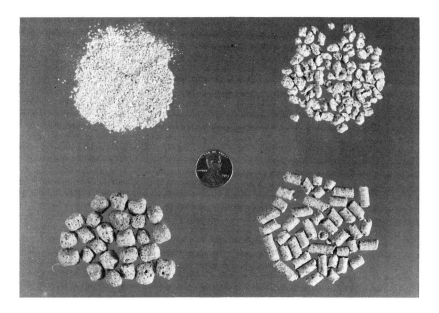

FIGURE 10.2.

Various types of catfish feed: a meal type for fry (upper left), "crumbles" for advanced fry and small fingerlings (upper right), and extruded and pelleted feeds for large fingerlings and adults.

sive, and have higher nutrient losses than those produced by steam pelleting. A typical extruded feed is shown in Figure 10.1.

Preparation of Meals and Crumbles

Small catfish require small particle feeds such as flours, meals, or crumbles (Fig. 10.2). Flour- or meal-type feeds are generally prepared by one of two methods: by reducing the particle size of pelleted feeds by grinding and then screening or by finely grinding feed ingredients to a particle size of less than 0.02 inch (0.5 millimeter) and mixing the ground ingredients to make the flour or meal. Crumbles are prepared by crumbling (crushing) feed pellets and screening for proper size. If flour or meal feeds are not pelleted, water-soluble nutrients are more likely to be lost to the water when the feed is fed. Supplemental fat sprayed on the surface of meal or crumbled feed improves water stability and floatability and reduces nutrient losses to the water.

FEEDING PRACTICES

Feeding catfish is more of an art than a science; thus an experienced feeder that is able to optimize consumption without excessive feed wastage is invaluable to the catfish producer. Feeding the most suitable feed in a manner that results in fast growth and efficient feed conversion results in more efficient production and increased profit. Uneaten feed costs the producer directly in decreased profits and indirectly in degradation of water quality. Since about one-half of the variable production cost in catfish production is feed cost, the catfish producer can reduce the overall cost of production by using the most appropriate feed for the specific system being used for culture and feeding it in the manner that most efficiently produces fish. To achieve this goal, the producer must be aware of the type of feed needed for various stages of growth as well as how much, when, and how to feed.

Feed Types

Most cultured catfish from fingerling to market size are fed a floating feed. Sinking feeds represent a small percentage of catfish feed. The floating feed is preferred by the catfish producer because of its management value. The feeder can observe the fish for signs of disease and general well-being. Floating feeds remain at the surface for considerable periods of time and thus are not subject to loss in bottom muds as is the sinking feed. Sinking feeds also disintegrate quickly in the water. If the fish are feeding slowly, considerable feed waste can occur when sinking feeds are used.

Feed particle size should be increased as fish increase in size (Table 10.9). Fry and fingerling catfish are fed finely ground meals or larger particle-size crumbles. Small fingerlings may be fed a crumble or a small-size floating pellet. Larger stocking-size fingerlings are generally fed a larger floating pellet throughout the growing season.

Two medicated feeds are available for use in treating bacterial diseases in catfish. Terramycin® and Romet® are approved for use on food fish. Terramycin® is generally available as a sinking pellet (some feed manufacturers spray Terramycin® on the outside of floating feed pellets), and Romet® is available as a floating pellet.

Feed Allowances

Feeding allowances for catfish may be expressed as a percentage of body weight or as pounds of feed fed/acre/day. Feeding is most affected

TABLE 10.9.
Optimal Feed Particle Size for Catfish

Fish Size (Inches)	Particle Size	Maximum Diameter (Millimeters)
Fry	Starter	0.5
1/2–1	No. 1	0.8
1–1 1/2	No. 2	1.2
1 1/2–2 1/2	No. 3	2.0
2 1/2–4	No. 4	3.3
4–6	3/32–1/8 inch	3.2
6 and larger	3/16-inch pellet	4.8

Source: Adapted from National Research Council (1983).

by fish size and water temperature. Smaller fish eat more feed in relation to their size and feed more frequently than larger fish. All fish eat less feed at colder water temperatures. Suggested feed allowances and feeding frequencies for catfish are presented in Tables 10.10 and 10.11.

If feeding allowance is based on a percentage of body weight, the amount of feed that should be fed changes daily. However, it is inconvenient (even with computer programs) to adjust feed allowance daily. Feed allowance should be adjusted weekly if one chooses to use a percentage of body weight. This may be accomplished by knowing the weight of fish at stocking and either assuming a feed conversion ratio (2.0 pounds of feed/1 pound of gain is a good assumption for fish reared on commercial catfish farms) or actually sampling the fish and basing biomass on sample weight. Although the second method is probably more accurate, sampling is problematic in that it is time-consuming and may stress the fish. To calculate the feed allowance using the information in Tables 10.10 and 10.11 the average fish weight is needed. Simply divide the total weight at stocking by the total number of fish stocked. Use that table to find the percentage body weight recommended for feeding fish of that size. Multiply the percentage body weight by the total weight of fish stocked to determine the amount that should be fed on that day. A new estimate should be calculated weekly on the basis of estimated growth. Using an assumed feed conversion ratio, estimate growth by dividing the total amount of feed (pounds) fed during the first week by the estimated feed conversion ratio. This will give the theoretical gain for the previous week. This gain should be added to the initial weight of fish to provide the theoretical weight to be used for calculating a new feeding allowance. Repeat the proce-

TABLE 10.10.

Daily Feed Allowances for Channel Catfish in Ponds in the Southeastern United States, April 15 to October 15

Water Temperature[a] (°F)	Fish Size (Pounds)	Feed Allowance/Day: Percentage of Fish Weight
68	0.04	2.0
72	0.07	2.5
77	0.11	2.8
80	0.15	3.0
82	0.22	3.0
84	0.92	3.0
85	0.35	2.8
85	0.42	2.5
86	0.59	2.2
86	0.75	1.8
82	0.90	1.6
79	1.00	1.4
73	1.10	1.1

[a] Mean temperature at 3-foot depth at feeding time.
Source: Adapted from Lovell (1989).

TABLE 10.11.

Suggested Maximum Feeding Rates and Feeding Frequencies at Different Water Temperatures

Water Temperature (°F)	Fry or Fingerlings		Food Size Fish	
	Feeding Frequency	Feeding Rate Percentage (Daily)[a]	Feeding Frequency	Feeding Rate Percentage (Daily)[a]
87 and above	2 times/day	2	1 time/day	1
80–86	4 times/day	6	2 times/day	3
68–79	2 times/day	3	1 time/day	2
58–67	1 time/day	2	1 time/day	2
50–57	Alternate days	2	Alternate days	1
49 and below	3rd to 4th day	1	3rd to 4th day	1/2

[a] Percentage of body weight.
Source: Dupree (1984).

dure for each period that feed allowance is adjusted. The following equations may be useful in adjusting feed allowance:

$$\text{Initial average fish weight} = \frac{\text{weight of fish stocked}}{\text{number of fish stocked}} \quad (10.1)$$

$$\text{feed allowance} = \frac{(\% \text{ from Table 10.10 or 10.11}) \times (\text{total fish weight})}{100} \quad (10.2)$$

$$\text{Theoretical weight gain} = \frac{\text{total pounds of feed fed}}{\text{feed conversion ratio}} \quad (10.3)$$

$$\text{Average theoretical fish weight} = \frac{\text{theoretical gain} + \text{previous weight}}{\text{total number of fish}} \quad (10.4)$$

As an example, assume that you stock 5,000 fish weighing 550 pounds. Initial average fish weight can be calculated:

$$\text{Initial average fish weight} = \frac{550}{5,000} = 0.11 \text{ pound} \quad (10.5)$$

According to the tables, feed allowance for fish of that size at a water temperature of 68°F (assuming that is your water temperature at stocking) is about 3 percent of body weight. Feed allowance can be calculated:

$$\text{Feed allowance} = \frac{3 \times 550}{100} = 16.5 \text{ pounds} \quad (10.6)$$

Thus on day one, 16.5 pounds of feed should be offered. Assuming daily feeding and feed allowance adjusted weekly, the total amount of feed for a 7-day period will be 115.5 pounds (16.5 × 7). Assuming a feed conversion ratio of 2:1, then theoretical weight gain can be calculated:

$$\text{Theoretical weight gain} = \frac{115.5}{2.0} = 57.75 \text{ pounds} \quad (\text{round to 58}) \quad (10.7)$$

Average theoretical fish weight is then calculated:

$$\text{Average theoretical fish weight} = \frac{58 + 550}{5,000} = 0.12 \text{ pound} \quad (10.8)$$

Total number of fish used to calculate average fish weight should be adjusted for mortalities (if known).

Feed allowance recommendations should be considered as guidelines, because the amount of feed that the pond can "metabolize" effectively influences feed allowance. Thus each producer must make decisions about feeding fish in individual ponds based on feed consumption and water quality. It is important to offer enough feed to attempt to satiate all fish without overfeeding. Uneaten feed results in waste and can be detrimental to water quality. Underfeeding results in greater size variation in harvested fish, because the more aggressive fish (usually the larger fish) consume a greater share of the feed.

Feeding Methods

Feeding methods differ for fry, fingerlings, food fish, and brood fish. Generally, the method used to feed at each stage of development should provide ample feed to supply essential nutrients and energy necessary for rapid growth, good feed conversion, and maintenance of good health or, in the case of brood fish, good reproductive capacity.

Feeding fry. Channel catfish fry are cultured inside in containers and outside in ponds. Culture in indoor troughs and tanks is usually for only 5 to 10 days and is used primarily to protect fry from predation but also to allow the producer to stock larger and perhaps hardier fry into nursery ponds. The fry are stocked in nursery ponds and grown to suitable size for stocking into growout ponds.

Fry live on nutrients stored in their yolk sac for 5 to 10 days, at which time the yolk sac is absorbed. The fry darken in color and swim to the water surface and appear to be seeking feed. Fry should be fed a finely ground feed containing 45 to 50 percent protein, which should be primarily from fish meal (Table 10.6). Frequent feeding is desired; and eight to ten feedings over a 24-hour period is not excessive. Fry are generally fed more than they can consume (25 percent body weight or more) to ensure that they are satiated. Excess feed must be removed from the tanks and troughs daily to prevent fouling and deterioration of water quality.

Fry stocked into earthen ponds should also be fed. The feed should be spread over a wide area to encourage consumption. Since the fry are not visible for several weeks until they reach a size of 1 or 2 inches, it is difficult to determine whether they are actually consuming the feed. The feed may simply serve as a fertilizer, thus stimulating production of natural food organisms. Natural food organisms are presumed to be the primary source of nutrition to catfish fry until they are capable of utilizing crumbled or pelleted feeds.

Feeding fingerlings. Techniques for feeding fingerlings stocked at relatively high densities are similar to those used for feeding fry in ponds. A 36 percent protein feed containing about 12 percent fish meal is adequate for feeding fingerlings (Table 10.7). The fish should initially be fed a crumble, and as they increase in size feed particle size should be increased accordingly. In practice, some producers feed fingerlings the same feed as they feed fish for growout, assuming that as the larger pellets soften and begin to break up in the water the fingerlings will be able to consume the feed. Although there is no experimental evidence, it is likely that fingerlings fed in that manner are more variable in size than fish fed different sizes of feed as they grow.

Feeding advanced fingerlings to harvestable-size fish. Typically, a 32 percent protein floating feed is given to large fingerlings to harvest. Sinking feeds can be used, but management is more difficult. Most catfish producers prefer a floating feed. The fish should be fed in the morning after dissolved oxygen levels have started to rise and no later than midafternoon, to allow digestion to occur during periods of relatively high dissolved oxygen. Fish of all sizes do not consume and assimilate feed efficiently when the oxygen levels are low.

On large commercial catfish farms, the feed is scattered over a wide area of the pond by using mechanical feeders that are either mounted on or pulled by vehicles. It is desirable to feed on all sides of the pond to increase opportunity for all fish to feed, but this may not be practical on large farms where fish in many ponds must be fed in a limited time. Also, under breezy conditions apply feed only along the upwind shore to prevent feed from washing ashore too quickly.

Feeding rates are affected by numerous factors, including standing crop, water quality, and water temperature. In the multiple stocking–multiple harvest program employed by many catfish producers the numbers of fish and the standing crop are usually high and fish size is quite variable. Thus to attempt to ensure adequate feed for all fish, the producer must offer ample feed but should be careful to add only as much as can be "metabolized" by the biological system. The amount of feed fed may be as high as 200 or more pounds/acre/day, but most catfish producers limit feed to 100 to 150 pounds/acre/day, and most feed once daily. Feeding twice daily is beneficial, but on large commercial catfish farms it is virtually impossible.

Catfish producers using cages, net pens, and raceways generally offer the amount of feed that can be consumed within 10 to 15 minutes. They may feed more than once a day and benefit from considerably faster growth rate and improved feed conversion.

Feeding brood fish. Feeding techniques recommended for feeding advanced fingerlings are also applicable to brood fish. Brood fish are normally fed the same feed used for growout of advanced finger-

TABLE 10.12.

Winter Feeding Schedule for Catfish

Temperature (°F)	Adults (1/2 Pound and Above)		Fingerlings	
	Percentage of Body Weight	Frequency	Percentage of Body Weight	Frequency
45–50	0.5	Weekly	0.5	3 days/week
51–55	1.0	2 days/week	1.0	Every other day
56–60	1.0	Every other day	1.0	Daily
61–65	1.5	Every other day	2.0	Daily
66–70	2.0	Every other day	2.5	Daily

lings, which appears sufficient to meet the needs of the brood fish. They should be fed at a rate of 1 to 2 percent of body weight daily. They may be fed a sinking or floating feed. Brood fish generally feed slowly, and if a sinking feed is used the pellets break up and may not be consumed. Although brood fish feed at the surface, some catfish producers prefer sinking feeds even though the water stability of sinking feeds is generally poor. Some producers provide forage fish for their brood fish.

Winter feeding. Feed consumption by catfish is directly related to water temperature. Optimal temperature for culturing catfish is about 86°F. As the water temperature decreases consumption declines. Feeding is inconsistent below about 70°F, and although catfish feed at temperatures as low as 50°F consumption is greatly reduced. However, some form of winter feeding appears to be beneficial to prevent weight loss and maintain health. Winter feeding of catfish is practiced by many producers, but the activity is often restricted because some catfish producers cannot drive on their pond levees during wet weather.

Several schedules for winter feeding have been suggested; generally, water temperature dictates feeding frequency. Typical winter feeding schedules are given in Tables 10.11 and 10.12.

REFERENCES

Dupree, H. K. 1984. Feeding practices. In *Nutrition and Feeding of Channel Catfish* (revised), Southern Cooperative Series Bulletin No. 296, eds. E. H. Robinson

References

and R. T. Lovell, pp. 34–40. College Station: Texas Agricultural Experiment Station.

Lovell, R. T. 1980. *Utilization of Catfish Processing Waste*, Bulletin 521. Auburn: Alabama Agricultural Experiment Station.

Lovell, R. T. 1989. *Nutrition and Feeding of Fish*. New York: Van Nostrand Reinhold.

National Research Council. 1983. *Nutrient Requirements of Warmwater Fishes and Shellfishes*. Washington D.C.: National Academy of Sciences.

Robinette, H. R. 1984. Feed formulation and processing. In *Nutrition and Feeding of Channel Catfish* (revised), Southern Cooperative Series Bulletin No. 296, eds. E. H. Robinson and R. T. Lovell, pp. 29–33. College Station: Texas Agricultural Experiment Station.

Robinson, E. H. 1989. Channel catfish nutrition. *Reviews in Aquatic Sciences* 1:365–391.

Robinson, E. H., and R. P. Wilson. 1985. Nutrition and feeding. In *Channel Catfish Culture*, ed. C. S. Tucker, pp. 323–404. Amsterdam: Elsevier.

Robinson, E. H., J. K. Miller, V. M. Vergara, and G. A. Ducharme. 1985. Evaluation of dry extrusion-cooked protein mixes as replacements for soybean meal and fish meal in catfish diets. *Progressive Fish-Culturist* 48:233–237.

CHAPTER 11
Infectious Diseases

Diseases of channel catfish are classified as noninfectious and infectious. Noninfectious diseases include nutritional disorders (Chapter 5) and toxicoses related to suboptimal environmental conditions (Chapter 4). Infectious diseases are caused by living organisms that parasitize or infect fish. Infectious diseases of channel catfish are caused by viruses, bacteria, fungi, protozoans, and metazoans.

The economic cost of infectious diseases to catfish producers is difficult to determine. The most obvious result of infectious disease is death, which represents a direct loss of investment. Making accurate estimates of the number of fish lost during disease outbreaks (called *epizootics*) is nearly impossible on large farms (Fig. 11.1). Furthermore, many epizootics are not reported to government or university laboratories because some farmers attempt to diagnose and treat the problem without consultation. Economic losses also occur as a result of treatment expenses and because diseased fish grow more slowly and convert feed inefficiently. Rough estimates of the economic impact of infectious diseases run from about 2 to over 6 percent of the farm gate value of the crop. These estimates are based on anecdotal reports. In terms of 1988 production and prices, this would amount to losses ranging from $5 million to over $20 million.

Regardless of the exact economic costs, farmers consider infectious diseases to be one of the major problems faced during production. Major losses to noninfectious diseases occur but are largely preventable

FIGURE 11.1.

Infectious diseases can cause large losses of fish. Making reliable estimates of the number of fish lost is difficult during fish kills like this one. (Photo: Mississippi Cooperative Extension Service)

with good management. For instance, loss of fish to dissolved oxygen depletion is a constant threat in raising fish in ponds, but if adequate aeration equipment is available and dissolved oxygen concentrations are monitored vigilantly, major losses occur only under extraordinary circumstances such as electrical power outages. In contrast, losses to infectious diseases are common in raising channel catfish, although various management practices can reduce the occurrence of disease and early diagnosis and proper treatment can often reduce losses.

Diagnosis of infectious diseases requires laboratory tests. The facilities and equipment required to conduct these tests are usually available only at laboratories specializing in fish disease diagnostics. Furthermore, interrelationships among environmental factors and various infectious diseases are often complex, and considerable experience and expertise are required to interpret test results and plan an efficacious treatment regimen. For these reasons, diagnosis of infectious diseases (including treatment recommendations) should be made by a qualified fish health specialist. However, all commercial catfish farmers should have a basic understanding of the major infectious diseases of channel catfish. A knowledgeable producer can aid in the diagnostic procedure by supplying pertinent information on environmental conditions, fish behavior, recent management activities, and other facts concerning the disease outbreak. Preventative measures, early detection, and effective implementation of treatments are also keys to minimizing economic losses to infectious diseases, and these aspects of disease management are the responsibility of the producer. A precise diagnosis by a fish health specialist is meaningless if the disease has already run its course or if the treatment is improperly administered.

The practical aspects of the major infectious diseases of channel catfish will be discussed in the remainder of this chapter. Readers interested in more detailed discussions of these diseases or information on diseases of lesser importance should consult MacMillan (1985) or Plumb (1985a).

ROLE OF ENVIRONMENTAL CONDITIONS

The mere presence of an infectious disease organism (pathogen) is often not sufficient to induce a diseased state in channel catfish because the natural disease defense system of the fish (see the section Immune Function) usually acts to prevent disease. Diseases occur when environmental conditions interact with the fish and pathogen. This relationship is illustrated in Figure 11.2. The environment can interact and lead to disease outbreaks in three ways: (1) environmental conditions, particularly water temperature, can affect the pathogenicity (ability to produce a disease) of the disease organism; (2) some pathogens are more abundant in the water under certain environmental conditions; and (3) exposure of fish to suboptimal environmental conditions can decrease their immunocompetence (ability to resist diseases).

Each fish disease organism has an optimal temperature range for growth and virulence. However, development of the disease (pathogenesis) and its outcome are related not only to the effects of temperature on the pathogen but also to the effect of temperature on the immune system of the fish. Farmers should be aware that certain diseases are more prevalent and cause higher rates of mortality at characteristic temperatures. This causes a definite seasonality of occurrence of each disease for pond-raised fish. Water temperatures cannot be controlled in ponds, but alteration of water temperature can be an effective method of controlling some diseases of catfish cultured in closed, water-recirculating systems.

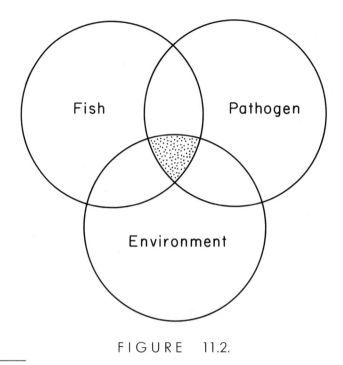

FIGURE 11.2.

Representation of the interaction among environment, pathogen, and host necessary to produce a diseased state. Diseases occur only under a certain combination of factors (stippled area).

Several important catfish disease organisms are commonly found free-living in the water and are called *facultative disease organisms*. Facultative disease organisms normally do not cause diseases unless the fish's defense system is compromised. Examples of facultative disease organisms include the bacteria *Aeromonas hydrophila* and *Flexibacter columnaris* and fungi such as *Saprolegnia* spp. When free-living, these organisms use dissolved organic compounds, organic detritus, or other living or dead microorganisms as food sources. Consequently, facultative disease organisms are usually more abundant in nutrient-rich, organic matter–laden waters than under less eutrophic conditions. Most waters used to raise catfish contain high concentrations of dissolved and particulate organic matter, and facultative disease organisms are always present. Warm water temperatures (>70°F) also favor the presence and growth in water of facultative bacterial and protozoan pathogens. Fungal pathogens grow better at lower water temperatures than do bacteria and are relatively more abundant in cooler waters.

The most important aspect of the interaction among environment, fish, and pathogen is the effect of adverse environmental conditions on the immunocompetence of the fish. Fish that are fed well, uncrowded, and living in a good environment are relatively resistant to most infectious diseases. The ability of fish to resist infectious diseases is decreased when they are harvested and transported (Chapter 12), fed an insufficient or nutritionally inadequate diet (Chapters 5 and 10), or exposed to water of poor or rapidly fluctuating quality (Chapters 4 and 9). These factors (stressors) elicit a series of physiological and behavioral responses (stress responses) that are an attempt by the fish to maintain or reestablish normal metabolism. When conditions are such that the capacity of the fish to adjust is temporarily exceeded, the fish become stressed. If the stress response can reestablish an adequate physiological state under the new conditions, adaptation to the stressor occurs. If adaptation does not occur, death is usually the result.

One function of the stress response is modulation of the animal's immune system. In effect, changes that enhance chances of survival in the short term occur. However, if stress is severe or prolonged, an undesirable consequence is suppression of certain aspects of the immune system. The exact mechanisms of stress-induced immunosuppression are not known, but it is well established that serious outbreaks of infectious diseases often follow in the days after exposure to a stressor. Also, if a disease outbreak has occurred, subsequent stress can result in a marked increase in the number of fish that ultimately die from the disease.

CLINICAL SIGNS OF FISH DISEASES

Sick fish exhibit a variety of behavioral and physical clinical signs (symptoms). Catfish farmers should know these signs so that disease outbreaks can be recognized early. Prompt diagnosis and treatment are essential for effective control of any disease.

Channel catfish are not readily observable during culture, so it is difficult to detect abnormal behavior or physical signs in individual fish during the initial stages of a disease outbreak. Usually, the first noticeable sign that fish are sick is a change in the behavior of the population within the culture unit. For instance, fish may feed less vigorously or behave strangely when dissolved oxygen concentrations are marginally low. Individual sick or dead fish are often not seen at this early stage. If abnormal behavior is noticed, measurements of dissolved oxygen, un-ionized ammonia, and nitrite concentrations should be made immediately to determine whether the cause is poor water quality. It is often difficult to verify the presence or absence of pathogens during the early stages of a disease outbreak because only part of the population is infected and representative fish samples are difficult to obtain. Usually, the first opportunity to verify the infectious nature of a disease is when individual fish exhibiting behavioral or physical signs are captured and submitted to a diagnostic lab.

Behavioral Signs

Healthy channel catfish are seen only during feeding and spend most of their time swimming near the bottom of the pond or culture unit. Any change in normal behavior should alert the farmer to potential problems.

Reduced feeding activity. Most cultured catfish are fed a floating feed (except during the winter), and the activity of the fish during feeding is easy to observe. The amount of feed consumed and the vigor with which it is consumed indicate the general health of the population. When water temperatures are above 70°F, feeding activity should be vigorous and a reasonable amount of feed consumed within 15 minutes or less. Although feeding activity varies from day to day, a sudden, large decrease in the amount or a steady decline in feed consumption over several days usually indicates the presence of an environmental stressor, an infectious disease, or both. A sudden drop in feeding activity is often related to an acute environmental problem, such as dissolved oxygen depletion; steady declines over several days are more typical of the early stages of an infectious disease outbreak.

Reduced feeding activity associated with diseases has important implications if the disease is due to a systemic (internal) bacterial infection. Antibiotic therapy is usually recommended, and the only practical method of delivering the antibiotic is by adding the drug to the feed. Obviously, if fish are not eating, this therapy is useless. This makes prompt diagnosis just that much more important: treatment must be started before fish stop feeding.

Intolerance of low dissolved oxygen concentrations. Certain bacteria or parasites growing on or within gill tissues cause lesions that reduce respiratory efficiency. These pathogens may destroy gill tissue, cause changes in gill tissue morphology, stimulate excessive mucus production, or physically impede diffusion of dissolved oxygen from the water into the blood by their presence. Under these conditions fish are intolerant of marginally low concentrations of dissolved oxygen. Fish with systemic viral or bacterial infections may also be intolerant of suboptimal dissolved oxygen concentrations because they are debilitated by the disease. Healthy channel catfish can survive for long periods at dissolved oxygen concentrations down to at least 2 ppm. If concentrations are greater than about 2 ppm but fish exhibit signs of respiratory distress (such as swimming at the surface gulping air) a complicating factor should be suspected (see the section Dissolved Oxygen).

Lethargy or erratic swimming. Healthy channel catfish swim away rapidly when startled. Sick fish often lie along the bank or swim sluggishly near the surface. Weak fish drift with currents and may accumulate along the leeward shore of ponds or at the downstream end of raceways or tanks. Lethargy is not specific to fish with infectious diseases; exposure to low concentrations of dissolved oxygen, high concentrations of nitrite, or other toxic substances can also cause lethargy or reduced vigor. Sick fish also may appear "nervous," swim erratically in spirals, exhibit loss of equilibrium, or swim in quick, sporadic bursts. This behavior is commonly associated with infectious diseases but can also be caused by exposure to certain toxicants, such as insecticides or un-ionized ammonia.

Physical Signs

In addition to the changes in behavior described, sick fish exhibit certain physical signs. Physical signs usually develop in the later stages of the epizootic and are often specific to certain infectious diseases.

Skin lesions. Ulcers, sores, blisters, or small red spots (petechial hemorrhages) are characteristic of several infectious diseases. White, pinhead-size spots covering the skin are extremely characteristic of heavy infestation of the parasite *Ichthyophthirius multifilis* ("Ich").

Eroded or reddish fins. Frayed or eroded fins, usually accompanied by hemorrhagic areas at the base of the fin, are particularly characteristic of bacterial infections.

Abnormal skin coloration. Abnormally pale skin can be caused by acute hypoxia (very low concentrations of dissolved oxygen), anemia, external bacterial infections, or parasitic infestations. Infestation of the skin with the parasite *Icthyobodo (Costia)* sometimes causes excessive mucus production, giving the skin a gray or bluish coloration.

Abnormal appearance of the gills. Brown- or chocolate-colored gills are associated with high concentrations of environmental nitrite (see the section Nitrite). Swollen gills, often with excessive production of mucus, can be caused by infectious diseases or exposure to environmental toxicants. Other gill lesions associated with infectious disease include frayed or eroded gill tips, discoloration (uniformly pale or blotchy red-brown), and areas of hemorrhage.

Swollen abdomen. A distended, swollen abdomen indicates that the body cavity is filled with fluid (check to make sure the swelling is not due to feed in the gut). The fluid in the body cavity may be cloudy, bloody, or clear. This sign is often associated with channel catfish virus disease or systemic bacterial infections.

Exophthalmia ("Popeye"). Bulging eyes can be caused by gas bubble trauma or by environmental stressors that affect osmoregulation. Usually, however, exophthalmia is caused by the presence of excessive amounts of fluid in the body cavities and is related to viral or systemic bacterial infections.

DISEASE DIAGNOSIS

Some diseases, particularly environmental diseases such as hypoxia or methemoglobinemia, can easily be diagnosed and treated by an experienced catfish farmer. However, if there is any reason to suspect that the disease is infectious, fish samples must be submitted to a laboratory for diagnosis. Confirmation of viral or bacterial diseases requires special equipment and skills. Furthermore, if the disease is caused by bacteria, antibiotic sensitivity tests should be conducted to verify the most efficacious treatment. Misdiagnosis and inappropriate treatment can actually cause greater losses than if the disease remains untreated. For example, assume that a fish has an infestation of gill parasites as well as a concurrent systemic bacterial infection (a very common occurrence). If the diagnosis is attempted in the field or with only the aid of a microscope, it is likely that only the more obvious problem (the gill parasites) will be identified. Gill parasites are usually treated with chemicals such as potassium permanganate, copper sulfate, or formalin. Exposure to these chemicals stresses fish. Also, their use in ponds often kills all or part of the phytoplankton bloom and causes water quality to deteriorate, further stressing the fish. The stress induced by the treatment decreases the fish's resistance to the bacterial infection, and mortality rates may increase dramatically. Additionally, the treatment may cause the fish to go off feed, making subsequent therapy with medicated feed impossible. The results of an accurate and complete diagnosis would have indicated that the first action should have been use of the proper medicated feed. Often, this therapy alone brings the problem under control, but if the parasites still require treatment, it is administered after the systemic bacterial infection has been controlled.

The procedure for submitting fish for disease diagnosis consists of obtaining a fish sample that offers a good chance of finding the cause of the disease and transporting that sample to the laboratory in a manner that minimizes changes in the condition of the fish. Water samples or other information that may aid the disease specialist in diagnosing the disease should also be collected and submitted with the fish sample. Fish disease diagnostic laboratories in the major catfish-producing areas of the United States are listed in the Appendices.

Collecting the Sample

The best samples consist of live fish that are behaving abnormally (lethargic, swimming erratically, and so on) or exhibiting any of the physical signs of disease discussed earlier. These fish can be scooped up in a long-handled dip net, caught in a cast net, or collected by any other convenient method. Try to catch at least two fish from the diseased population; a sample of four to six is ideal.

Samples of recently dead fish are less suitable, and the probability of identifying the cause of death decreases the longer the fish has been dead. Tissues rapidly begin to break down after a fish dies, especially in warm water. A wide variety of bacteria rapidly colonize dead tissues, making identification of the pathogen difficult. Cloudy eyes, pale gills, fishy odor, or pale, blotchy skin indicates that the fish has been dead for some time and is totally unusable as a sample.

Fish taken randomly from the population (such as a group captured by seining) or fish captured by angling are not good samples for diagnosis of infectious diseases. The probability of finding infected fish from a random sample is low in the initial stages of the epizootic because the prevalence of the disease among the fish is low. At later stages, good samples of live fish exhibiting behavioral or physical symptoms are easily obtained and should be used as samples. Fish caught by angling are poor samples regardless of the stage of the epizootic because sick fish feed poorly and are unlikely to bite; fish caught by angling are usually healthy.

Transporting and Shipping the Sample

Before submitting the sample, notify the disease specialist so that the sample will be expected. This is especially important if the sample is to be shipped.

If the diagnostic laboratory is nearby, place the fish in an insulated ice chest about half full of clean water. Do not use tap water because it contains chlorine that will kill the fish and many of the parasites or bacteria on the external surface of the fish. A small battery-powered aquarium-type aerator can be used to keep the water oxygenated. A few pieces of ice can also be added to keep the water cool. To prevent cross-contamination of samples, do not place samples of fish from different culture units in the same water.

It is usually more convenient to transport samples on ice than in water. Clip off the fish's spines and place the fish in a sealable plastic bag. Preferably, double-bag the sample. Be sure to include a label, written in pencil on paper, in the bag and also label the bag with a permanent marker. Place the bag on ice in an insulated ice chest and cover with more ice. Fish can be held in good condition in this manner for up to 1 day.

If fish must be shipped rather than taken to the laboratory in person, pack them on ice as described, but be sure the container (a styrofoam cooler is commonly used) is well sealed. The fastest way to send the sample is usually by bus. Tape an envelope containing all pertinent information to the top of the container. Address the envelope to the diagnostic laboratory and write on the outside of the container:

Call on arrival: (phone number)
Perishable scientific specimen

The phone number can be the laboratory phone number or the diagnostician's home phone number, depending on what arrangements have been made.

Additional Information

The diagnostic process usually involves more than just laboratory analyses. Other information will greatly aid the disease specialist. Be prepared to supply (or send) the following information.

When sick fish were first noticed. The time when illness was first observed indicates how far the disease has progressed. Also, point out the time of day that sick fish were first observed and whether sick fish are seen more often at particular times of the day. This may indicate the involvement of environmental variables; for instance, if low dissolved oxygen concentrations are the primary cause of the problem, more fish will be seen in distress in the early morning.

Number of fish lost each day. Information on the daily loss of fish is necessary to make a rational decision as to the appropriate treatment. All disease treatments are expensive and, depending on the disease, may not be warranted if only a few fish are dying each day. The pattern of fish lost over time also indicates whether the problem is getting worse or is in remission.

Water quality information. If possible, provide measurements of water temperature and concentrations of dissolved oxygen, nitrite, and ammonia for the week or so before sick fish were first noticed as well as measurements made since then. This information may indicate whether poor water quality is the cause of the problem or is a predisposing factor for the infectious disease. Some fish disease specialists may also request that a water sample from the affected culture unit be provided with the fish sample.

Type and size of culture unit. Tell the diagnostician whether fish are from a tank, net pens, a raceway, or a pond. Also provide the dimensions or volume (if known) of the unit. This information is necessary to determine the appropriate therapy and treatment rate. It also gives a better perspective on the number of fish that are dying.

Fish stocking density. Stocking density information also helps put mortality rates into perspective. It may also indicate the basis for certain problems: water quality problems (and hence stress-related infectious diseases) are more common in densely stocked culture units. Stocking density may also give some indication of the probable progression of the disease because infectious diseases spread more rapidly when fish densities are high.

Recent chemical treatments. Tell the diagnostician what chemicals (herbicides, disease treatments, etc.) have been used in the last month or so, how much was used, and the reason for their use. Many commonly used chemicals are somewhat toxic to fish, even when properly used. Moreover, many chemicals are toxic to the plankton bloom and their use can cause poor water quality. Be sure to indicate whether attempts to control the disease have been made and what the results were. This may aid in the decision as to the most appropriate new course of action.

Other information. Provide any other pertinent information, such as whether fish were recently stocked or harvested and whether other culture units on the farm also contain sick fish.

VIRAL DISEASES

Viruses are small infectious agents that do not carry out independent metabolic functions. Virus particles (virions) can only be seen with an electron microscope. Virions carry viral genetic material (nucleic acid: either deoxyribonucleic acid [DNA] or ribonucleic acid [RNA], depending on the type of virus) from the cell in which the virion was produced to another cell where the viral nucleic acid is introduced. The viral nucleic acid redirects the metabolic machinery of the host cell to produce more viral nucleic acid and other components of the virus. Viruses are thus obligate intracellular parasites.

Viral diseases are difficult to treat because the virus resides primarily within the host cells. Viral diseases of higher vertebrates are controlled by use of *vaccines,* dead virus or virus rendered nonpathogenic (attenuated) to illicit a long-lasting immune response that will increase resistance to disease from subsequent exposures to the virus. Viral diseases of channel catfish cannot be controlled by chemicals or drugs, and vaccines are not currently available.

There are two viruses known to infect channel catfish: channel catfish virus and channel catfish reovirus. Channel catfish virus disease is an important infectious disease of young catfish and can cause devastating losses. Channel catfish reovirus has only recently been discovered and has not yet been confirmed as the cause of large epizootics. Plumb (1985b, 1988a) and Wolf (1988) provide excellent reviews of viral diseases of channel catfish.

Channel Catfish Virus Disease

Channel catfish virus disease (CCVD) was discovered in 1968 and has since been widely reported in the southeastern United States and other areas where channel catfish are cultured. The disease primarily affects channel catfish less than 6 inches long. Outbreaks are most common from June through September, when large numbers of young fish are present on farms and water temperatures are most conducive to rapid spread of the disease within a population.

Viral Diseases **329**

Channel catfish virus disease epizootics are widespread but sporadic in occurrence, and relatively few cases are reported each year. However, when it does occur, losses to the individual farmer can be devastating.

Causative agent. Channel catfish virus is a herpesvirus, icosohedral in shape and about 200 nanometers in diameter including the envelope. The virus persists in dead fish for 2 or 3 days at 70°F, but for at least 2 weeks in fish stored on ice, and over 6 months in frozen fish. Channel catfish virus survives for several days in pond water at 80°F and for at least a month at 40°F, but it is quickly killed under dry conditions and will not live more than 1 or 2 days on dry surfaces or seines. The virus is rapidly inactivated in pond sediments because clay particles firmly adsorb it.

Channel catfish virus is quite host-specific, and channel catfish are the only species known to sustain natural epizootics. However, susceptibility to CCVD varies greatly among different strains of channel catfish (see, for example, Table 6.1). The disease can be induced in blue catfish, but natural epizootics apparently do not occur. Channel catfish × blue catfish hybrids are as susceptible to CCVD as the channel catfish parental line (Plumb and Chappell 1978).

Clinical signs. Fish with CCVD feed poorly and swim erratically, usually in a spiral pattern. In the late stages of the disease, they may float head-up in the water. The most obvious physical signs are a distended abdomen and exophthalmia. The body cavity is usually filled with a clear, yellowish fluid. The gills may be pale, and frequently there is petechial hemorrhage at the base of the fins. Internally, the liver and kidney are usually pale and the spleen is swollen and dark red. The musculature often contains diffuse petechial hemorrhage.

Characteristics of the disease. The course and outcome of CCVD are strongly affected by age and size of fish. Fish less than 1 month old are very sensitive, and mortality rates can exceed 90 percent in less than 1 week. Older fish are more resistant and mortalities extend over a longer time. Fish larger than about 8 inches are considered resistant to CCVD.

Mortality and rates of development of clinical signs are related to water temperature. Fish do not develop CCVD at temperatures below 60°F; occurrence of the disease is irregular and mortalities are usually low at temperatures between 70 and 80°F. At 70°F, up to 10 days after exposure to the virus may be required before clinical signs appear. At temperatures above 80°F, clinical signs can develop within 2 to 4 days after exposure and mortality rates may be high. Highest mortality rates occur at temperatures around 85°F.

Although CCVD epizootics are often acute, at times the progression of the disease is slow, with relatively few fish dying over a period of weeks, even when water temperatures are optimal for the disease. Environmental stressors appear to be an important factor in triggering acute outbreaks of CCVD. Handling, seining, or exposure to poor environmental conditions (particularly low dissolved oxygen concentrations) can cause sudden, massive mortality in populations with CCVD.

Transmission. Development of CCVD in fry may be the result of transmission of virus from parents to offspring via reproductive products. This mode of transmission has long been suspected, and with the development of sensitive methods of detecting CCV genetic material in adult fish (Wise and Boyle 1985), it has apparently been confirmed (Wise et al. 1988). The exact mode of transmission is not known, however. Also, it appears that many adult fish on commercial catfish farms in Mississippi carry the viral genetic material in a latent state (Wise, Bowser, and Boyle 1985). This then raises the question of why CCVD is not more prevalent in commercial stocks of fry and fingerlings. It may be that many fish are asymptomatic carriers of the virus and that outbreaks of CCVD occur after fish are stressed.

During outbreaks of CCVD, virus is easily transmitted from fish to fish through the water. Virus is shed into the water by infected fish and enters others through the gill or intestinal epithelium. Virus can be spread to other culture units in the effluent from a unit holding fish with CCVD or by equipment (nets, seines, etc.) that has not been properly disinfected after being used to handle fish with CCVD. Also, fish-eating birds, reptiles, or mammals may possibly spread the disease by carrying infected fish from one culture unit to another.

Diagnosis. Channel catfish virus diseases should be suspected when fry or fingerlings suddenly begin dying in large numbers when water temperatures are above about 70°F. Behavioral and physical signs can also aid in the diagnosis, but the clinical signs of CCVD do not differ greatly from those associated with systemic bacterial diseases, particularly enteric septicemia of catfish and motile aeromonad septicemias. An accurate diagnosis of CCVD is imperative because misdiagnosis may result in an inappropriate treatment that stresses fish and causes even greater losses. Diagnosis of viral diseases requires special laboratory tests because viruses can propagate only within living cells. Fish disease diagnostic laboratories maintain cultures of fish cells known to be susceptible to the virus. Tissues from fish suspected of being infected with CCV are pulverized and then passed through a filter that removes bacteria and cell debris but allows viruses to pass through. The filtrate is inoculated into the fish cell culture, which is then allowed to incubate for a week. If virus is present, the fish cells change size and shape in a characteristic fashion (usually within 2 days). These changes, called *cytopathic effects* (CPEs), are easily seen with a light microscope. The development of syncytia (multinucleate masses caused by the merging of cells) is the characteristic CPE of CCV infection of susceptible cells. Positive confirmation that the CPE was in fact due to a specific virus requires further testing (normally using serum neutralization tests), although this is not usually required during routine diagnosis. Virus can be isolated only from infected fish during an epizootic or at most for 1 or 2 days afterward. Fish samples should preferably be alive, but virus can be isolated from freshly dead samples held for several days on ice. After mortality stops, it is usually not possible to isolate CCV from survivors, even though they may still carry the virus in a latent form.

Control. Presently there is no cure for CCVD, but certain management practices can minimize the incidence and severity of epizootics. These practices include avoiding CCV-carrier brood fish, minimizing stress in susceptible fry or fingerling populations, using adequate hygienic practices to minimize the potential spread of the disease, and manipulating water temperature (in culture systems where it is feasible). Use of antibiotics during outbreaks of CCVD may have some benefit because secondary bacterial infections are common and contribute to overall mortality. In the future, CCVD may be controlled by vaccination or use of disease-resistant strains or hybrids.

There is sound evidence that CCV is transmitted from brood stock to progeny, so the use of CCV-free brood fish should help prevent the disease. However, latent virus is apparently widespread in commercial stocks of adult channel catfish, and this practice is probably not possible except in areas somewhat remote from the major catfish-producing regions in the southeastern United States. Potential CCV-carrier fish can be identified by the presence of CCV-neutralizing antibodies in the serum of adult fish or by the presence of CCV genetic material within catfish cells.

Because stress is a strong contributing factor in the incidence and severity of CCVD epizootics, the best current practice for controlling the disease is to use accepted management practices and maintain good environmental conditions in the brood pond, hatchery, and fry nursery culture units. Avoid wild spawning (see the section Methods of Propagation) in brood ponds because fry densities cannot be controlled and may become excessively high. Overcrowding hastens the spread of CCVD during epizootics and also leads to poor water quality, which stresses the fish. Fry in wild-spawn ponds may also contract CCVD from adult fish in the pond. The risk of losses to CCVD in hatcheries can be minimized by using a water supply of good quality and maintaining optimal environmental conditions within the hatchery (Chapter 7). Hatchery water temperature is particularly important because problems with CCVD are more common when temperatures are above about 80°F. Transport fry and fingerlings only when water temperatures are below 85°F and avoid all unnecessary stress during transport. Channel catfish virus disease occurs more frequently and losses are greater in densely stocked nursery ponds, so stock fish at moderate densities, preferably less than 150,000 fry/acre. Monitor water quality in nursery ponds and provide adequate aeration; try to maintain dissolved oxygen concentrations above about 4 ppm. Never seine or transport fish with active CCVD; the stress of handling and the crowded conditions during transport can result in nearly 100 percent mortality, particularly if water temperatures are above 80°F. Sick fry or fingerlings should always be checked for CCVD before any treatment is administered. Therapeutant chemicals used to treat external parasites or bacterial infections can stress fish, causing great losses if there is a concurrent CCV infection.

Good sanitation practices in hatcheries are important in preventing the spread of CCVD as well as generally improving environmental conditions in hatching and rearing tanks. Tanks should be cleaned and disinfected between batches of fry. If CCVD occurs in a group of fry in the hatchery, the entire group should be carefully removed from the hatchery and destroyed. Avoid contaminating other tanks with water from the tank with infected fish or with equipment that has been in contact with either the water or infected fish. Thoroughly disinfect the tank and all possibly contaminated equipment with solutions of 5 percent formalin, 40 ppm calcium hypochlorite, or 1,000 ppm benzalkonium chloride. If CCVD occurs in a pond population of fish, quarantine that pond to the extent possible. Do not let effluent from the pond flow into any other pond. Any equipment (seines, aerators, etc.) used in the pond with infected fish should be thoroughly dried or disinfected before use in another pond. Promptly remove dead fish from any culture unit containing fish with CCVD. The dead fish may serve as a reservoir for the virus and may be moved from pond to pond by animals. Susceptible fish should not be stocked into ponds with a history of CCVD unless the ponds have been drained and allowed to dry to the point where no puddles exist before refilling. It is not necessary to dry the pond bottom completely because mud inactivates CCV and the virus will not be transmitted to subsequent batches of fish.

Antibiotics have no effect on CCV, but their use during CCVD epizootics may decrease overall mortality. The prophylactic use of antibiotics is usually not recommended, but bacterial infections are so common during outbreaks of CCVD that antibiotic therapy is often warranted. As mentioned previously, never use therapeutant chemicals to treat concurrent disease problems of fish with CCVD.

Lowering the water temperature to less than 70°F significantly reduces mortality during CCVD epizootics. This therapy is limited to small laboratory culture units and some water-reuse systems where temperature control is possible.

A vaccine using live attenuated virus has been found to protect fish against CCVD under experimental conditions (Noga and Hartman 1981). The vaccine has not been licensed by the United States Department of Agriculture and is not available commercially. Furthermore, it appears unlikely that a commercial vaccine for CCVD will soon be available because the market is relatively limited and several important basic and practical aspects of the use of the vaccine need to be investigated (Plumb 1988a).

The use of CCVD-resistant strains or hybrids holds some promise for the future, but it is unlikely that these fish will be widely available anytime soon. Although it is clear that CCVD-resistant fish exist, studies must be conducted at various locations to determine correlations between this trait and other important commercial traits: breeding CCVD-resistant fish is counterproductive if the fish also grow slowly or are susceptible to other diseases.

Channel Catfish Reovirus

Channel catfish reovirus (CRV) was discovered in 1982 during routine virological examination of juvenile channel catfish on a farm in California (Amend, McDowell, and Hedrick 1984). The virus and the pathology of the disease have not been well characterized. It appears that CRV is not highly pathogenic and does not cause high rates of mortality. Specific clinical signs have not been described; hyperplasia (an increased number of cells, causing swelling) of the gill lamellae in fish from which the virus was isolated was reported, but this may not have been directly attributable to CRV. Optimal temperature for infection is about 77°F. The virus can be isolated on the same cell lines used to isolate channel catfish virus, and the cytopathic effect produced on these cells is similar to that produced by channel catfish virus. The potential impact of CRV on commercial catfish culture is unknown but appears to be minimal. Channel catfish reovirus has not been reported outside southern California.

BACTERIAL DISEASES

Bacteria are a diverse group of procaryotic, single-cell organisms that reproduce asexually by binary fission (one cell splitting into two identical daughter cells). Hundreds of species of bacteria are found in catfish pond waters. Most aquatic bacteria are free-living and perform beneficial functions such as the decomposition of organic matter. A few species are opportunistic pathogens and cause disease only when the immune system of the fish is suppressed. Only one of the common bacterial diseases of channel catfish, enteric septicemia of catfish, is caused by an obligate pathogen. The bacterium causing that disease, *Edwardsiella ictaluri*, normally is not found in water in the absence of channel catfish.

Bacterial diseases account for well over half the farm-raised channel catfish lost to infectious diseases. Antibiotics are available to treat bacterial disease of channel catfish, but effective use requires prompt diagnosis and initiation of treatment. Consult a competent fish disease specialist whenever a bacterial disease is suspected because many bacterial diseases are caused by mixed infections (two or more pathogens) and certain strains of bacteria may be resistant to a particular antibiotic. Failure to diagnose the problem fully can result in ineffective treatment.

Enteric Septicemia of Catfish

Enteric septicemia of catfish (ESC) is the most serious infectious disease problem of farm-raised channel catfish. The disease was first described in 1976 (Hawke 1979) for channel catfish in Alabama and Georgia but has now been found throughout the United States wherever channel catfish are commercially cultured. The disease can affect all sizes of channel catfish, but it is primarily a problem in populations of fry and fingerlings. Epizootics of ESC are highly seasonal, occurring primarily in spring and fall, when water temperatures are between 72°F and 82°F.

Causative agent. Enteric septicemia of catfish is caused by *Edwardsiella ictaluri*, a gram-negative bacterium in the enteric group (Enterobacteriaceae). Hawke (1979) describes the biochemical and physical properties of *E. ictaluri*. Bacteria survive for less than 2 weeks in pond water but may survive for at least 3 months in pond muds at 70 to 85°F.

Edwardsiella ictaluri is fairly host-specific for channel catfish, although it has also been isolated from walking catfish, blue catfish, white catfish, brown bullhead, and two species of tropical aquarium fish. Experimental infections have been established in blue tilapia (Plumb and Sanchez 1983), chinook salmon, and rainbow trout (Baxa

Bacterial Diseases 335

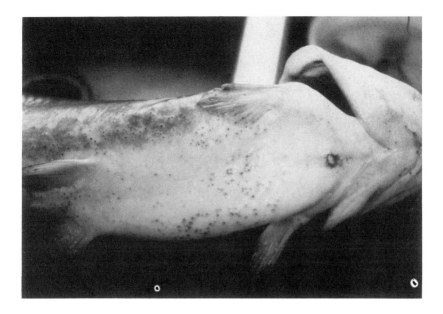

FIGURE 11.3.

Rashlike areas of hemorrhage on the belly of a channel catfish with enteric septicemia of catfish. (Photo: A. J. Mitchell)

and Hedrick 1989), but natural infections in these fishes have not been described.

Clinical signs. The array of clinical signs exhibited by fish with ESC varies considerably, depending on the size of fish, water temperature, and course of the disease. Clinical signs may resemble those associated with other systemic bacterial infection, and in small fish they are often similar to those associated with CCVD. Gross external signs may be absent in acute cases.

Fish with ESC do not feed well and may be listless, swimming slowly at the surface with their tails hanging down, or they may swim erratically, often in circles in a spiraling fashion. Fish may have petechial hemorrhages around the mouth and on the underside of the fish. Small red or white ulcerlike spots may also be present on the sides or belly of the fish (Fig. 11.3). The gills are often pale. The presence of an ulcerative lesion on the top of the head between the eyes (Fig. 11.4) is considered the most characteristic clinical sign of the disease and gives it one of its common names, "hole-in-the-head disease." The lesion varies from a soft, pale area or raised bump to a large open ulcer extend-

336 Infectious Diseases

FIGURE 11.4.

An ulcerative lesion on the top of the head is characteristic of enteric septicemia of catfish. (Photo: A. J. Mitchell)

ing through the frontal bones into the brain cavity. Infected fish often have a distended abdomen and exophthalmia; fluid in the body cavity is usually clear or straw-colored. The muscle, visceral fat, intestine, and inner walls of the body cavity may have areas of petechial hemorrhage. The kidney and spleen are usually enlarged, and the liver is usually covered with white necrotic spots.

Characteristics of the disease. The occurrence of ESC is highly temperature-dependent. Epizootics are most common and mortalities are greatest when water temperatures are between 72 and 82°F. Within this temperature range the disease is usually acute: fish rapidly reduce feeding activity and large numbers of sick or dead fish are soon observed with little or no initial period of low-grade mortalities. At marginal water temperatures (65 to 72°F or 82 to 85°F) losses are usually chronic and mortality rates are low unless the fish are severely stressed. Secondary infection by other bacteria or parasites is common at these marginal temperatures. Fish losses to ESC are rare when water temperatures are consistently below 65 or above 85°F.

Channel catfish of all ages and sizes are susceptible to ESC, but fingerlings account for most of the fish lost to ESC. Epizootics frequently occur during the fingerling's first fall, when water temperatures begin falling into the 72 to 82°F range, or in the following spring, when water temperatures rise and again become conducive for the disease.

Recurrence of the disease in a population that has survived an ESC epizootic can occur, but fish often develop an apparent immunity and are able to survive subsequent encounters with the bacterium. Fish that survive ESC may become asymptomatic carriers of the bacterium. If these fish are transferred to a culture unit containing fish that have not been previously exposed, an epizootic involving the naive fish often develops when water temperatures become conducive for the disease. The reverse situation may also occur, wherein naive fish are stocked into a culture unit containing fish that have survived a previous ESC epizootic. These situations are common in ponds managed under the multiple-batch production system (see the section Food-Fish Production). In this production system, ponds contain several cohorts of fish and each cohort has a unique history with respect to previous exposure to infectious diseases. The common use of the multiple-batch production system has probably contributed to the rapid spread of ESC since it was first identified.

Transmission. Edwardsiella ictaluri is highly virulent and is easily spread from fish to fish. Fish with active infections or asymptomatic carrier fish shed the bacteria into the water with excremental products. Contaminated water or mud may also serve as a source of bacteria. Water-borne bacteria may enter fish through the nares or the intestinal lining (Shotts, Blazer, and Waltman 1986). Bacteria entering the fish through the nares migrate along the olfactory nerve and establish an infection in the brain. Involvement of the brain explains the unusual spiraling or erratic swimming behavior in fish with ESC. Also, the hole-in-the-head lesion is associated with the establishment of an infection in the brain. Fish can also become infected by cannabilizing infected carcasses. Bacteria in the food enter through the intestinal lining and rapidly spread throughout the body. The brain is often not involved when fish are infected only via the intestinal lining (Shotts, Blazer, and Waltman 1986).

The disease may be spread from one culture unit to another by water-borne bacteria in effluents. Bacteria may also be spread by using contaminated nets or seines, although this has not been demonstrated. Fish-eating birds or other animals possibly spread the disease by transporting infected carcasses or by eating infected fish and then shedding the bacteria in feces.

Diagnosis. A tentative diagnosis can be made on the basis of clinical signs, particularly if the characteristic hole-in-the-head lesion is observed and water temperatures are conducive to the disease. Because the disease frequently progresses rapidly with sudden cessation of feeding, a strong tentative diagnosis can be sufficient reason to begin antibiotic therapy. The diagnosis should subsequently be confirmed by laboratory isolation and identification of bacteria from sick fish. A complete diagnosis is necessary not only to confirm the cause of the disease but also to detect any mixed infections that may require later attention and verify that the bacteria are susceptible to the antibiotic being used. Laboratory diagnostic methods are outlined by Lewis and Plumb (1985).

Control. Prompt treatment is critical to successful control of ESC. Because the occurrence is highly seasonal, ponds should be closely monitored during the spring and fall for any sign of sick fish. Pay particular attention to fry nursery and fingerling ponds. Begin antibiotic therapy as soon as fish are diagnosed as having ESC. Two antibiotics are presently available: oxytetracycline (Terramycin®) and sulfadimethoxine potentiated with ormetoprim (Romet®). Both antibiotics are made to be formulated into medicated feeds. The use of these two antibiotics is discussed in the section Disease Treatments. Commercial catfish farmers report variable treatment success with either antibiotic. Poor success is probably related to delayed initiation of treatment and subsequent failure to achieve the necessary levels of antibiotic in fish because of poor feeding response.

Edwardsiella ictaluri is highly pathogenic and can cause disease even in otherwise healthy fish. Nevertheless, mortality is greatest when fish with ESC are stressed. To the extent possible, maintain good water quality and avoid seining fingerlings when water temperatures are conducive to ESC. Never seine or transport fingerlings with active ESC. Because the disease is usually more severe in ponds stocked at high fish densities, use conservative production practices. The spread of ESC from carrier to naive fish argues against the use of the multiple-batch production system; if this system is used and ESC causes large losses year after year, consider using the single-batch system unless other economic factors are deemed more important.

Fish with ESC often have concurrent infestations of external parasites or external infections of *Flexibacter columnaris* bacteria. Normally, these problems should not be treated until the ESC is brought under control by medicated-feed therapy because the chemicals used to treat these problems may stress the fish or cause them to decrease feeding activity. Occasionally, however, initial treatment for the external parasites or bacteria may be warranted. Extremely high levels of these organisms may so debilitate the fish that feeding response is nil. Chemical treatment may then be required to increase feeding activity.

This treatment regimen is rather risky and should only be undertaken after consultation with an experienced fish disease specialist and a complete diagnosis of the problem.

In some types of culture systems, ESC can be controlled by manipulation of water temperature. Optimal growth of channel catfish occurs at water temperatures of 80 to 85°F. Maintenance of temperatures at the high end of this range, or even slightly higher, should decrease the incidence and severity of ESC outbreaks. Temporarily decreasing the water temperature to below 65°F will decrease or stop losses during outbreaks of ESC, but mortality rate may simply increase again when water temperatures are subsequently increased.

Under laboratory conditions, mortality rates of channel catfish infected with *E. ictaluri* decrease with increasing levels of vitamin C in the diet (Li and Lovell 1985). High-C feeds have been used by some catfish farmers in Mississippi to decrease losses to ESC, but the results are not clear.

Vaccines to prevent ESC are not currently available, but this approach has promise. The vaccine should have considerable market potential because the disease is serious and widespread. This should attract interest by established vaccine manufacturers. Vaccination will probably involve immersion of fry or fingerling fish in a dilute solution of vaccine. An oral booster consisting of the vaccine incorporated into feed may then be administered shortly before water temperatures are expected to move into the 72 to 82°F range. The use of vaccines against *E. ictaluri* is discussed in detail by Plumb (1988b).

Columnaris

Columnaris is the second most common bacterial disease of farm-raised channel catfish. It occurs throughout the warmer months of the year, particularly in late spring and early fall. Outbreaks of the disease are usually associated with stressful conditions.

Causative agent. Columnaris is caused by *Flexibacter columnaris* and perhaps other closely related species. *Flexibacter columnaris* is a long, thin gram-negative rod in the family Cytophagaceae. The bacterium is motile by a gliding or flexing action and forms characteristic columns or "haystacks" that can be observed microscopically on wet mounts of gills or skin scrapings from infected fish.

Flexibacter columnaris is considered a saprophytic bacterium that is a normal inhabitant of most natural waters, especially those with high calcium and organic matter content. The bacteria live on decomposing organic matter and do not require fish as a host.

340 Infectious Diseases

FIGURE 11.5.

Columnaris lesions on channel catfish. Lesions may vary from areas of depigmentation (top) to ulcers (bottom) with exposure of the muscles and bone. (Photo: Mississippi Cooperative Extension Service)

Columnaris is a common fish disease worldwide and affects virtually all freshwater fish species. Different strains of *F. columnaris* affect warm-water and cold-water fish (Pyle and Shotts 1980), and strains also differ in virulence.

Clinical signs. Columnaris usually begins as an infection of the external surfaces of the fish that may progress to an internal, systemic bacteremia. Lesions are most common on the gills but may occur on the body, head, or fins or inside the mouth. Infection of gills results in tissue death with yellow-brown areas of necrosis beginning at the distal (free) end of the gill filaments, progressing to the basal (attached) end. Infections of the body surface often begin on the fins, which become ragged and frayed. Skin lesions begin as areas of depigmentation or discoloration that may enlarge to become shallow, grayish ulcers with reddened areas around the periphery (Fig. 11.5). Masses of the yellow-pigmented bacteria may cover infected areas of the skin or gills. In severe cases, the skin may be destroyed, exposing the underlying muscles

Bacterial Diseases 341

and bone. External lesions are often invaded by fungi if water temperatures fall below about 60°F.

Internal infections often follow the external infections, but internal infections may also be present in the absence of external lesions. Fish with internal columnaris often show little internal evidence of the disease.

Characteristics of the disease. Columnaris rarely occurs in catfish populations unless fish are stressed. Poor water quality and handling of fish in warm water favor the development of the disease. Columnaris is most common when water temperatures are 75 to 85°F. If fish are severely stressed within this temperature range, columnaris epizootics are often acute and progress rapidly. Fish rapidly reduce feeding activity and large numbers of fish may be lost within a few days. At lower water temperatures, the disease is chronic and losses are usually lower than at warmer temperatures. All ages and sizes of channel catfish can be affected. Recurrence of the disease is common, particularly if the conditions predisposing fish to the initial infection are not addressed.

Transmission. Flexibacter columnaris is present in most waters and fish are continuously exposed to the bacterium. During outbreaks of columnaris, bacteria may also be transmitted from fish to fish by physical contact or through water. Transmission is facilitated by crowded conditions. Because F. columnaris is ubiquitous and usually requires some predisposing stressful condition to initiate disease, transmission of the disease from one culture unit to another by using contaminated equipment or by animals moving dead fish from pond to pond is less likely than with obligate, highly virulent pathogens such as channel catfish virus or E. ictaluri. Nevertheless, use good judgment and try to quarantine culture units containing fish with columnaris.

Diagnosis. Observation of bacteria in scrapings from external lesions is usually considered diagnostic for columnaris. The slender rods can be seen under a microscope at magnifications of 200× or greater. If wet mounts of scrapings are allowed to stand for a few minutes, F. columnaris may form easily observed "haystacks" of bacteria (Fig. 11.6). Internal columnaris can only be confirmed by laboratory isolation. Hawke and Thune (1988) describe appropriate methods for its detection.

Control. The occurrence of columnaris epizootics can be minimized by preventing stressful situations. Try to maintain good environmental conditions in culture units and avoid handling fish when water

FIGURE 11.6.

"Haystacks" of *Flexibacter columnaris* on a wet mount of tissue. These haystacks are about 25 micrometers high. (Photo: A. J. Mitchell)

temperatures are high. When fingerlings must be seined and moved in warm water, complete the operation as quickly as possible and avoid crowding the fish excessively.

Treatment for columnaris depends on whether the infection is external or internal. For a purely external infection, treatment with potassium permanganate is usually effective (see the section Disease Treatments). It is important to confirm that the infection is purely external and that a mixed infection with another bacterium or channel catfish virus does not exist because treatment with potassium permanganate stresses fish and may increase mortality.

Internal columnaris requires antibiotic therapy with either Terramycin® or Romet®. When both internal and external columnaris are present, antibiotic therapy alone may control both infections.

Motile Aeromonad Septicemia

Motile aeromonad septicemia (MAS) is caused by a heterogeneous group of bacteria that are commonly found in aquatic habitats and

Bacterial Diseases **343**

among the normal microflora of the catfish intestinal tract. The bacteria that cause MAS are opportunistic pathogens that produce disease when fish are stressed. They are also common secondary invaders during other diseases. Motile aeromonad septicemia can be an acute or chronic disease, and epizootics are most common during the spring and fall months, when water temperatures are between 65 and 85°F.

Causative agents. Motile aeromonad septicemias of channel catfish are caused by Aeromonas sobria and A. hydrophila. These bacteria are motile gram-negative rods in the family Vibrionaceae. The motile aeromonads are a diverse group of bacteria, and strains differ greatly in virulence. Strains isolated from water or mud are usually less virulent than strains of the same species isolated from diseased fish (De Figuerredo and Plumb 1977). Also, strains isolated from diseased fish often lose virulence quickly when successively cultured on artificial media in the laboratory. Biochemical characteristics of the motile aeromonads are provided by Popoff and Lallier (1984).

All species of freshwater fish are susceptible to the disease, but MAS is most common in intensively cultured warm-water fish and in cold-water fish species, such as trout and salmon, held at marginally high water temperatures.

Clinical signs. Clinical signs of MAS vary greatly. The signs and severity of the disease depend on the interrelationships among the kind and degree of predisposing stress, general health of the fish, and virulence of the strain of Aeromonas involved in the infection.

In chronic MAS, mortality rate is low and feeding activity may not be significantly affected. Usually a small percentage of the population is affected, but affected fish may have severe external lesions. Lesions include frayed fins with hemorrhage at their base, petechial hemorrhages over the body surface, or shallow, round, open ulcers. In severe cases the underlying muscle may be exposed.

In acute MAS, losses may be great and a greater percentage of the population affected. Feeding activity usually decreases noticeably with onset of the disease. Petechial hemorrhages on the body surface are often the only external lesions. Commonly, the abdomen is distended and the eyes bulge. The body cavity may be filled with bloody fluid and the liver may be soft and enlarged. The lower intestine is often swollen, hemorrhagic, and filled with bloody or yellowish mucus.

Characteristics of the disease. Motile aeromonad septicemia is usually associated with a predisposing stressor such as low dissolved oxygen concentrations, high concentrations of nitrite or un-ionized ammonia, sudden changes in water temperature, or improper handling of fish. Epizootics can occur anytime during the year but are most common in the spring and fall, when water temperatures are between 65 and 85°F. The disease is particularly common after fish are seined or transported when water temperatures are above about 80°F. Infection by motile aeromonads is often the final insult to fish already debilitated by other bacterial infections or infestations of parasites. Fish grown at high densities in cages, raceways, or tanks are particularly prone to MAS, and under these conditions the disease can spread rapidly and losses may approach 100 percent. All sizes of catfish are susceptible to MAS, but the disease is more common in fingerling populations. Recurrence of the disease within a population is common if the underlying stressor is not removed.

Transmission. Little is known of the modes of pathogen transmission during outbreaks of MAS. Motile aeromonads are always present in the culture environment, but the role that routine environmental isolates play in outbreaks of the disease is not known with certainty. In the past it was assumed that bacteria from the environment infected fish when they became weakened by some type of stress. However, recent work has shown that bacteria isolated from the environment are less virulent and differ genetically from those isolated from fish with MAS (DeFiguerredo and Plumb 1977; Ford and Thune 1988). Thus, the factors involved in the initiation and spread of the disease may be less straightforward than previously believed. During outbreaks of acute MAS, highly virulent bacteria are transmitted from fish to fish through the water, and it is possible that the disease can be transferred from one culture unit to another in effluents or on contaminated equipment.

Diagnosis. Clinical signs alone cannot be used to diagnose MAS because they vary among epizootics and often resemble those of other systemic bacterial infections. Diagnosis must be based on identification of bacteria in laboratory cultures isolated from the internal organs of sick fish. A complete laboratory diagnosis is imperative because *Aeromonas* is frequently involved with mixed infections, and all infectious agents must be identified to treat the problem properly. Laboratory identification of motile aeromonads is discussed by Cipriano, Bullock, and Pyle (1984). Antibiotic-sensitivity tests should also be performed as part of the diagnostic procedure because *Aeromonas* bacteria are notorious for developing resistance to commonly used antibiotics.

Control. The key to minimizing the incidence and severity of MAS is the use of good cultural practices. Culture units should be stocked at conservative densities and every effort made to maintain good environmental conditions. Do not handle fish in warm water, but if they must be handled, try to minimize stress by conducting operations quickly and preventing overcrowding.

Antibiotic therapy, using either Terramycin® or Romet®, is used to treat MAS. In general, MAS is easier to control than other bacterial diseases if the following protocol is used:

1. Identify and remove the predisposing stressor. Often this alone controls the disease.
2. Identify any other infectious agents; in particular, confirm the presence or absence of *Edwardsiella ictaluri*. Secondary infection by *Aeromonas* is common during outbreaks of ESC; if this is the case, treatment should usually be directed at the potentially more serious ESC problem.
3. Identify the antibiotic to which the bacteria are sensitive.
4. Initiate antibiotic therapy early, before fish go off feed.
5. Do not stress fish during an outbreak of MAS; they should not be seined or transported until well after the disease has been controlled.

FUNGAL DISEASES

Fungi are important disease-causing organisms of channel catfish eggs and occasionally cause disease in fry, fingerlings, or adult fish. Species of *Saprolegnia* and *Achlya* are the most common fungi affecting channel catfish. These fungi, commonly called *water molds*, are all similar in appearance, and even experts may disagree on identification and classification. For convenience, the term *saprolegniasis* is used to describe external fungal infections caused by these water molds.

Water molds are present at all times in fresh water and normally are saprophytic on decaying organic matter. Saprolegniasis occurs only after fish are stressed or injured. Fungi are also common secondary invaders after other infectious diseases. Although saprolegniasis can occur at any time, the disease is much more common in channel catfish when water temperatures are below about 60°F.

The infective stage of water molds is the asexual reproductive spore (zoospore). The spores are free-swimming and attach to dead or injured tissues, where they germinate to produce filaments called *hyphae*. The hyphae grow and invade surrounding tissue to form a colony (mycelium). Infections can occur on any epidermal surface, including fins, barbels, and gills. In severe infections the physical integrity of the fish is affected to such a degree that loss of vital body fluids causes death. Saprolegniasis of channel catfish is limited to external infections; systemic infections of channel catfish by water molds have not been reported.

Conditions responsible for the development of saprolegniasis in channel catfish are not fully understood, although it is clear that some factor must predispose the fish to the disease and allow initial infection by zoospores. Physical injuries or lesions caused by other diseases are often sites for initial fungal colonization. Physical injuries may occur as a result of seining, handling, crowding, or fighting among fish during spawning. Saprolegniasis may also develop after exposure to stressful environmental conditions. Saprolegniasis is usually a chronic disease with a low mortality rate over a relatively long period of time. However, fungi are usually associated with "winter kill" (see the section Diseases of Uncertain Origin) and mortality rates may be high. But it is not known whether fungi are the primary cause of death in this poorly characterized syndrome.

In many respects, saprolegniasis can be considered the cold water counterpart of external columnaris infections. Both diseases are caused by ubiquitous aquatic microorganisms, and some predisposing factor is required for initiation of infection. External columnaris lesions are frequently invaded by fungi as water temperatures drop, and at moderate water temperatures (55 to 70°F) mixed infections of fungi and *Flexibacter columnaris* are common. At colder temperatures only fungi may be found. Likewise, lesions initially caused by fungi at low water temperatures may be invaded by *F. columnaris* as water temperatures rise.

Saprolegniasis can be tentatively diagnosed by the presence of white to brownish, cottony patches of fungus on any external surface of the fish. Patches may be small focal spots or larger. Areas of depigmentation with loss of the mucus layer may be present before masses of mycelia are obvious. The diagnosis can be confirmed by microscopic examination of scrapings from the lesion. Hyphae are long, continuous cells that occasionally branch. Cells do not have cross walls (septae) except at the terminal cells, which are sporangia containing round zoospores. A complete diagnosis should be performed to detect any possible mixed infections involving bacteria or parasites.

The occurrence of saprolegniasis can be reduced by maintaining good environmental conditions and otherwise minimizing stress during periods of cool or cold water temperatures. Episodes of high concentrations of environmental nitrite and ammonia are the major water quality problems in culture ponds during the cooler months of the year (see Chapter 9). Stress due to elevated nitrite concentrations can be minimized by monitoring nitrite concentrations and applying common salt to provide chloride, which decreases the toxicity of nitrite. Problems with high ammonia concentrations are difficult to manage in large ponds, but problems with ammonia (and nitrite) are less frequent and less severe in ponds stocked at moderate fish densities. In general, the occurrence of saprolegniasis (like that of most infectious disease) increases as stocking density increases.

Small batches of fish with saprolegniasis can be treated with a 1- to 2-minute dip in a 3 to 5 percent solution of salt (NaCl). Fish in larger culture units can be treated with potassium permanganate or formalin. The decision to treat with potassium permanganate or formalin should be carefully considered on the basis of the number of fish being lost and the expected benefit of the treatment. Treatment of large ponds with potassium permanganate or formalin is expensive, and fungi will simply reinfect the fish unless the underlying stressor that predisposed fish to the initial infection is identified and removed.

PROTOZOAN PARASITES

The *protozoans* are a heterogeneous group of microscopic, single-celled animals. Protozoans may be free-living, commensal, symbiotic, or parasitic; solitary or colonial; motile or sessile. Some protozoans, notably *Ichthyophthirius multifilis*, are obligate fish parasites and require fish as a host. Other protozoans are commonly found in or on fish and the association is considered normal; when present in low numbers they are regarded as commensal and cause insignificant or no damage to the fish. Under certain conditions, however, the relationship between the fish and protozoan is altered and the abundance of the protozoan increases dramatically to the detriment of the fish. The factors regulating fish parasitism are poorly understood, but epizootics involving certain protozoans often occur when poor environmental conditions, poor nutrition, or overcrowding diminishes the fish's natural resistance to parasites.

Most protozoan parasite infestations of channel catfish can be treated relatively effectively with one of the chemicals registered for use in food-fish production. The use of these chemicals is discussed in the section Disease Treatments. The decision to treat must, however, be carefully evaluated because the benefits of treatment often do not justify the cost. Consultation of a qualified fish disease specialist is highly recommended. Four factors should be considered before treating: First, the parasite must be accurately identified. The expected course of the epizootic and possible treatments depend on the particular parasite causing the problem. Second, most problems with parasites are the result of some adverse condition that predisposes the fish to infestation. Unless the underlying stress is addressed, treatment success may be only temporary. Third, mortality rates and expected fish loss must justify the expense of treatment. Some chemical treatments are expensive, particularly when used in large ponds, and there may be no economic justification for treatment if only a few fish are dying or if only a small percentage of the fish in the culture unit are affected. Only experience can indicate whether the number of fish lost in the early stages of an epizootic will escalate. Finally, many protozoan infestations are secondary to bacterial or viral infections. Some chemical treatments for parasites stress fish and may actually increase fish loss. The stress of treatment may also cause fish to go off feed, reducing the effectiveness of antibiotic therapy using medicated feed. A complete diagnosis, including bacteriology and virology, is necessary before a safe and effective treatment regimen can be formulated.

Five of the important protozoan parasites of channel catfish are briefly discussed in the following sections. Another important disease of channel catfish, proliferative gill disease, is probably caused by a protozoan, but this disease is discussed in the section Diseases of Uncertain Origin. Readers interested in more detailed information on protozoan parasites should consult Hoffman (1967), Hoffman and Meyer (1974), Post (1983), MacMillan (1985), and Rogers (1985).

Ichthyophthirius Multifilis

Ichthyophthirius multifilis ("Ich") is the most devastating parasite affecting channel catfish. Although epizootics are relatively uncommon, infestations of Ich spread rapidly under crowded culture conditions and entire populations of fish can be killed within a few days when conditions are optimal for the disease. Also, Ich is difficult to control, particularly under pond culture conditions.

Ich is the largest protozoan found on fish. Cells of the adult stage, called trophozoites, are oval to round and may measure up to 1/32 inch (1 mm) long. Large trophozoites can be seen with the naked eye. Trophozoites are uniformly ciliated around the body and a C-shaped nucleus is clearly visible in mature cells (Fig. 11.7). Trophozoites cause great tissue destruction as they migrate throughout the epidermis feeding on tissue fluids and cell debris. The trophozoite leaves the fish after maturing and settles on some substrate, such as pond bottom muds. The cell encysts and divides to produce 1,000 to 2,000 infective cells called tomites. These small (30 to 50 micrometers [μm]), ciliated, pear-shaped cells actively seek a new fish host. Once a host is contacted, the tomite burrows into the skin, where it matures to a trophozoite and the cycle begins again (Fig. 11.8).

Protozoan Parasites **349**

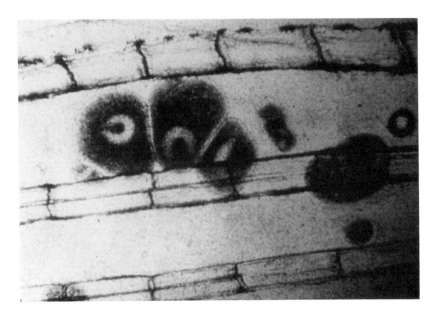

FIGURE 11.7.

Trophozoites of *Ichthyophthirius multifilis* with the characteristic C-shaped nucleus. (Photo: F. P. Meyer)

Ich is an obligate fish parasite. Tomites die if they do not locate a fish host within about 2 days after the cyst ruptures. The optimal temperature for reproduction is about 75°F; at this temperature the entire life cycle can be completed in 3 or 4 days. Maturation and reproduction are slower at cooler temperatures. At 45°F, the life cycle requires about 1 month to be completed.

Epizootics occur most frequently in the spring, when water temperatures are 70 to 80°F, although some outbreaks occur in fall and winter. Ich is extremely rare during the summer months, when water temperatures are consistently above 80°F. Fingerlings are particularly susceptible because they are usually held at high densities that enhance the spread of the disease.

The most obvious clinical sign of Ich is the presence of many raised, pinhead-sized spots on the skin or gills. These spots are areas of inflammation and tissue damage where tomites have burrowed into the skin. Fish may have a "rough" appearance due to sheets of mucus that slough off the skin. Fish behavior depends on the severity of the infestation. In the early stages of an epizootic when infestation is light, no changes in behavior will be noticed. When infestations become

350 Infectious Diseases

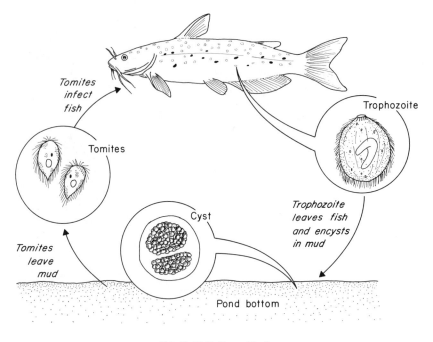

FIGURE 11.8.

Life cycle of *Ichthyophthirius multifilis*.

heavy, fish will be lethargic and may "flash" as they writhe in the water, apparently in response to the irritation caused by the parasite. Heavy parasite infestations on the gills can hinder gas exchange, and fish may swim at the surface gasping air, even when dissolved oxygen concentrations are normally adequate.

Tentative diagnosis is based on the presence of the white spots described previously, but this is not specific to this disease. Microscopic observation of mature trophozoites from one or more of the white spots is conclusive.

Ich can be extremely difficult to control because the parasite resides beneath the epithelium of the skin. Epizootics are controlled by breaking the life cycle by killing the parasite once it has left the fish as a trophozoite or before it infects the fish as a tomite. Two different treatment regimens have been used; both are based on treatments that are repeated until the disease is controlled. One treatment regimen consists of treatments of copper sulfate repeated every other day. The amount of copper sulfate applied at each treatment depends on the total alkalinity of the water (see the section Disease Treatment). Five to seven treatments is usually sufficient, but fish should be sampled

Protozoan Parasites **351**

throughout the treatment to verify success. More than seven treatments may be necessary in some instances. The other treatment regimen involves alternating treatments of potassium permanganate, formalin, and copper sulfate. The treatment rate for potassium permanganate depends on the amount of organic material in the water (see the section Disease Treatment). Formalin is used at 15 ppm; copper sulfate is applied at a rate that varies with total alkalinity. This treatment is more expensive than use of copper sulfate alone but is generally more effective and may only require one cycle of three treatments (one treatment with each chemical). The interval between successive treatments with the three different chemicals should vary with water temperature because the life cycle of Ich slows as water temperatures decrease. The following schedule is suggested:

Water temperature	Treatment schedule
80°F	Every day
70°F	Every other day
60°F	Every third day
50°F	Every fourth day
40°F	Once a week

Every effort should be made to prevent introduction of the disease to the farm and spread of disease once an outbreak starts. Never transfer fish with Ich into a culture unit with other fish. Wild fish in the water supply or in the culture unit can serve as a reservoir for the parasite. Ich can be transferred from one culture unit to another through effluents or on boots, seines, boats, or other equipment, so the culture unit containing fish with Ich should be quarantined to the extent possible.

Trichodina Species

Trichodina species are saucer-shaped parasites with an inner ring of interlocking denticles (Fig. 11.9). Several species of Trichodina can infest fish at the same time, and two "forms" (possibly several different species) are common: small cells found primarily on the gills and larger cells found primarily on the body. Trichodinids on the gills are more detrimental than those on the body. Heavy infestations of the parasite irritate gill surfaces, causing excessive mucus production, which decreases respiratory efficiency. Fish may suffocate at normally adequate concentrations of dissolved oxygen.

Trichodinids are common on normal, healthy channel catfish and cause problems only when they become excessively abundant. Epizoot-

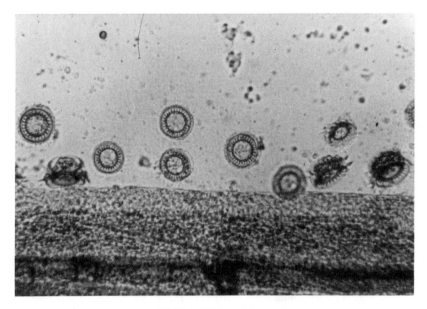

FIGURE 11.9.

Trichodina species on gills. Parasites are about 25 micrometers in diameter. (Photo: A. J. Mitchell)

ics are invariably associated with a predisposing stressor, usually poor water quality. Secondary infestation by Trichodina is also common after fish are debilitated by a bacterial disease. Epizootics involving Trichodina occur throughout the year but are most common in the spring and fall.

The abundance of Trichodina infesting fish can be effectively reduced by treating with copper sulfate. Potassium permanganate and formalin are also effective. Unless the condition that predisposed the fish to the initial infestation by Trichodina is also addressed, reinfestation will occur.

Ambiphrya (Scyphidia) Species

The Ambiphrya (formerly Scyphidia) protozoan is barrel-shaped with a ring of cilia on the distal (unattached) end of the cell (Fig. 11.10). Ambiphrya is common on the skin and gills of healthy catfish and becomes a problem only when numbers increase dramatically under poor environmental conditions. Epizootics seem to be most common in

Protozoan Parasites **353**

FIGURE 11.10.

Ambiphrya species on gills of channel catfish. Parasites are about 25 micrometers in diameter. (Photos: G. L. Hoffman, left, and A. J. Mitchell, right)

nursery ponds where fingerlings are held under highly crowded conditions (>200,000 fish/acre). Heavy infestations on gills cause excessive mucus production, which reduces respiratory efficiency. Also, Ambiphrya may become so abundant on gills that the layer of parasites physically impedes gas exchange and the fish suffocates. Epizootics involving Ambiphrya occur throughout the year but are most common in the fall, winter, and spring. Maintenance of good water quality reduces the incidence and severity of Ambiphrya epizootics. Ambiphrya is usually controlled effectively with copper sulfate. Potassium permanganate, formalin, and concentrated salt (NaCl) solutions also can be used.

354 Infectious Diseases

FIGURE 11.11.

Trichophrya species. Parasites are about 35 micrometers in diameter. (Photo: Roger Dexter)

Trichophrya Species

Trichophrya adults are orange or brown, dome-shaped cells with tentacles that are used for gathering food (Fig. 11.11). The organism is commonly found in low numbers on channel catfish gills, often residing in the interlamellar troughs. At low numbers they are considered commensals and of little consequence to fish health. However, Trichophrya can be a serious parasite of channel catfish when present on gills in moderate to high numbers. Increased abundance of Trichophrya usually occurs after fish are stressed. Epizootics are most common in spring and fall.

Heavy infestations of Trichophrya cause swelling and erosion of the gill tissues as well as excessive production of mucus. The gills look ragged and fish may become anemic. Diagnosis is based on observation of the organism in slide preparations of gill tissue. Copper sulfate is the most effective chemical treatment.

Ichthyobodo (Costia)

Ichthyobodo necator (formerly Costia necatrix) is an obligate fish parasite that can cause extensive losses of cultured channel catfish. Ichthy-

Protozoan Parasites **355**

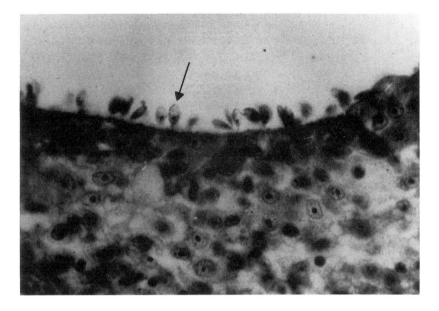

FIGURE 11.12.

Icthyobodo necator (arrow). Parasites are about 2 micrometers long. (Photo: H. S. Davis)

obodo is a small, teardrop-shaped parasite that attaches to the gills or skin (Fig. 11.12). The parasite is small, about the size of a catfish red blood cell, and can be difficult to see under a microscope. After *Ichthyobodo* attaches to a host cell, small tubules penetrate the cell and the cell contents are sucked out as food. *Ichthyobodo* infestations can cause considerable tissue damage and loss of vital fluids. Excess mucus production associated with gill infestations of *Ichthyobodo* may also decrease respiratory efficiency.

Epizootics associated with *Ichthyobodo* usually occur after some predisposing stress; catfish crowded in cages, for example, often carry heavy infestations. *Ichthyobodo* can be found on fish at any time of the year, but most epizootics occur during the cooler months of the year. Epizootics may be chronic or acute, and losses can be considerable.

Fish infested with *Ichthyobodo* may become listless and refuse to feed. The skin and gills of affected fish may become covered with copious, thick mucus. Occasionally the mucus takes on a bluish coloration, giving the disease its common name, "blue slime disease." Diagnosis is based on observation of the parasites on microscope slide preparations of gill or skin scrapings. The parasite is most easily observed when alive and motile; on gill preparations it may be seen "fluttering" like a tree leaf in a breeze.

Maintenance of good environmental conditions decreases the incidence and severity of the disease. Epizootics associated with *Ichthyobodo* can usually be treated effectively with potassium permanganate or formalin.

METAZOAN PARASITES

Metazoans are multicellular organisms that develop from embryos. The metazoan parasites of channel catfish include certain crustaceans, trematodes (flukes), cestodes (tapeworms), nematodes (roundworms), and leeches. These parasites are common in channel catfish, particularly those captured from the wild. Their effect on fish health is usually minimal. With the exception of the gill flukes (discussed later), metazoan parasites are not known to cause significant problems in populations of cultured channel catfish. Information on metazoan parasites of fish can be found in Post (1983) and Plumb (1985a).

Gill flukes in the genus *Ligictaluridus* (formerly *Cleidodiscus*) are the only metazoan parasites of significance to commercial channel catfish culture. Gill flukes are trematodes that have a single host in their life cycle (monogenetic). Adult gill flukes have a posterior organ of attachment (haptor) bearing two pairs of hooks. The haptor is embedded into the gill epithelium and causes mild inflammation. The presence of four eyespots is characteristic of *Ligictaluridus* spp. Gill flukes possibly interfere with respiration when present in large numbers. The greatest risk associated with the presence of gill flukes is secondary infection by bacteria. Generally, gill flukes are a problem only for small fingerlings. The adult flukes produce eggs from which small, ciliated larvae hatch. The larvae then actively seek a new host. Proliferation of gill flukes is usually associated with poor environmental conditions and overcrowding. Potassium permanganate or formalin can be used to reduce the numbers of gill flukes affecting fish, but it is difficult to eliminate them completely.

DISEASES OF UNCERTAIN ORIGIN

The causes of three important diseases of cultured channel catfish are somewhat obscure. One of these diseases, called "winter kill," is a leading cause of fish loss in pond-raised channel catfish in the southeastern United States. The two other diseases, proliferative gill disease and "no blood disease," are relatively uncommon, although large numbers of fish can be lost when they occur. Proliferative gill disease appears to be caused by a highly infectious and pathogenic protozoan parasite. The causes of the other two diseases are not known and may be environment- or nutrition-related.

Winter Kill

The term *winter kill* has historically referred to the loss of fish to dissolved oxygen depletion in northern lakes covered with ice for long periods of time. The same term has been colloquially coined to refer to a serious disease problem affecting pond-reared channel catfish during the late fall, winter, and early spring when water temperatures are below about 55°F. Other terms used to describe this problem include "winter mortality," "winter fungus," and "winter dwindles."

The condition is characterized by loss of mucus; dry, whitened patches of skin; and sunken eyes (endophthalmia). External fungal infections are common. Occasionally large numbers of fish die with little external evidence of the disease. The disease is usually chronic, and total losses within a pond population may be high. Losses appear to be greatest after periods of ice cover but are not related to dissolved oxygen depletion. Although all sizes of fish can be affected, larger fish (over 0.5 pound) are most likely to be lost.

It is currently believed that winter kill is caused by some predisposing condition that decreases the ability of the fish to adapt to cold or rapidly fluctuating water temperatures. Failure to adapt then affects the normal function of the skin, gills, or immune system. Several factors, including previous exposure to poor environmental conditions, history of infectious disease, and inadequate nutrition, have been suggested to predispose fish to winter kill. However, the cause and pathogenesis of the disease remain largely unknown.

There is no cure for winter kill once it has started. Treatment with potassium permanganate sometimes removes the external fungus temporarily, but because the fish's immune system is suppressed at low water temperatures, fungi often reinvade the lesions. Control of winter kill must therefore be directed to prevention. Use conservative fish stocking densities and try to reduce standing crops of fish by harvesting those ponds with excessive standing crops of market-sized fish before the onset of winter. To the extent possible, maintain good water quality throughout the year, especially in the fall, when water quality problems are common. Make sure that any infectious disease problem is

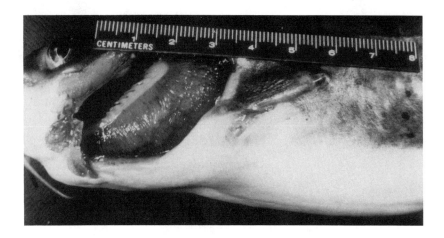

FIGURE 11.13.

Proliferative gill disease of channel catfish. The operculum has been removed to show the swollen, mottled gills. (Photo: A. J. Mitchell)

quickly identified and effectively treated. Also, maintain the best plan of nutrition possible by feeding fish throughout the winter according to an accepted schedule (see the section Feeding Practices). If winter kill is noticed in a pond, try to minimize losses by harvesting the fish. If the disease is identified early, only a small percentage of the fish may be affected and the remainder will be completely acceptable for processing.

Proliferative Gill Disease

Proliferative gill disease (PGD) is an acute, devastating disease of pond-raised channel catfish. The gills of fish with PGD become severely swollen and fragile (Fig. 11.13). The gills may bleed if they are touched. Areas of the gills become necrotic and brown. The raw, mottled appearance of the gills gives the disease its common name, "hamburger gill disease." The severe changes in gill structure reduce respiratory efficiency, and fish with PGD behave as if they are suffocating, even when dissolved oxygen concentrations are normally adequate. Fish may swim lethargically at the surface gasping for oxygen or may lie quietly along the bank. The disease can affect all sizes of channel catfish.

Epizootics of PGD are relatively uncommon, but losses can be tremendous in affected populations. The occurrence of the disease is temperature-dependent: epizootics occur when water temperatures are consistently between 60 and 70°F. In Mississippi, PGD occurs in mid-April to mid-May and, less frequently, in October. The disease appears to occur most frequently in newly constructed ponds or those that have been drained, dried for an extended period, and then refilled.

Several causes of PGD have been postulated, including various protozoan parasites and water-borne toxic substances such as un-ionized ammonia and algal toxins. It now appears the PGD is caused by a myxosporean parasite, probably a species of *Sphaerospora* or a myxosporean closely resembling *Sphaerospora* (Groff, McDowell, and Hedrick 1989; MacMillan, Wilson, and Thiyagarajah 1989). The organism invades the gills and causes a marked inflammatory response. The life cycle of the parasite and the mode of transmission of the disease are unknown. Fish can become infected when exposed to water or mud from a pond containing channel catfish with PGD, but fish-to-fish transmission does not appear to occur readily.

Proliferative gill disease can be tentatively diagnosed from the appearance of the gills and the occurrence of the disease at conducive temperatures, but the gross clinical signs are not restricted to PGD. Microscopically, distinct breaks in the gill cartilage, appearing as clear areas between darker areas of cartilage, are indicative of proliferative gill disease. Confirmation requires microscopic examination of gill tissues prepared by special techniques.

There is no cure for PGD, nor is there any known way to prevent it. Treatment with chemicals is ineffective and may further irritate the gills, making conditions worse. The only effective treatment is to maintain dissolved oxygen concentrations as near saturation as possible by using supplemental aeration. Some farmers have reported that mortality due to PGD decreases when water from an adjacent pond is pumped into the pond with PGD-affected fish, but this practice has no logical basis unless the pumped water provides additional oxygenated water. Because so little is known about PGD, make every effort to avoid the possibility of transferring the disease from pond to pond. Do not transfer fish with PGD to another pond containing unaffected fish, and do not use equipment that has been in contact with water, mud, or fish with PGD in other ponds until it has been disinfected or allowed to dry thoroughly.

"No Blood Disease"

"No blood disease" is a trivial name given to a disorder characterized by severe anemia. The packed red blood cell volumes (hematocrits) of affected fish may be less than 5 percent. Packed cell volumes of normal channel catfish range from about 25 to 45 percent. The disease is relatively rare, although at least 70 cases were reported in both 1983 and 1984 from Alabama and Georgia (Butterworth, Plumb, and Grizzle 1986). The disease is usually chronic, but many fish can be killed if dissolved oxygen concentrations fall to low levels in the pond with affected fish.

The gills and internal organs of affected fish are pale to white, and the blood is pink to straw-colored rather than the usual bright red. Often, the intestine is telescoped upon itself (intussusception). In some cases only the larger fish in the pond are anemic and the smaller fish continue to feed vigorously and appear unaffected. In other instances, the severity of the anemia within the population is variable but does not seem size-related. The anemia seems to be tolerated until dissolved oxygen concentrations fall to marginal levels (less than 4 to 5 ppm), at which time affected fish show signs of respiratory distress: fish become listless and may lie motionless along the bank. Severe anemia can occur at any time of the year but is usually reported most frequently in the warmer months. In Mississippi the disease occurs most frequently in early spring and late fall (MacMillan 1985); in Alabama and Georgia outbreaks usually occur in late spring and summer (Klar, Hanson, and Brown 1986). Whether these differences in seasonal incidence are related to different causes of the anemias is not known.

Anemias can be caused by a variety of agents and disorders, and this characteristic complicates the diagnosis of this disease. Anemias, usually mild in nature, are associated with many infectious diseases. A generally mild but variable anemia also develops after exposure to high concentrations of nitrite (Tucker, Francis-Floyd, and Beleau 1989). However, most cases of severe anemia in farm-raised channel catfish appear to be feed-related (Butterworth, Plumb, and Grizzle 1986; Klar, Hanson, and Brown 1986). Anemia develops when fish consume feed that is contaminated with microorganisms capable of converting folic acid to pteroic acid. Pteroic acid is a folic acid antagonist and causes cessation of red blood cell maturation. If the anemia is determined to be feed-related (usually by ruling out other possible causes), using a fresh batch of feed often results in recovery. Feed-related anemias are more common on small farms where large purchases of feed are made once or twice a year and feed is stored for long periods; problems can be prevented by not using old, moldy feeds. If possible, store feeds under cool dry conditions or purchase limited amounts so it can be completely used in a relatively short period. Regardless of the cause, dissolved oxygen concentrations should be maintained as near saturation as possible if fish are found to be anemic.

DISEASES OF EGGS

Channel catfish eggs in hatcheries can become infected with fungi or bacteria. If accepted hatchery management practices are used and the hatchery water supply is of good quality, diseases are uncommon and no more than an occasional nuisance. Losses can be great, however, if eggs are mishandled or environmental conditions are poor. Severe disease problems in hatcheries affect egg hatchability, and any fry produced may be weakened and survival rate in nursery ponds is poor.

The three most common causes of disease problems in hatcheries are improper water temperature, inadequate aeration, and poor sanitation. Optimal conditions for egg development include water temperatures of 78 to 82°F and minimum dissolved oxygen concentration of 5 ppm. Environmental requirements and cultural practices for hatchery operation are discussed in Chapter 7.

Fungal Diseases

Water molds, such as *Saprolegnia,* are the most common microorganisms infecting channel catfish eggs. These fungi are common in most waters and are introduced into the hatchery on egg masses brought into the hatchery. Hatchery waters are also continually inoculated with airborne propagules of fungi, so it is nearly impossible to prevent their introduction into the hatchery. Water molds live on dead organic matter and are more abundant in tanks and troughs in which large amounts of egg debris or dead fry have accumulated as a result of infrequent cleaning. Fungal growth in egg masses usually begins on infertile or dead eggs. Once the colony is established, fungi may then invade and kill healthy eggs. Fungal colonies appear as white to brown cottonlike growths on the egg mass.

Problems with fungi are most common in hatcheries using cool water. Fungal infections of egg masses occur most frequently if the hatchery water temperature is below 75°F; problems are relatively rare when temperature is above 78°F. Eggs with fungal infections are also common in the early part of the spawning season when water temperatures in brood ponds are marginal for spawning activity (70 to 75°F). Not only are these temperatures conducive to fungal infection but early-season spawns often contain many infertile eggs that can serve as foci for colonization by fungi.

If only a few egg masses within the hatchery are infected, the spread of fungal colonies throughout individual egg masses can be controlled by inspecting the masses frequently and pinching off and discarding the small portions that are infested. If the problem is more widespread and many egg masses are infected, treatment with formalin kills the fungus. Turn the water off during treatment but leave the paddles in the hatching trough turning. Add formalin evenly throughout the tank to achieve a concentration of 100 ppm. After 15 minutes, flush the trough completely. Do not treat eggs with formalin if eggs are within 1 day of hatching because they may be killed.

Bacterial Egg Rot

Bacteria of the genera *Aeromonas*, *Flavobacter*, and *Acinetobacter* can infect and kill eggs. These bacteria are common inhabitants of aquatic environments and cause problems only when eggs are mishandled or poorly aerated or when water temperatures are above 82°F. Problems with bacterial egg rot are also more common when troughs are cleaned infrequently. The first indication of bacterial infection is a milky white patch of dead eggs, often on the underside or center of the egg mass. The gelatinous matrix between the eggs is dissolved as the infection progresses. There is no effective registered therapeutant for bacterial egg rot. The disease can usually be prevented by using good sanitation, maintaining dissolved oxygen concentrations above 5 ppm, and keeping water temperatures between 78 and 82°F. If bacterial egg rot is noticed on an occasional egg mass, try to excise and discard the affected portion. Large egg masses with bacterial egg rot should be split after excision into two smaller masses. This will help assure better oxygenation of the eggs in the center of the mass.

DISEASE TREATMENTS

Prevention, rather than treatment, should be the goal in fish health management. Many diseases occur because stress reduces the natural ability of fish to ward off disease organisms. The incidence of disease can be reduced by maintaining good water quality, ensuring good nutrition, and preventing stresses related to overcrowding and improper handling. With the possible exception of providing good nutrition, it is not always possible to avoid these stressors in commercial catfish culture. For catfish farming to be economically feasible, fish must be grown at unnaturally high densities that lead to poor environmental conditions. Thus, disease problems are relatively common in commercial catfish culture, and occasionally corrective treatment is necessary.

All therapeutants used in catfish farming act as supplements to the fish's natural defense system. They are not "magic bullets" that cure the disease simply because they are applied. Therapeutants retard the growth or kill most of disease-causing organisms, but ultimately, the fish's defense system must function to overcome the disease. Failure of treatment with therapeutants to resolve disease problems can often be traced to some underlying factor, such as poor water quality, that prevents natural defense mechanisms from operating effectively. Treatment may temporarily reduce the abundance of bacteria or parasites, but the disease recurs because fish are still weakened. Identification and treatment of predisposing stressors must therefore be the first steps in fish health management. The efficacy of subsequent treatment, if required, then depends on a prompt, accurate diagnosis of the disease; evaluation of the benefits of treatment relative to costs and risks; and selection of the proper treatment regimen. These considerations have been mentioned in previous sections of this chapter and are also discussed by MacMillan (1985) and Wellborn (1985). The importance of consultation of an experienced, qualified fish disease specialist cannot be overstated.

Relatively few therapeutant compounds are legal for use in the production of catfish for human consumption. The limited options available for disease treatment and the inconsistent results sometimes obtained from their use often tempt catfish farmers to use unregistered compounds. Admittedly, there is a pressing need for additional registered therapeutic agents; however, the use of unregistered compounds is illegal and the catfish farmer is ethically and morally responsible for ensuring the proper use of therapeutants and production of a wholesome food product. Also, most infectious diseases can be treated effectively with registered compounds. As mentioned, inconsistent treatment results are often attributable to failure to address all aspects of the disease properly. Meyer and Schnick (1989) discuss the registration process and list the compounds currently registered for use in aquaculture by the United States Food and Drug Administration or the Environmental Protection Agency.

Antibiotics

Oxytetracycline (Terramycin®) and sulfadimethoxine plus ormetoprim (Romet®) are the only two antibiotics currently registered for use by the U.S. Food and Drug Administration for use in catfish grown for human consumption. Terramycin® is marketed by Pfizer, Inc.; Romet® is marketed by Hoffman-LaRoche, Inc. Both drugs are for use as treatments for bacterial diseases and can only be used in the manufacture of medicated feeds.

Medicated feeds are formulated to deliver a certain amount of drug per given weight of fish per day. Standard units of treatment are often given as grams of drug/100 pounds of fish/day. The amount of drug delivered thus depends on the concentration of drug in feed and the feed intake by fish.

Farmers often become frustrated when medicated feed therapy fails to stop mortality within 1 or 2 weeks. Poor results can usually be attributed to one or more of the following factors: (1) failure of the fish to eat the feed, (2) bacterial resistance to the drug used, or (3) factors other than bacterial infection contributing to mortality.

Oral application of antibiotics is the only feasible method of treating populations of fish in large culture units, but it is sometimes difficult to achieve the proper dosage because sick fish usually feed poorly. In fact, one of the first indications of a disease problem is reduced feeding activity. The disease must be promptly diagnosed and treatment initiated as soon as possible before feeding activity is reduced to the point where inadequate amounts of drug are consumed by the fish. When using medicated feed, make every effort to coax fish into consuming the proper amount. Feed in the afternoon when dissolved oxygen concentrations are highest and fish are most likely to feed well. Feeding more than once a day and applying the feed over a large area rather than in one spot may also help increase feed consumption.

Possible resistance of bacteria to Terramycin® or Romet® must be determined as soon as possible to ensure that the drug used will be effective. Antibiotic sensitivity tests can be performed at most government or university fish disease diagnostic laboratories in areas where catfish are grown commercially. Personnel at these laboratories should also be able to provide a complete diagnosis and identify all factors contributing to mortality. If antibiotic therapy fails to stop mortality within 7 to 14 days, additional fish should be examined to determine whether the cause of the disease or antibiotic resistance patterns have changed.

Oxytetracycline (Terramycin®).

Oxytetracycline is a broad-spectrum bacteriostatic drug that inhibits protein synthesis in many gram-positive and gram-negative bacteria. The drug is registered for treatment of infections by *Aeromonas hydrophila*. Oxytetracycline-medicated feed has also been used to treat other *Aeromonas* infections, enteric septicemia of catfish, and systemic columnaris infections. Medicated feed containing oxytetracycline is formulated and then fed at a rate to deliver dosages of 2.50 to 3.75 grams oxytetracycline/100 pounds of fish/day. The most common feed formulation currently used contains 50 pounds of TM-100®/ton of finished feed (TM-100® is a Terramycin® premix containing 100 grams of oxytetracycline/pound of premix). The finished feed contains 2.5 grams oxytetracycline/pound of feed. When fed at 1.0 percent of body weight per day, the dosage is 2.50 grams oxytetracycline/100 pounds of fish/day. Feeds containing oxytetracycline are usually formulated as sinking pellets because the drug is heat-labile and is destroyed during the manufacture of extruded, floating feeds. Some feed manufacturers produce a floating medicated feed by spraying a solution of oxytetracycline onto the surface of floating feed pellets.

Oxytetracycline-medicated feed is fed for 7 to 10 days. There is a mandatory 21-day withdrawal period before slaughter.

Sulfadimethoxine plus ormetoprim (Romet®). Romet® contains a 5:1 mixture of sulfadimethoxine and ormetoprim. Sulfadimethoxine is a broad-spectrum sulfonamide antibiotic whose antibacterial activity is potentiated by ormetoprim. Romet® is registered for control of enteric septicemia of catfish caused by *Edwardsiella ictaluri*. The drug is also effective in the treatment of motile aeromonas septicemias caused by susceptible strains of *Aeromonas hydrophila* and *A. sobria* and systemic columnaris infections caused by susceptible strains of *Flexibacter columnaris*.

Romet®-medicated feed is administered to deliver 2.3 grams active ingredients/100 pounds fish/day. The most common commercial formulation contains 33.3 pounds of Romet-30® premix/ton of finished feed. Romet-30® premix contains 113.5 grams (25 percent) sulfadimethoxine and 22.7 grams (5 percent) ormetoprim/pound of premix. This feed delivers the required dosage rate when fed at 1 percent of body weight daily. Feeds containing Romet® should be formulated with higher levels of fish meal to improve the palatability of the feed (Robinson et al. 1990). The palatability of Romet®-medicated feeds is reduced to catfish by the bitter taste of the ormetoprim portion of the drug (Wilson and Poe 1989). High levels of fish meal (at least 16 percent) partially mask that taste. Romet® is heat-stable and is usually formulated in extruded, floating pellets.

Romet®-medicated feeds should be fed for 5 consecutive days. If mortality continues, additional sick fish should be examined to reestablish the cause of mortality and susceptibility of the bacteria to the drug. An additional period of medicated feed therapy may be necessary. A 3-day withdrawal period before slaughter is mandatory.

Chemical Therapeutants

External parasite infestations and some external bacterial infections must be treated by applying chemicals to the water. The most commonly used chemical therapeutants are potassium permanganate, copper sulfate, formalin, and salt (sodium chloride). With the exception of salt, these compounds are rather toxic to fish and the concentration of chemical required for an effective treatment may be only slightly lower than the concentration that will kill fish. Also, treatment rates with potassium permanganate and copper sulfate vary depending on the quality of water in which the chemical will be used. Inexperienced catfish farmers should consult a fish disease specialist before using chemical therapeutants.

Three methods can be used to treat diseases with chemical therapeutants: dip treatments, bath treatments, and indefinite treatments. Choice of treatment depends largely on the type of culture unit being used.

Dip treatments are used when relatively small numbers of fish are to be treated. They usually involve placing fish in a net and dipping them into a strong solution of chemical for a brief period (usually less than 60 seconds). Dip treatments are seldom used in commercial catfish culture because of the labor involved, the stress imposed on the fish from excessive handling, and the potential danger involved in exposing fish to concentrated chemicals.

Bath treatments involve adding the chemical directly to the culture unit and exposing fish to the chemical for a specified time, usually 1 hour or less. The chemical is then quickly flushed from the culture unit with fresh water. The chemical must be applied evenly throughout the unit to prevent "hot spots" of excessively high chemical concentrations that could kill fish. Fish should be observed throughout the treatment period to make sure they are tolerating it. Fry, small fingerlings, and sick fish of any size often cannot tolerate recommended dosages. Flush the chemical from the unit at the first sign of distress. Bath treatments are used for fish in raceways, tanks, or cages (see Chapter 13).

Indefinite treatments are used in ponds. A relatively low concentration of the chemical is applied and allowed to dissipate naturally. Although this is the only feasible method for treating fish in ponds, large amounts of chemical are required and treatment may be expensive. Also, potassium permanganate, copper sulfate, and formalin are toxic to plants, and their use may result in depletion of dissolved oxygen or other water quality problems. Adequate aeration equipment must be readily available any time these chemicals are used in ponds.

Chemicals used in indefinite pond treatment must be quickly and evenly applied. A boat specially equipped for chemical application should be available on large catfish ponds. If such a boat is not available, dry chemicals (potassium permanganate or copper sulfate) can be weighed into burlap bags and towed behind, letting the chemical dissolve into the water. Liquids, such as formalin, can be siphoned or pumped from the container into the propeller wash of the outboard motor.

Potassium permanganate, copper sulfate, and formalin are potentially toxic to humans and may irritate skin or mucous membranes. These chemicals should be applied only by people experienced in handling dangerous chemicals. Applicators should avoid contact with the chemical and should wear protective clothing, goggles, masks, and gloves during applications. Be sure to read and carefully observe all label directions and precautions.

Potassium permanganate. Potassium permanganate ($KMnO_4$) is a strong oxidizing agent that is registered for use as an oxidizer and detoxifier in waters used to raise food fish. The chemical is a dry powder that must be completely dissolved during treatment. For purposes of treatment rate calculations it is considered 100 percent active.

Potassium permanganate has been used to control external protozoan parasites, gill flukes, and external fungal and bacterial infections of catfish. The chemical is particularly effective for treatment of infestations of *Trichodina* or *Ambiphrya* and external columnaris infections. Potassium permanganate is also used as part of a treatment regimen with copper sulfate and formalin to treat infestations of *Ichthyophthirius multifilis* (Ich). The protozoan parasite *Trichophrya* is relatively resistant to potassium permanganate treatments. The extent of mortality and the prognosis for the disease must be carefully considered before using potassium permanganate because it is one of the more expensive chemicals used in catfish culture. Potassium permanganate is also quite toxic to catfish, and considerable care must be exercised whenever it is used. The chemical is phytotoxic (toxic to plants) and may kill the phytoplankton bloom and cause a dissolved oxygen depletion when used in ponds.

Potassium permanganate can be used as a bath treatment at 10 ppm for 20 to 30 minutes. Fish must be watched carefully during treatment because the toxicity of the chemicals varies widely with fish size, general condition of the fish, and amount of organic matter in the water. Dip treatments using concentrated solutions of the chemical are not recommended because strong potassium permanganate solutions are extremely caustic.

Potassium permanganate is consumed in reactions with organic material present in pond water, and the amount of chemical needed for an effective indefinite treatment of fish in ponds increases as the organic content of the water increases. In relatively clear ponds with sparse phytoplankton blooms, diseases can be controlled with treatments of 2 to 4 ppm. Treatment rates exceeding 12 ppm may be needed in ponds with dense phytoplankton blooms because so much of the chemical is destroyed in reactions with organic matter. The amount of potassium permanganate needed to treat ponds is usually determined by applying the chemical in increments of 2 to 4 ppm until a wine red color, characteristic of residual permanganate, persists in the pond for several hours. As the chemical reacts with organic matter in the water, the color changes from wine red to reddish brown to a dirty brown color. Judging the correct color requires considerable experience and it is easy to undertreat or overtreat ponds. If too little chemical is added, the treatment is ineffective and will have to be repeated. If too much is added fish may be killed.

The amount of potassium permanganate needed for an effective, but safe, indefinite treatment of fish in ponds can be estimated from a simple test called the "15-minute potassium permanganate demand test." This test determines the amount of potassium permanganate destroyed by organic matter during a 15-minute period. This amount of potassium permanganate is the 15-minute potassium permanganate demand of the water. The amount of chemical needed to treat a particular pond effectively is roughly 2.5 times the 15-minute potassium permanganate demand of a sample of water from that pond (Tucker 1989).

The 15-minute potassium permanganate demand test is easy to perform, but requires some laboratory glassware and an accurate balance, so it is best conducted in a fish disease diagnostic laboratory. The test is conducted as follows:

1. Prepare a 1,000 ppm potassium permanganate solution by dissolving 1.000 gram of potassium permanganate and diluting to 1,000 milliliters with distilled water. This solution must be prepared the day of the test.
2. Measure eight portions of 1,000 milliliters of pond water into a series of glass beakers. Pond water samples should be obtained from several locations and combined into a single sample. A total of about 2 1/2 gallons of pond water will be needed.
3. Make potassium permanganate treatments of 1, 2, 3, 4, 5, 6, 7, and 8 ppm by adding 1.0, 2.0, 3.0, 4.0, 5.0, 6.0, 7.0, and 8.0 milliliters of the stock 1,000 ppm potassium permanganate solution, respectively. Immediately stir the solutions to mix. You now have a series of eight samples whose colors range from light pink to purple-red.
4. After 15 minutes determine the lowest concentration of potassium permanganate that still has a faint pink color. This is the 15-minute potassium permanganate demand. Differences in the series of colors will be subtle, and the tendency is to over-

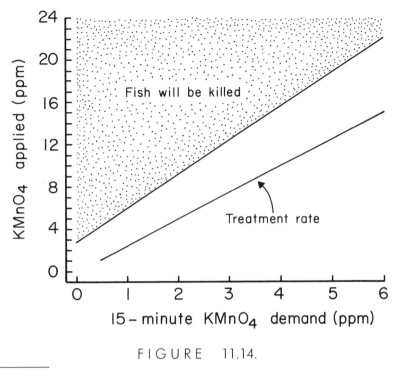

FIGURE 11.14.

Relationships between the 15-minute potassium permanganate ($KMnO_4$) demand of water and (1) the disease treatment rate estimated by multiplying the 15-minute demand by a factor of 2.5; (2) concentrations of potassium permanganate toxic to channel catfish (adapted from Tucker 1989).

estimate the demand by looking for an obvious pink color. The appropriate choice will contain only a hint of pink color.

The estimated pond treatment rate is then calculated by multiplying the 15-minute potassium permanganate demand by 2.5. The estimated rate should only be used as a guide because there may be other factors that affect the behavior of potassium permanganate in pond water. The chemical should still be applied in increments, and as the total amount applied approaches the estimated rate, the applicator must use judgment as to the correct treatment.

The presence of organic material in water decreases the toxicity of potassium permanganate to channel catfish, so a margin of safety is provided even at the high treatment rates necessary in organically laden pond water. In Figure 11.14, estimated potassium permanganate treatment rates at various 15-minute demand values are compared to laboratory-derived estimates of potassium permanganate concentrations that will kill fish.

Copper sulfate. Copper sulfate ($CuSO_4 \cdot 5H_2O$) is registered for food-fish use as a herbicide and algicide. It has also been used as a parasiticide and is usually the treatment of choice for *Ichthyophthirius* and *Ambiphrya*. Copper sulfate is available as various-sized granules or as a powder. The powder form should be used because it dissolves quickly. Copper sulfate is considered 100 percent active when calculating treatment rates.

The efficacy of copper sulfate as a herbicide or parasiticide (and its toxicity to fish) varies in a complex fashion with the pH, total alkalinity, and total hardness of the water. In acidic waters of low total alkalinity and hardness, copper sulfate is extremely toxic to fish. To simplify calculations, copper sulfate treatment rates are usually based only on the total alkalinity of the water. Total alkalinity is the major factor modifying the toxicity of copper in most waters used to grow catfish, and pH and total hardness usually vary with total alkalinity. Never use copper sulfate unless the total alkalinity of the water is known. It is best to measure total alkalinity immediately before treatment rather than assume a value based on past measurements or experience. Although total alkalinity usually changes only slowly in ponds, heavy rainfalls can dilute the water and significantly decrease the total alkalinity.

Indefinite copper sulfate treatment rates for ponds are calculated by the following formula:

$$\text{parts per million } CuSO_4 \cdot 5H_2O = \text{total alkalinity} \div 100 \quad (11.1)$$

Total alkalinity is expressed in units of ppm as $CaCO_3$. Copper sulfate should not be used if the total alkalinity of the water is less than about 50 ppm as $CaCO_3$ because it may kill fish. Copper rapidly forms insoluble precipitates in waters of high alkalinity, and treatment efficacy is greatly reduced in waters with total alkalinities greater than about 300 ppm as $CaCO_3$. Copper sulfate is phytotoxic and often causes depletion of dissolved oxygen when used in ponds, particularly if water temperatures are high. Copper sulfate should not be used as a dip or bath treatment because little is known of the relationship between water quality and the toxicity of concentrated copper solutions.

Formalin. Formalin is registered for food-fish use as a parasiticide. It has been used to treat infestations of *Ichthyophthirius*, *Ichthyobodo*, *Trichodina*, gill flukes, and certain other external parasites. It is also used to control fungal infections on eggs and fish. The chemical is ineffective against bacterial infections of fish. Formalin is a liquid containing 37 percent formaldehyde by weight and 15 percent methanol. For purposes of calculating treatment rates, formalin is considered 100 percent active. Methanol is added to inhibit the formation of paraformaldehyde. Paraformaldehyde may be present in formalin stored for long periods or exposed to temperatures below 40°F. Fresh formalin is clear; the presence of a white precipitate indicates that paraformaldehyde is present and the solution should be discarded. Paraformaldehyde is extremely toxic to fish.

Fungal infections of eggs can be controlled by treating with 100 ppm formalin for 15 minutes. Formalin can be used as a bath treatment for fish in raceways or tanks by treating for up to 1 hour at concentrations of 100 to 250 ppm, depending on water temperature. At water temperatures below about 70°F, use 150 to 250 ppm formalin; at water temperatures above 70°F, use 100 to 150 ppm. Make sure adequate aeration is available and watch the fish closely during treatment.

Formalin is used at 15 to 25 ppm as an indefinite treatment for ponds. Organic material in the water decreases the effectiveness of formalin. Treatment rates of 15 to 20 ppm are usually effective in clear ponds with sparse phytoplankton blooms, but treatments of 20 to 25 ppm are more effective in most catfish ponds. Formalin is phytotoxic and may cause dissolved oxygen depletions in ponds, particularly when used in warm water.

Salt. Salt (NaCl) is registered for food-fish use as an osmoregulatory enhancer. Salt has also been used as a treatment for certain external parasites and as a treatment for nitrite-induced methemoglobinemia (brown-blood disease; see the section Nitrite). Concentrated salt solutions increase mucus flow on the gills and skin, and parasites may slough off along with the excess mucus. Salt solutions may also be directly toxic to certain parasites. Salt is not used as a pond treatment for parasites because unreasonable quantities are required to achieve the high concentrations required for parasite control.

As a dip treatment, salt is used at 2 to 3 percent (20,000 to 30,000 ppm). Fish are dipped into the solution for 60 seconds or less if they show signs of stress. Prolonged bath treatments with 1,000 to 2,000 ppm salt are occasionally used when fish are temporarily held in tanks or when they are transported in hauling tanks. Channel catfish can survive indefinitely at these salt concentrations. The treatment is usually employed to reduce osmoregulatory stress when fish are handled or transported but may also provide some benefit in controlling external parasites.

TABLE 11.1.

Weights per Unit Volume of Water Equivalent to a Concentration of 1 Part per Million

2.7 pounds/acre-foot	= 1 ppm
1.23 kilograms/acre-foot	= 1 ppm
0.0283 gram/cubic foot	= 1 ppm
28.35 grams/1,000 cubic feet	= 1 ppm
1.0 ounce (avoirdupois)/1,000 cubic feet	= 1 ppm
3.78 milligrams/gallon	= 1 ppm
0.0038 grams/gallon	= 1 ppm
3.78 grams/1,000 gallons	= 1 ppm
0.13 ounce (avoirdupois)/1,000 gallons	= 1 ppm
1.00 milligram/liter	= 1 ppm

TREATMENT RATE CALCULATIONS

Chemical treatment rates in the preceding section were given in units of parts per million (ppm). This unit of measure means that one weight unit of chemical is added to 999,999 weight units of water: for instance, 1 pound of chemical added to 999,999 pounds of water gives a 1 ppm solution of the chemical. In scientific literature, metric units are used and the common expression for concentration is milligrams/liter (mg/L). For solutions in fresh water, mg/L and ppm are equivalent units (1 ppm = 1 mg/L) because 1 liter of fresh water weighs 1,000,000 milligrams. It is seldom possible to weigh the water being treated, but the weights of given water volumes are known or can be calculated, so treatment rate calculations are usually determined from water volumes. For instance, pond volumes are usually expressed in acre-feet of water (1 acre-foot is the volume of water covering 1 acre to a depth of 1 foot). An acre-foot of water weighs about 2.7 million pounds, so 2.7 pounds of a chemical added to 1 acre-foot of water gives a 1 ppm solution. The weights of chemical needed to give 1 ppm solutions in various volumes of water are given in Table 11.1. Once the volume of the water to be treated is known, use of the conversions in Table 11.1 to calculate treatment rates is straightforward for the dry chemicals potassium permanganate, copper sulfate, and salt. When using formalin or other liquids, the weight of a given volume of chemical must be known to calculate the number of gallons (or other volumes) of chemical that must be used.

Treatment Rate Calculations   **373**

The treatment rate calculations discussed in the following are not limited to determining disease treatment rates. Application rates for aquatic herbicides and salt treatments for methemoglobinemia are similarly calculated once the desired concentration (in ppm) is known. Readers are urged to consult the excellent guide to calculations of treatment rates compiled by Jensen (1988). That booklet also illustrates many other calculations used in catfish culture.

Determination of Volumes

The correct volume of water to be treated must be known within an error of not more than about ±10 percent. Never try to guess culture unit volumes for a chemical treatment because incorrect estimates can result in ineffective treatment or dead fish. Volumes of troughs, vats, or tanks can usually be determined with a high degree of accuracy because most have a regular geometric shape. Pond volumes are more difficult to determine accurately because they are not one standard shape and the depth is not constant over the entire pond.

Noncircular troughs, tanks, and other containers. Troughs and tanks are usually rectangular. Volume is determined from inside length and width and average depth. If the tank has a sloped bottom, measure depth at three or four places in the middle along the long axis to determine the average depth. Volume of rectangular solids is calculated by the following formula:

$$\text{Volume} = \text{length} \times \text{width} \times \text{depth} \quad (11.2)$$

Volumes are expressed in cubic feet, cubic meters, or the units in which the linear measurements were made. Volume in cubic units can then be converted to gallons or liters using the conversion factors in Table 11.2.

Example: A hatching trough is 8 feet long and 2 feet wide and filled to an average depth of 10 inches (0.83 foot). What is the volume in gallons?

$$\begin{aligned}\text{Volume (feet}^3) &= 8 \text{ feet} \times 2 \text{ feet} \times 0.83 \text{ foot} \\ &= 13.28 \end{aligned} \quad (11.3)$$

$$\begin{aligned}\text{Volume (gallons)} &= 13.28 \text{ feet}^3 \times 7.48 \text{ gallons/feet}^3 \\ &= 99.33, \text{ or about } 99.3 \text{ gallons}\end{aligned} \quad (11.4)$$

Circular tanks. The volume of these units is calculated from the formula for the volume of a cylinder:

TABLE 11.2.

Some Useful Volume and Weight Conversions

To Convert Column 1 to Column 2, Multiply by	Column 1	Column 2	To Convert Column 2 to Column 1, Multiply by
1	Cubic centimeter	Milliliter	1
0.001	Cubic centimeter	Liter	1,000
0.000264	Cubic centimeter	Gallon	3,785
1,000	Cubic meter	Liter	0.001
35.31	Cubic meter	Cubic foot	0.02832
264.2	Cubic meter	Gallon	0.003785
0.00081	Cubic meter	Acre-foot	1,233.6
16.39	Cubic inch	Cubic centimeter	0.0610
0.0164	Cubic inch	Liter	60.98
0.00058	Cubic inch	Cubic foot	1,728
0.00433	Cubic inch	Gallon	231
28.32	Cubic foot	Liter	0.0353
7.48	Cubic foot	Gallon	0.1337
2.3×10^{-5}	Cubic foot	Acre-foot	43,560
325,850	Acre-foot	Gallon	3.07×10^{-6}
2,711,000	Acre-foot	Pounds of water[a]	3.70×10^{-7}
8.32	Gallon	Pounds of water[a]	0.12
62.23	Cubic foot	Pounds of water[a]	0.0161

[a] For a water temperature of 77°F.

$$\text{Volume} = \pi \times r^2 \times \text{depth} \quad (11.5)$$

where: $\pi = 3.14$

$r^2 =$ the square of the radius, r. The radius is half the diameter of the tank.

Example: A circular fish-holding tank has a diameter of 8.5 feet and is filled to an average depth of 40 inches (3.33 feet). What is the volume in gallons?

$$\begin{aligned}\text{Volume (feet}^3) &= 3.14 \times (4.25 \text{ feet})^2 \times 3.33 \text{ feet} \\ &= 188.87 \end{aligned} \quad (11.6)$$

Volume (gallons) = 188.87 feet³ × 7.48 gallons/feet³
= 1,412.71, or about 1,413 gallons   (11.7)

Ponds. Pond volumes are usually expressed in acre-feet of water and are determined by the following formula:

volume (acre-feet) = area (acres) × depth (feet)   (11.8)

Pond surface areas are usually determined at the time of construction; ask for a copy of the blueprints and keep them on file. If the area is unknown or the pond shape or size has changed (from erosion of the banks, for example), the best way to determine surface area is to have the ponds surveyed. Local Soil Conservation Service personnel can help make arrangements for a survey. Alternatively, the area of ponds with regular shapes (square, rectangular, triangular, or trapezoidal) can be estimated by measuring distances along the banks and using geometric formulas to calculate area. Be sure to measure distances as accurately as possible: using repeated measures of a tape length (called "chaining") is more accurate than "pacing." Methods of measuring distance and calculating areas are given by Jensen (1988).

A simple method of measuring depth is to make transects across the pond and take depth readings at regular intervals. Readings can be taken with a long pole marked off in inches. Lower the pole into the water until it just touches the surface of the mud. Record all measurements, then determine average depth by totaling all the measurements and dividing by the number of measurements taken. The more measurements taken, the more accurate is the estimate of average depth.

Calculations

The weight of chemical needed for a treatment is calculated using this basic formula

Weight of chemical = $V \times CF \times ppm \times (100 \div \%AI)$ (11.9)

where: V = the volume of water to be treated

CF = the appropriate conversion factor to give the weight of chemical that will equal 1 ppm in one unit volume of water

ppm = the desired treatment concentration

$100 \div \%AI$ = 100 divided by the % active ingredient (AI) of the chemical

Potassium permanganate, copper sulfate, formalin, and salt are all considered 100 percent active, so this term becomes 1 and is dropped from the calculation. The percent active ingredient (% AI) of other aquaculture chemicals is usually listed on the label.

The following two examples illustrate how this formula is used to calculate treatment rates with dry chemicals.

Example: A pond with a surface area of 17.2 acres, with an average depth of 4.2 feet, is to be treated with copper sulfate to control the parasite *Trichophrya*. The total alkalinity of the pond water is 175 ppm as $CaCO_3$. How many pounds of copper sulfate ($CuSO_4 \cdot 5H_2O$) will be needed for the treatment?

1. First determine the ppm of copper sulfate needed. Remember (page 370) that the treatment rate is

    ppm $CuSO_4 \cdot 5H_2O$ = total alkalinity ÷ 100
    = 175 ÷ 100
    = 1.75

2. Next determine the pond volume, V:

    volume (acre-feet) = area (acres) × depth (feet)
    = 17.2 × 4.2
    = 72.24

3. The appropriate conversion factor (CF) to give pounds from volume measured in acre-feet is 2.7 pounds/acre-foot/ppm (Table 11.1).

4. Copper sulfate is considered 100 percent, so the term 100 ÷ %AI becomes 1 and is dropped.

5. The pounds of copper sulfate needed is determined by substituting the appropriate values into the formula:

    Weight of chemical = V × CF × ppm
    = 72.24 × 2.7 × 1.75
    = 341.3 pounds, or about 340 pounds

Example: How many grams of potassium permanganate must be added to a hauling tank to achieve a concentration of 10 ppm potassium permanganate? The hauling tank is 8 feet long and 4 feet wide and filled with water to a depth of 3 feet.

1. Determine the volume (V) of the tank:

    Volume (feet^3) = 8 feet × 4 feet × 3 feet
    = 96

2. The appropriate conversion factor to give grams from volume measured in cubic feet is 0.028 grams/cubic foot/ppm (Table 11.1).

3. The desired treatment rate is 10 ppm.
4. Potassium permanganate is 100 percent active so the term 100 ÷ %AI is dropped.
5. Grams of potassium permanganate needed
   = 96 × 0.028 × 10
   = 26.88, or about 27 grams

Notice that the treatment rate was calculated in grams rather than pounds. This is common (and easiest) when working in tanks and vats because there are no convenient small English units to use. Every catfish farm should have an accurate gram balance to weigh out chemicals for treatment of small volumes of water. Also note that the volume of the tank could be calculated in gallons (96 cubic feet × 7.48 gallons/cubic foot = 718 gallons) and a different conversion value (0.0038 grams/gallon/ppm) used. The resulting answer (718 × 0.0038 × 10 = 27.2 grams) is essentially the same. The slight difference in the answers is due to small errors in rounding off the conversion values.

When treating with formalin, the answer (in weight units) provided by the preceding formula is usually converted to volume units to make measurements easier. Use the following conversions:

$$1 \text{ gram of formalin} = 0.93 \text{ milliliter}$$

$$1 \text{ pound of formalin} = 0.11 \text{ gallon}$$

Example: How much formalin is required to obtain a 10 ppm solution in a hatching trough measuring 8 feet × 2 feet × 10 inches?

1. Determine the volume of the tank:

   Volume (cubic feet) = 8 feet × 2 feet × 0.83 foot
   = 13.28

2. The appropriate conversion factor to give grams from volume measured in cubic feet is 0.028 grams/cubic foot/ppm (Table 11.1).

3. The desired treatment rate is 100 ppm.

4. Formalin is 100 percent active so the term 100 ÷ %AI is dropped.

5. The grams of formalin needed = 13.28 × 0.028 × 100
   = 37.18

6. The volume of formalin needed is

   37.18 grams × 0.93 milliliter/gram = 34.58, or about 35 milliliters

Inexperienced catfish farmers should seek assistance when making treatment rate calculations. If a fish disease specialist is not available, ask someone good at mathematics to check the calculations. With experience, you will be able to tell whether a calculation is correct just by the magnitude of the answer. For instance, after awhile you will know that a 1 to 2 ppm treatment in a 15- to 20-acre pond that is about 4 feet deep will require an amount of chemical in the approximate range of 150 to 500 pounds. If your calculation gives an answer in the thousands of pounds, you will recognize it as wrong.

REFERENCES

Amend, D. F., T. McDowell, and R. P. Hedrick. 1984. Characteristics of a previously unidentified virus from channel catfish (Ictalurus punctatus). Canadian Journal of Fisheries and Aquatic Sciences 41:807–811.
Baxa, D. V., and R. P. Hedrick. 1989. Two more species are susceptible to experimental infections with Edwardsiella ictaluri. FHS/AFS Newsletter 17:4.
Butterworth, C. E., J. A. Plumb, and J. M. Grizzle. 1986. Abnormal folate metabolism in feed-related anemia of cultured channel catfish. Proceedings of the Society for Experimental Biology and Medicine 181:49–58.
Cipriano, R. C., G. L. Bullock, and S. W. Pyle. 1984. Aeromonas hydrophila and Motile Aeromonad Septicemias of Fish. Fish Disease Leaflet 68. Washington D.C.: United States Department of the Interior.
DeFigueirrido, J., and J. A. Plumb. 1977. Virulence of different isolates of Aeromonas hydrophila in channel catfish. Aquaculture 11:349–354.
Ford, L. A., and R. L. Thune. 1988. Aeromonad virulence factors: a field study. Abstract of paper read at the International Fish Health Conference, 19–21 July 1988, at Vancouver, British Columbia, Canada.
Groff, J. M., T. McDowell, and R. P. Hedrick. 1989. Sphaerospores observed in the kidney of channel catfish (Ictalurus punctatus). FHS/AFS Newsletter 17:5.
Hawke, J. P. 1979. A bacterium associated with disease of pond-cultured channel catfish, Ictalurus punctatus. Journal of the Fisheries Research Board of Canada 36:1508–1512.
Hawke, J. P., and R. L. Thune. 1988. Evaluation of a medium for the selective isolation of Flexibacter columnaris from diseased channel catfish. Abstract of paper read at the International Fish Health Conference, 19–21 July 1988, at Vancouver, British Columbia, Canada.
Hoffman, G. L. 1967. Parasites of North American Freshwater Fishes. Berkeley: University of California Press.
Hoffman, G. L., and F. P. Meyer. 1974. Parasites of Freshwater Fishes. Neptune City, N.J.: TFH Publications.
Jensen, G. 1988. Handbook of Common Calculations in Finfish Aquaculture. Baton Rouge: Louisiana Agricultural Experiment Station.
Klar, D. W., L. A. Hanson, and S. W. Brown. 1986. Diet-related anemia in channel catfish: case history and laboratory induction. Progressive Fish-Culturist 48:60–64.
Lewis, D. G., and J. A. Plumb. 1985. Bacterial diseases. In Principal Diseases of Farm-Raised Channel Catfish, ed. J. A. Plumb, pp. 13–21. Auburn: Alabama Agricultural Experiment Station.

Li, Y., and R. T. Lovell. 1985. Elevated levels of dietary ascorbic acid increase immune responses in channel catfish. *Journal of Nutrition* 115:123–131.

MacMillan, J. R. 1985. Infectious diseases. In *Channel Catfish Culture*, ed. C. S. Tucker, pp. 405–496. Amsterdam: Elsevier.

MacMillan, J. R., C. Wilson, and A. Thiyagarajah. 1990. Experimental induction of proliferative gill disease in specific pathogen free channel catfish. *Journal of Aquatic Animal Health* (in press).

Meyer, F. P., and R. A. Schnick. 1989. A review of chemicals used for the control of fish diseases. *Reviews in Aquatic Sciences* 1:694–710.

Noga, E. J., and J. X. Hartman. 1981. Establishment of walking catfish *(Clarias batrachus)* cell lines and development of a channel catfish *(Ictalurus punctatus)* virus vaccine. *Canadian Journal of Fisheries and Aquatic Sciences* 38:925–930.

Plumb, J. A., ed. 1985a. *Principal Diseases of Farm-Raised Catfish*. Southern Cooperative Series Bulletin 225. Auburn: Alabama Agricultural Experiment Station.

Plumb, J. A. 1985b. Viral diseases. In *Principal Diseases of Farm-Raised Channel Catfish*, ed. J. A. Plumb, pp. 9–12. Auburn: Alabama Agricultural Experiment Station.

Plumb, J. A. 1988a. Vaccination against channel catfish virus. In *Fish Vaccination*, ed. A. E. Ellis, pp. 216–223. London: Academic Press.

Plumb, J. A. 1988b. Vaccination against *Edwardsiella ictaluri*. In *Fish Vaccination*, ed. A. E. Ellis, pp. 152–161. London: Academic Press.

Plumb, J. A., and J. Chappell. 1978. Susceptibility of blue catfish to channel catfish virus. *Proceedings of the Annual Conference of Southeastern Associations of Fish and Wildlife Agencies* 32:680–685.

Plumb, J. A., and D. J. Sanchez. 1983. Susceptibility fo five species of fish to *Edwardsiella ictaluri*. *Journal of Fish Diseases* 6:261–266.

Popoff, M., and R. Lallier. 1984. Biochemical and serological characteristics of *Aeromonas*. In *Methods in Microbiology*, Vol. 16, ed. T. Bergan, pp. 127–145. London: Academic Press.

Post, G. W. 1983. *Textbook of Fish Health*. Neptune City, N.J.: TFH Publications.

Pyle, S. W., and E. B. Shotts. 1989. A new approach for differentiating flexibacteria isolated from coldwater and warmwater fish. *Canadian Journal of Fisheries and Aquatic Sciences* 37:1040–1042.

Robinson, E. H., J. R. Brent, J. T. Crabtree, and C. S. Tucker. 1990. Improved palatability of channel catfish feed containing Romet-30®. *Journal of Aquatic Animal Health* 2:43–48.

Rogers, W. A. 1985. Protozoan parasites. In *Principal Diseases of Farm-Raised Channel Catfish*, ed. J. A. Plumb, pp. 24–32. Auburn: Alabama Agricultural Experiment Station.

Shotts, E. B., V. S. Blazer, and W. D. Waltman. 1986. Pathogenesis of experimental *Edwardsiella ictaluri* infections in channel catfish *(Ictalurus punctatus)*. *Canadian Journal of Fisheries and Aquatic Sciences* 43:36–42.

Tucker, C. S. 1989. Method for estimating potassium permanganate disease treatment rates for channel catfish in ponds. *Progressive Fish-Culturist* 51:24–26.

Tucker, C. S., R. Francis-Floyd, and M. H. Beleau. 1989. Nitrite-induced anemia in channel catfish, *Ictalurus punctatus* Rafinesque. *Bulletin of Environmental Contamination and Toxicology* 43:295–301.

Wellborn, T. L. 1985. Control and therapy. In *Principal Diseases of Farm-Raised Channel Catfish*, ed. J. A. Plumb, pp. 50–67. Auburn: Alabama Agricultural Experiment Station.

Wilson, R. P., and W. E. Poe. 1989. Palatability of diets containing sulfadimethoxine, ormetoprim, and Romet-30® to channel catfish. *Progressive Fish-Culturist* 51:226–228.

Wise, J. A., and J. Boyle. 1985. Detection of channel catfish virus in channel catfish: use of a nucleic acid probe. *Journal of Fish Diseases* 8:417–424.

Wise, J. A., P. R. Bowser, and J. Boyle. 1985. Detection of channel catfish virus in asymptomatic adult channel catfish, *Ictalurus punctatus* (Rafinesque). *Journal of Fish Diseases* 8:485–493.

Wise, J. A., S. F. Harrel, R. L. Busch, and J. A. Boyle. 1988. Vertical transmission of channel catfish virus. *American Journal of Veterinary Research* 49:1506–1509.

Wolf, K. 1988. *Fish Viruses and Fish Viral Diseases*. Ithaca, N.Y.: Cornell University Press.

CHAPTER 12
Harvesting and Transporting

Regardless of the culture system that is used, channel catfish must be captured and transported several times before they are slaughtered for processing. Commonly, fish are moved (as eggs) from brood ponds to the hatchery, again as fry from the hatchery to a nursery facility, as fingerlings to a growout facility after a period of growth in the nursery facility, and then finally to the processing plant after growout to foodfish size. In some production schemes, fish may be handled even more often before final harvest. Also, some fish are "live-hauled" to special markets such as restaurants, fee-fishing lakes, or sport-fishing ponds.

During harvest and transport, fish are confined in a small volume of water and water quality can deteriorate rapidly. Catfish that are crowded may receive abrasions, cuts, or puncture wounds that become infected. Stress caused by poor environmental conditions or crowded conditions suppresses the immune system and predisposes fish to a variety of infectious diseases.

Stress associated with harvest and transportation cannot be prevented; however, the use of proper techniques can minimize stress and subsequent mortality. These techniques will be discussed in this chapter. Harvest practices for ponds will be emphasized in this chapter; techniques for harvesting fish from other culture systems will be mentioned in Chapter 13. Transportation of fish is similar regardless of culture system.

It should be pointed out that little systematic research has been conducted on harvest and transportation of channel catfish under commercial culture conditions. Many of the techniques used on catfish farms were developed by trial and error and may not be applicable to all situations. The best general advice is to use common sense and avoid situations that will excessively stress the fish.

CONSIDERATIONS BEFORE HARVEST

The harvest process must proceed smoothly to minimize stress on the fish. The following advice applies regardless of the type of culture system from which fish are to be harvested:

> Develop a plan before harvesting. All arrangements for harvest, transport, and sale should be made days or even weeks before harvest. You must know approximately how many fish will be harvested and where they will be delivered. Arrangements must be made with processing plants or whoever will be buying the fish. Make sure that all necessary equipment and labor will be on hand when needed. Describe the plan to all personnel involved in the harvest so that everyone will know what to do.
>
> Keep all equipment in good working order. Equipment breakdowns during harvest cause delays and fish are more likely to be stressed. Seines and nets should be routinely inspected and kept in good repair. Tractors, seine reels, boom trucks, boat, aerators, transport trucks, and all other equipment used in harvest should be inspected and serviced regularly.
>
> If food-sized fish are to be harvested, check for off-flavors before harvesting. Most processing plants require flavor checks before accepting fish for processing, but all fish sold for food should be checked, even if it is not required. Poor-tasting fish in the marketplace may cause future sales problems.
>
> Do not harvest sick fish. Harvesting and restocking fingerlings with an infectious disease can spread the disease to other ponds. Also, many fish may die after harvest and transport because sick fish may not withstand the stress of handling. Note, however, that harvest of food-sized fish for slaughter may be the best way to prevent certain disease-related losses if the disease is detected early and only a small percentage of the population is affected. The decision to harvest fish in the early

stages of an epizootic should be made after deliberation with a fish disease specialist.

Do not feed fish within 24 hours of harvest. Fish that are full of feed do not withstand the stress of handling as well as those with empty stomachs. Fish disgorge feed recently consumed, and the disgorged feed can foul the water in holding tanks. Also, most processors deduct from the fish purchase price if fish have noticeable amounts of feed in their stomachs.

Take special care when harvesting fish in hot weather. Fish handle poorly when water temperatures are above about 85°F. To the extent possible, try to schedule most fish harvests in the cooler months of the year. If fish must be harvested in the summer, be sure to have adequate aeration equipment available. In particular, avoid handling fry or small fingerlings when water temperatures are above 85°F because the stress may trigger an outbreak of channel catfish virus disease (see the section Viral Diseases).

Do not harvest fish if water quality is poor. The combined stressors of poor water quality and handling can easily kill fish. Do not harvest fish from waters with low concentrations of dissolved oxygen or if fish are stressed by high concentrations of un-ionized ammonia or nitrite. Dissolved oxygen concentrations in ponds can be quite low right after dawn, so try to schedule pond harvests later in the day when dissolved oxygen concentrations are higher.

Conduct the harvest as efficiently as possible. A considerable amount of time (several hours at least) is required to harvest fish from large ponds, but make every effort to make the operation proceed smoothly from start to finish without unnecessary delays.

HARVESTING FISH FROM PONDS

The method of harvest used in ponds depends on the type and shape of the pond and the production strategy. Levee ponds (Chapter 8) are relatively easy to harvest by seining, and harvest methods have become somewhat standardized. Although some watershed ponds are constructed to allow efficient fish harvest by seining, many are too deep or too irregular to be seined effectively and a variety of alternatives have been used in attempts to harvest fish.

In the major catfish-producing areas of the southeastern United States, farmers have the option of harvesting their own fish or using custom harvesting services. The decision must be based on an economic analysis of the individual farm situation. Custom harvesters currently charge about 3 cents/pound of fish harvested. They usually supply seines, a boom truck for loading fish, a seining crew, a boat, and other necessary equipment. The farmer usually must supply tractors for pulling the seine and aeration equipment. Generally, the convenience and costs of custom harvesting are attractive for small farms where harvests are relatively infrequent and labor may be in short supply. Sufficient labor is usually available on large farms to justify purchasing seining equipment and conducting harvests using available personnel. The type of production system used on the farm must also be considered when deciding whether to use custom harvesting services. If ponds are managed as single-batch production systems (see the section Food-Fish Production), farms should have their own harvesting crews and equipment because this cropping system usually requires more harvests of each pond per fish crop than ponds managed as multiple-batch systems. Also, custom harvesters are often unwilling to conduct several consecutive harvests in an attempt to capture all the fish in a pond. Relatively few fish are captured after a pond has been seined once or twice, and because custom harvesters are paid by the weight of fish captured, the expense of the last attempts at harvest is not justified by the returns.

Harvesting Fish from Levee Ponds

Harvesting fish from levee ponds by seining is relatively simple because most levee ponds have regular shapes (usually rectangular) and are shallow with smooth bottoms devoid of snags. Levees should be free of trees or obstacles and are usually wide enough to permit the use of heavy equipment, such as tractors and large fish-hauling trucks. Also, levee ponds are usually supplied with a dependable source of high-quality water that can be used to maintain good water quality in the harvest and loading area of the pond.

Seines used to harvest catfish ponds are purchased custom-made from commercial sources. The dimensions of the seine and, to some extent, the materials used in construction are based on the needs of the individual farmer or fish harvester. Seines should be about 1.5 times the maximum depth of the ponds to be seined. Seines of 8- to 10-foot depth are used in most levee ponds. Seines should be at least 1.5 times as long as the width of the pond; a typical rectangular pond of about 17 acres will require a seine at least 1,000 feet long. Seines are made

FIGURE 12.1.

A mud line made by rolling fine mesh netting around the lead lines of a seine.

from knotted polyethylene or nylon netting. Nylon netting requires a coating of tar-based or plastic "net coat" to protect the netting from sunlight, which causes nylon to deteriorate. The net coat also helps prevent fish spines from snagging in the netting. Nylon seines should be treated every 1 or 2 years; polyethylene netting does not require coating.

Seines have a float line with foam or plastic floats placed every 18 to 24 inches and a weighted lead line with 2-ounce lead weights spaced 12 to 18 inches apart. Bare lead lines dig into the bottom muds to such an extent that seining is usually impossible unless the pond bottom is unusually hard. Seines are almost always equipped with a "mud line" of some sort to make the weighted bottom line slide over the mud rather than digging in. Three types of mud lines are available: rope "many-ends" mud lines, rolled-netting mud lines, and "mud rollers." The many-ends mud line is made by binding 10 to 15 strands of 1/2- to 5/8-inch nylon rope around the lead line to form a bundle roughly 3 inches in diameter. The rolled-netting mud line (Fig. 12.1) is made from fine-meshed knotless nylon netting rolled and bound around the lead line to form a bundle about 3 inches in diameter. Mud rollers use rubber-coating cylindrical weights (rollers) spaced about 3 feet apart, with regular lead weights at 1-foot intervals between the rollers.

Seines may also be built with a tapered tunnel about 100 to 200 feet from one end of the seine. The tunnel is equipped with a drawstring closure that can be connected to the metal frame of a live car

TABLE 12.1.
Net Square Mesh Size for Grading and Harvesting Channel Catfish

Square Mesh Size (Inches)	Approximate Minimum Size Range of Fish Retained
2	1 3/4–2 pounds
1 3/4	1 1/4–1 1/2 pounds
1 5/8	3/4–1 pound
1 1/2	1/2–3/4 pound
1 3/8	1/4–1/2 pound
1	7–9 inches
3/4	6–7 inches
5/8	5–6 inches
1/2	4–5 inches
3/8	3–4 inches
1/4	2–3 inches

(also known as fish grader nets, holding nets, or socks). Live cars are used to hold fish in confinement after seining.

The netting used to construct seines or live cars is available in a variety of mesh sizes. Mesh sizes are usually specified as *square mesh size*, which is the knot-to-knot distance. The mesh should be sized to capture the smallest fish desired. Fish smaller than that minimum size will pass or "grade" through the netting. Sizes of fish retained by netting of various square mesh sizes are listed in Table 12.1.

Use seines with the largest mesh size commensurate with the size of fish to be harvested; seines with small mesh tend to "mud down" because bottom sediments are scooped up and do not readily pass through the small mesh openings as the seine is pulled through the pond. For example, if you want to harvest fish of 1 pound and above from a food-fish growout pond, you could use a 1/4-inch square mesh fingerling seine to capture all fish in the pond, then grade fish with a 1 5/8-inch square mesh live car to retain fish of roughly 1 pound and above. However, harvest would be slow and tedious as the fine mesh seine continually mudded down. Harvest would be easier with a 1 1/2-inch square mesh seine. Fish captured with the 1 1/2-inch mesh seine could then be graded with a 1 5/8-inch square mesh live car to retain fish of the desired size.

A 1 1/2-inch square mesh seine is a good general-purpose seine for food-sized catfish. Seines of 3/8- or 1/2-inch square mesh are com-

FIGURE 12.2.

A seine reel used to store and beach seines. The reel operates off the hydraulic system of the tractor.

monly used as general-purpose fingerling seines. Of course, the appropriate mesh size for seines varies, depending on the production goals of the individual farm. Likewise, desired net mesh sizes for live cars vary from farm to farm. On farms producing only food-sized fish, live cars with mesh sizes of 1 1/2 or 1 5/8 inches are popular. Fingerling producers often have live cars of several mesh sizes to grade fingerlings into different size classes.

Two tractors are needed to harvest ponds larger than about 0.75 acre. One tractor pulls a hydraulically operated seine reel (Fig. 12.2) mounted on a trailer. The reel serves as a seine storage unit and as a winch during final stages of seining. The actual process of seining varies somewhat with the configuration of the pond. Ponds are usually seined from the deep end to the shallow end. The shallower water makes final harvest operations easier, and the water inlet is usually located at the end. The seine is released from the reel and placed along the bank at the deep end of the pond. The amount of seine laid out should be roughly 1.5 times the width of the pond. The loose end of the seine is attached to the second tractor and the seine is pulled slowly (0.5 to 1 mile/hour) up the length of the pond. One or two people should stand on the seine mud line in shallow water near the bank to prevent the seine from rising off the bottom and allowing fish to escape (Fig. 12.3). The float line of the seine should be closely watched during seining. If the floats in a section disappear beneath the water surface, this indicates that the seine has become snagged on some obstacle or, more commonly, that the seine has dug into the mud. A boat equipped

FIGURE 12.3.

One or two people stand on the mud line of the seine to prevent it from rising off the pond bottom.

with a "seine catcher" (Fig. 12.4) is used to catch the float line and push the seine forward to dump mud as it accumulates (Fig. 12.5).

The ends of the seine are brought together at the shallow end where the seine will be beached. Several different methods can be used to beach the seine. If two seine reels are used, the reels can be positioned next to each other and both ends of the seine slowly taken up. If only one seine reel is available, it can be positioned on a corner of the pond and the other tractor can simply drag the seine up the bank and then along a levee to crowd the fish. Other methods have also been used, and the farmer should adopt a method best suited to the particular situation.

Two methods are commonly used to complete the harvest. One method uses live cars to hold fish temporarily and the other uses a "cutting seine" to crowd fish for final harvest.

Live cars are open-topped bags made of netting (Fig. 12.6). One end of the live car is fashioned into a tunnel terminating in a metal frame. A special drawstring closure in the seine fits around the frame to connect the seine to the live car. The live car is not attached to the seine during the initial stages of seining because it increases the ten-

Harvesting Fish from Ponds **389**

FIGURE 12.4.

A boat equipped with a seine catcher.

FIGURE 12.5.

Mud can be dumped by catching the float line with the seine catcher and pushing it forward.

FIGURE 12.6.

A live car used to hold fish for final harvest.

dency of the seine to mud down and the weight of mud carried along in the live car offers considerable resistance while the seine is pulled up the pond. The live car is usually attached after the ends of the seine have been brought together and some of the seine has been beached. As the last portion of the seine is beached, the crowded fish are funneled into the live car (Fig. 12.7). When harvesting in warm water, the fish should be crowded slowly. If fish are crowded too rapidly they will not have time to funnel into the live car before using most of the dissolved oxygen in the harvest area. This can kill or severely stress the fish. In cold water, fish can be crowded more rapidly without ill effect; in fact, when water temperatures are below about 50°F, catfish become lethargic and must be crowded into an extremely dense mass and forced to "flow" into the live car. The final harvest area should be at least 3 feet deep to allow easy passage of fish into the live car and should be near the water inlet so fresh pumped well water can be used to maintain good water quality while fish are crowded. This is especially important if water temperatures are high.

After the live car is filled to approximately the recommended capacity (Table 12.2), it is detached and closed by wrapping a closure line tightly around the neck of the live car tunnel. Another live car can then be attached to receive additional fish. Several live cars may be needed to hold fish when large ponds are seined. Live cars can then be positioned and secured where a gentle flow of aerated water from the

Harvesting Fish from Ponds **391**

FIGURE 12.7.

Fish crowded in the seine are funneled into the live car.

well or an aerator can pass through the live car. Fish are held in live cars until they have graded or are loaded onto transport trucks.

No guidelines exist for the length of time required for adequate grading of fish in live cars. Grading efficiency is affected by the degree of crowding, water temperature, and size range of fish. Only an hour or two may be needed to grade food-sized fish in warm water adequately. Fingerling producers often let fish grade overnight in warm water to ensure good grading. Regardless of fish size, fish may not grade effectively in cold water even after 24 hours.

Loading fish onto transport trucks from live cars is simple. The live car is positioned near the bank and the fish are crowded into one end of the car by slowly gathering up the other end. The crowded fish are scooped into a loading net attached to a hydraulic boom (Fig. 12.8). The loading net is then lifted (Fig. 12.9), and the fish are dumped into holding tanks on the transport truck. In-line scales are used to record the weight.

The alternative method of completing the harvest uses a "cutting seine" rather than live cars to crowd fish before loading onto transport

TABLE 12.2.

Approximate Capacities of Fish That Can Be Held in Live Cars for up to 24 Hours in Water 70 to 85°F

Live Car Dimensions (Feet)		Capacity (Pounds)
Width	Length	
8	20	7,000
8	30	10,000
8	40	14,000
10	20	10,000
10	30	15,000
10	40	20,000
10	50	25,000
10	60	30,000
10	70	35,000
10	80	40,000

Note: Capacities can be increased by factors of 1.5 to 2 in cooler waters or when fish will be held for less than 2–4 hours.

FIGURE 12.8.

Fish crowded into one end of the live car are loaded onto transport trucks by scooping fish into a loading net suspended from a boom.

Harvesting Fish from Ponds **393**

FIGURE 12.9.

Lifting a loading net to transfer fish to a transport truck.

trucks. The use of a cutting seine may be appropriate when the harvest area is shallow, when a very large harvest is expected, or when live cars are not available. The pond is seined as described and the seine is beached until fish are confined to an area of roughly 0.2 to 0.5 acre, depending on the size of the pond and number of fish. The fish should not be excessively crowded at this point. The mud line of the seine is then securely anchored to the bottom by metal stakes. The stakes should also support the float line 1 foot or so above the water to prevent fish escape. Oxygenated water should flow through the confinement area, and fish can be allowed to grade for some desired period, perhaps overnight. A smaller seine, the *cutting seine*, is then pulled by hand through the enclosed area to crowd fish for loading. Further grading can be accomplished by using cutting seines of smaller mesh size than the harvest seine. Fish crowded into the cutting seine can be loaded by manually dip-netting fish into the loading net or by swinging the loading net through the fish to scoop them up.

The efficiency (measured as percentage of harvestable fish actually harvested) of seining levee ponds varies considerably. Efficiencies are highest in new ponds with firm smooth bottoms, in cold water, and when an experienced crew conducts the harvest. Under these conditions, 80 to over 90 percent of the harvestable fish can be captured in one seine haul. Seining efficiency is lower in older ponds because the activities of fish and the currents produced by aeration tend to produce depressions in the pond bottom that allow fish to escape under the seine. Also, old ponds usually have a deep layer of soft mud that has built up over time. This mud must be dumped if it is scooped up in the seine, and some fish may escape during the dumping process. Harvest efficiencies are highest in cold water because fish are lethargic; in warm water, fish actively seek to escape around or under the seine. Finally, skill in harvesting fish, like any other endeavor, improves with experience. An experienced seining crew is more likely to capture a higher percentage of harvestable fish and also less likely to stress the fish while doing so. Regardless of the conditions, however, it is nearly impossible to harvest all the fish in a pond by seining. Under the best conditions, 95 to 99 percent of the fish can be captured by seining the pond three or four times. Complete harvest is possible only by draining the pond and picking up the remaining fish by hand. This process, called "scrapping out," is probably the least desirable job on a fish farm.

Harvesting Fish from Watershed Ponds

Watershed ponds constructed with smooth, relatively flat bottoms and a maximum depth of no more than about 8 feet can be harvested by seining as described for levee ponds. In fact, watershed ponds for large-scale commercial culture should, if at all possible, be constructed to allow seining without draining the pond. Fish in watershed ponds that cannot be seined in the traditional manner can be harvested by partially or completely draining the pond or by trapping.

Most watershed ponds can be efficiently harvested by discharging water until the maximum water depth is 4 to 6 feet. The pool that remains may represent from 20 to over 50 percent of the original pond area, depending on the shape of the basin. Fish in the pool are seined toward the deepest water (usually toward the dam) with a seine of appropriate mesh size. The seine may have to be pulled by hand because tractors or other equipment will sink into the soft mud of the exposed pond bottom. The method used to load fish on transport trucks will depend on the height of the dam. If the dam is low enough, a hydraulic boom can be used to reach out and lower a fish loading net into the harvest area. In deep ponds, booms may not be long enough to reach the harvest area when the water level is lowered. Fish can be moved up the dam by constructing a system of drag buckets mounted on temporary wood or metal rails. Fish are dip-netted into the drag bucket and the bucket (which may hold several hundred pounds of fish) is then slid or pulled up the dam along the rails by using a winch.

Some watershed ponds are equipped with external or internal catch basins to facilitate harvest. External catch basins are preferred because internal basins can become filled with sediment while the pond is full. External basins are essentially concrete tanks built at the end of the drain pipe outside the pond. A weir at the lower end of the catch basin prevents fish escape. Generally it is not possible to build a basin large enough to hold all the fish in the pond, so the pond is partially drained and seined. The remaining fish are then allowed to pass through the drain pipe into the catch basin, where they can easily removed.

Various trapping methods can be used when only a portion of the fish in the pond are needed or when it is impossible to drain the pond for harvest. Several trapping methods are discussed by Huner, Dupree, and Greenland (1984); all rely on baiting fish into traps or trapping areas with feed. The easiest trapping method is called *corral seining*. A 100- to 200-foot seine, 6 to 10 feet long, is placed parallel to the bank about 50 feet from the shore. Ropes attached to the ends of the seine are led to the shore. Fish are trained to feed in the area between the seine and shore by offering feed for several days. They are trapped by offering feed to draw fish into the area and then rapidly pulling the ropes to enclose the fish. Corral seining cannot be used more than once a week because fish become wary after trapping. Trapping is not dependable during the colder months because fish feed poorly.

TRANSPORTING FISH

Commercial culture of any fish usually requires transporting (or "hauling") fish several times at different stages of growout. One of the advantages of culturing a hardy species such as channel catfish is that the fish withstand the rigors of transport fairly well at all life stages. Nevertheless, losses of catfish during and after transport are common because large numbers or weights of fish must be transported in a relatively small volume of water to make transportation efficient. The resulting crowded conditions together with the poor water quality that can occur in hauling tanks may damage or stress fish, leading to immediate loss of fish during transport or loss to stress-related infectious diseases shortly after they are moved to another culture unit.

Proper procedures are critical during transport of eggs, fry, and fingerlings because poor survival subsequent to transport decreases production efficiency and profits. "Live hauling" of food-sized catfish to specialty markets also demands care during transport because fish are expected to arrive alive and in good condition. Almost all food-sized channel catfish are transported live to processing plants. These trips are usually of short duration and, although fish should arrive at the plant alive, fish are transported at higher densities with less attention given to environmental conditions because fish are normally held only briefly after transport before slaughter.

The following sections outline some guidelines for transporting channel catfish. Additional information is provided by McCraren and Millard (1979), Dupree and Huner (1984), and Busch (1985).

Equipment

Fish hauling tanks are usually rectangular and constructed of aluminum, steel, or Fiberglas. Relatively small aluminum or Fiberglas tanks of 100- to 500-gallon capacity* are used on farms for hauling eggs and fry or small weights of larger fish. These tanks are carried on medium- or heavy-duty pickup trucks (1/2- to 1-ton trucks) and can be removed when not needed for hauling fish to free the truck for other duties on the farm. Larger weights of fish are hauled on larger double-axle or trailer rigs holding several tanks. Total capacity of individual tanks usually ranges from 500 to 800 gallons.

Transport tanks can be insulated or not; insulated tanks are useful if transit time is long, particularly in hot weather. Large tanks should have internal baffles to minimize sloshing of water within the tanks. Fish are loaded through doors in the top of the tank. Doors should be at least 2.5 by 2.5 feet to facilitate transfer of fish from loading nets to tanks. Fish are unloaded through drainage ports on the side or end of the tank. Ports are usually sealed with a cam-locking door that is removed when fish are unloaded. An interior sliding door prevents fish from escaping when the exterior door is removed for unloading. Drainage ports should be at least 10 by 10 inches so fish can be rapidly unloaded. When fish are to be unloaded, a chute is attached to the drainage port and the exterior door is removed. The sliding interior door is eased away and fish flow with hauling water into the receiving body of water.

---

*Capacities are given as total capacity of the tank. Tanks are usually filled one-half to two-thirds full with water to allow displacement by fish. About 12 gallons of water is displaced by 100 pounds of fish.

Three general types of aeration system are used to maintain adequate dissolved oxygen concentrations in hauling tanks: surface agitators, diffused air, or diffused oxygen. The type of aeration system used depends on the size of the hauling tank and its intended use.

Small "spin" agitators are used on the relatively small hauling tanks carried on pickup trucks. These agitators have a 12-volt, 1/20-horsepower motor connected in-line to a shaft with a stirring blade. The motor rests on top of the hauling tank, and the shaft protrudes through the tank top with the blade submerged in the water. The blade is covered by a screen to prevent damage to the fish. Water in the tank must be maintained at a level that allows the blade to be submersed at an optimal depth for aeration. The rapidly spinning blade effectively aerates shallow transport tanks. These agitators are relatively inexpensive, dependable, and easily connected to the electrical system of the truck. Two 1/20-horsepower agitators can aerate up to about 150 gallons of water at the fish loading rates described later.

Larger hauling tanks that are permanently mounted on trucks or trailers are aerated with diffused air or oxygen. Most large transport units hauling market-sized fish from growout ponds to processing plants use diffused air. Diesel-powered air blowers supply large volumes of air to a diffuser near or on the bottom of the tank. The diffuser (usually just a steel pipe with many small holes) releases bubbles, which transfer oxygen as they rise through the water. Diesel air blowers are expensive, but daily operational costs are low. The units are subject to mechanical failure, but with regular maintenance they are handy for routine daily use. The large volumes of air released in the tanks and the resulting turbulence are undesirable when transporting fry or small fingerlings. Also, water temperatures tend to rise over time in tanks aerated with diffused air because the air is heated as it is compressed in the blower. This can be a serious disadvantage when transit time is long.

Aeration systems using diffused oxygen are increasingly popular, particularly for hauling units used around the farm for moving fry and fingerlings. Oxygen is available as compressed oxygen gas or as liquid oxygen. Compressed oxygen gas can be used when small volumes of oxygen are needed for short-term hauling of fish in small (50- to 100-gallon) hauling tanks. A standard compressed oxygen cylinder holds about 250 cubic feet of oxygen and weighs about 200 pounds. Liquid oxygen stored in special vacuum-insulated containers (cryogenic or dewar tanks) is preferred for routine use on commercial farms. Capacities of cryogenic containers range from about 3 to about 6 cubic feet of liquid oxygen. A 5.6-cubic-foot container weighs about 750 pounds and provides about 4,500 cubic feet of gaseous oxygen. This is equivalent to about 18 standard compressed oxygen cylinders, so liquid oxygen is considerably more convenient when large amounts are required. Liquid oxygen is less expensive per cubic foot, but the containers are expensive and are often rented rather than purchased. Because oxygen is slowly vented from cryogenic containers to maintain safe internal pressures, liquid oxygen cannot be stored indefinitely. Check the tank gauge periodically, especially in hot weather, to make sure sufficient oxygen is available. It is good practice to refill the tank immediately before extended use.

Oxygen is delivered from cryogenic containers to the hauling tank as a gas. A delivery manifold is connected to a single-stage regulator on the container. Supply lines from the manifold to each hauling tank should have a flow meter to register flow and allow adjustment of flow rates. Oxygen is introduced into the tank through diffusers as fine bubbles. Diffusers are constructed of fused minerals (similar to airstones), porous plastic tubing, finely woven hoses, or other materials.

Oxygen delivered from cryogenic containers tends to cool the water. Also, turbulence in the tank is minimal because relatively small volumes of gas are introduced as fine bubbles. This characteristic makes liquid oxygen aeration systems ideal for transporting fry or small fingerlings and for transporting larger fish over long distances.

Procedures

Water quality. Clear well water should be used to fill hauling tanks. Well water is free of fish pathogens and, unlike most pond waters, does not contain plankton or other organisms that can consume oxygen.

The temperature of the water in hauling tanks should be within 5°F of the water from which fish are harvested. The temperature of water from shallow wells (100 to 200 feet deep) is usually about 65 to 75°F. Water of this temperature may be too cool to use in the summer, when pond water temperatures are 80 to over 85°F, and it is too warm to use in the winter, when water temperatures are usually below 55°F. Water can be warmed or cooled by letting the transport unit with water sit for some period; overnight is often convenient. The water temperature in hauling units can be decreased by adding ice; 0.5 pound of ice lowers the temperature of 10 gallons of water about 10°F. Also, problems with temperature differences in the pond and hauling tank can be reduced by loading fish in the early morning in the summer (when water temperatures in the pond are lowest) or loading fish in the afternoon in the winter (when pond water temperatures are highest). Always temper fish from one water to another if the water temperatures differ by more than 5°F; *tempering* involves changing the temperature of the holding water at less than 1°F/minute until the water temperatures are equal. Change the temperature even more slowly (<0.5°F/minute) if temperatures differ by more than 20°F.

The stress of handling causes osmoregulatory disturbances in fish. Dissolved calcium helps fish maintain osmotic balance. Calcium hardness should be at least 50 ppm as $CaCO_3$ in hauling tank waters, particularly if transit time will be greater than 1 hour. Even higher calcium hardness is beneficial, and some recommendations call for minimum calcium hardness values of 150 ppm in hauling water. Calcium can be added to deficient water as calcium chloride ($CaCl_2$ or $CaCl_2 \cdot H_2O$) or calcium sulfate (gypsum, $CaSO_4 \cdot 2H_2O$). The approximate amounts of these compounds needed to raise the calcium hardness (expressed as ppm as $CaCO_3$) of 100 gallons of water by 50 ppm are as follows: $CaCl_2$, 20 grams; $CaCl_2 \cdot H_2O$, 25 grams; $CaSO_4 \cdot 2H_2O$, 35 grams. Common salt (NaCl), without iodide, is also used to reduce osmoregulatory distress during hauling. Salt may also provide some benefit in controlling external parasites (see the section Disease Treatments). Salt is added to the hauling tank water to give solutions of 0.1 percent (1,000 ppm) to 0.3 percent (3,000 ppm). Adding 3.8 grams of NaCl/gallon (0.8 pound of NaCl/100 gallons) provides 1,000 ppm of salt. Note that many groundwaters are already highly mineralized and may not require addition of calcium or salt. For instance, shallow groundwaters in west-central Mississippi have calcium hardness values of 150 to 400 ppm as $CaCO_3$, and the total dissolved salt (all salts, not just NaCl) often exceeds 500 ppm. These waters are suitable for use in hauling tanks without additives.

During transport, fish consume oxygen and produce carbon dioxide and ammonia. Adequate supplies of dissolved oxygen are important regardless of the duration of transport because dissolved oxygen is depleted within minutes at the high densities of fish normally present in transport tanks. Dissolved oxygen concentrations should never fall below 5 ppm during transport. The detrimental effects of carbon dioxide and ammonia (Chapter 4) are prevented by reducing fish loading rates in proportion to transit time rather than attempting to remove the substances by chemical or physical means.

Factors affecting amounts of oxygen consumed by fish are discussed in the section Dissolved Oxygen. Briefly, oxygen consumption rates by fish increase as fish size decreases, as water temperature increases, after fish are fed, and when fish are excited. Each of these factors has important implications for fish transport.

Small fish consume more oxygen than large fish for the same total weight. Because oxygen transfer by aeration usually cannot be varied in transport tanks, loading rates must decrease in proportion to fish size.

Lowering the water temperature decreases metabolic activity and reduces oxygen consumption. The degree to which water temperatures can be manipulated is limited because large changes in temperature can stress fish. Nevertheless, it is recommended that water temperatures be reduced to 60 to 70°F when hauling fish long distances in the summer. Lower water temperatures also benefit fish by reducing carbon dioxide and ammonia production. Water temperatures can be decreased after fish are loaded by adding ice made from unchlorinated water. Remember, however, that fish may have to be slowly tempered to a higher water temperature when they are unloaded. Direct transfer of fish hauled at 65°F to water at 85°F can result in large losses from thermal shock.

Fingerling and adult fish should not be fed for at least 24 hours before transport. Fish with an empty digestive tract consume less oxygen than recently fed fish. Also, partially digested food that is regurgitated or voided from the intestine during transport fouls the hauling water. Subsequent decomposition of the food or fecal matter in the hauling tank consumes oxygen and produces ammonia and carbon dioxide.

Excitation and activity increase oxygen consumption by fish. Oxygen consumption rates are highest in the first 15 minutes after fish are loaded into the hauling tank. Fish loading rates and aeration capability must account for this initial high oxygen removal rate. As mentioned, lowering the water temperature to less than 70°F can reduce fish activity and oxygen consumption rates during long trips.

Transport practices. Fish should be in good health before transport. Fish having infectious diseases or stressed from exposure to poor water quality may not withstand the rigors of transport. Withhold feed for 1 or 2 days before transport and follow recommended practices during harvest to prevent excessively stressing the fish before transport. Make sure the temperature of the water in the hauling tank is within 5°F of water from which fish will be harvested.

Loading rates for channel catfish are presented in Table 12.3. These rates must be regarded as rough guidelines because safe loading rates vary, depending on the type of equipment used, initial water quality, condition of the fish, and other factors. In general, the loading rates listed in Table 12.3 can be increased by about 25 percent if the water temperature is below 55°F and should be reduced 25 percent if the water temperature is between 75 and 85°F. Reduce the loading rate by 50 percent if the water temperature is above 85°F.

Make sure that all aspects of the trip have been carefully planned so that unnecessary delays are prevented. Have all necessary permits if fish are to be transported across state lines. Check all equipment and

TABLE 12.3.
Estimated Pounds (Nearest 0.5 Pound) of Channel Catfish/Gallon of Water That Can Be Transported at 65°F

Size of Fish		Transit Period (Hours)		
Number/Pound	Pound/1,000 Fish	8	12	16
1	1,000	6.0	5.5	5.0
2	500	6.0	5.0	3.5
4	250	5.0	4.0	3.0
50	20	3.5	2.5	2.0
125	8	3.0	2.0	1.5
250	4	2.0	1.5	1.5
500	2	1.5	1.5	1.0
1,000	1	1.0	1.0	1.0
10,000	0.1	0.5	0.5	0.5

Source: Adapted from McCraren and Millard (1979).

carry replacement parts for any critical piece of equipment. Always carry an oxygen meter and thermometer when hauling fish long distances. Check fish frequently during transit to make sure there are no problems. Channel catfish should not be seen right at the water surface after they have adjusted to conditions in the hauling tank; if fish are seen at the surface, they may be in distress. Measure dissolved oxygen concentrations immediately and take remedial action if required.

Transporting Eggs and Fry

The need for transporting eggs or fry depends on the type of fry-production system used (Chapter 8). Commonly, fertilized eggs are moved from spawning containers in the brood fish pond to the hatchery. The fry produced in the hatchery are then moved to a nursery pond after the yolk sac has been absorbed and the fry have begun to consume commercial feeds.

Channel catfish eggs can be transported short distances in insulated coolers or ice chests. About 1 pound of eggs/gallon of water can be transported if the duration of transport is less than 15 minutes. If eggs are to be held longer, the amount of water should be increased and the water should be aerated by a small compressed oxygen cylinder and diffusers. Larger quantities of eggs can be transported in a small hauling tank (100 to 400 gallons) carried on the bed of a pickup truck.

These tanks are usually insulated and aerated with 12-volt agitators or, preferably, diffused oxygen. Small wire baskets can be fashioned to hold egg masses during transport. Regardless of the method used to transport eggs, dissolved oxygen concentrations in the hauling water should never fall below 3 ppm and preferably should exceed 5 ppm. The temperature of the hauling water should match that of the pond water from which the eggs are obtained. If the temperature of the hauling water differs from that in the hatchery by more than 5°F, slowly temper the eggs to the hatching water temperature by running water from the hatchery into the hauling unit so that the temperature changes less than 1°F/minute until the water temperature in the tank equals that in the hatchery.

Small numbers of fry can be transported short distances in aerated buckets or tubs. About 1 pound of fry (approximately 7,500 to 20,000 fry, depending on age) can be held no more than an hour in 5 to 10 gallons of aerated water. Commercial fry and fingerling producers usually transport fry in small tanks specifically designed to haul them. The tanks should have a sloped floor leading to a 2- to 3-inch drain. A length of hose is attached to the drain so that fry can be stocked into nursery ponds without handling the fish. Tanks are aerated with 12-volt agitators covered with small-mesh screen or, preferably, with bottled oxygen delivered through diffusers. Fry can be transported for at least 12 hours at a rate of about 0.50 pound of fry/gallon of water (McCraren and Millard 1978). Do not feed fry for 1 or 2 hours before transport. Maintain at least 5 ppm of dissolved oxygen during transport and temper fish at less than 1°F/minute if water temperatures in the hatchery, hauling tank, or nursery pond differ by more than 5°F.

Never use aeration systems using large air blowers to aerate water when transporting eggs or fry. Blower units deliver large volumes of air through perforated pipe or tubing. The large bubbles that are created cause excessive water turbulence that can break apart and damage egg masses or cause overexertion and fatigue in fry as they are rolled around in the hauling tank.

REFERENCES

Busch, R. L. 1985. Harvesting, grading, and transporting. In *Channel Catfish Culture*, ed. C. S. Tucker, pp. 543–567. Amsterdam: Elsevier.

Dupree, H. K., and J. V. Huner. 1984. Transportation of live fish. In *Third Report to Fish Farmers*, ed. H. K. Dupree and J. V. Huner, pp. 165–176. Washington D.C.: U.S. Fish and Wildlife Service.

Huner, J. V., H. K. Dupree, and D. C. Greenland. 1984. Harvesting, grading, and holding fish. In *Third Report to Fish Farmers*, ed. H. K. Dupree and J. V. Huner, pp. 158–164. Washington D.C.: U.S. Fish and Wildlife Service.

McCraren, J. P., and J. L. Millard. 1979. Transportation of warmwater fish. In *Manual of Fish Culture, Section G: Fish Transportation*, ed. Anonymous, pp. 43–88. Washington D.C.: U.S. Fish and Wildlife Service.

CHAPTER 13
Alternative Culture Systems

The variety of systems proposed for rearing fish appears to be limited only by the imagination. In theory, a number of systems may be applicable to catfish culture; in practice, however, commercial culture is typically conducted in large earthen ponds because presently this is the most economical method of production. The adage "Don't fix it if it isn't broken" doesn't seem to apply to aquaculture, for which complicated and expensive systems are often devised for high-density fish culture at production costs greatly exceeding those normally incurred in pond culture. Certainly high-density aquaculture is desirable and has a place, but only if it is profitable. Profit associated with various intensive systems developed for culture of catfish is questionable because of very high production costs coupled with excessive start-up costs (Losordo, Easley, and Westerman 1989). When considering promotional material concerning various systems for intensive culture of fish, be cautious, exercise common sense, and be aware that production data presented for a particular system are often extrapolated from small-scale production trials and do not necessarily reflect production in commercial-size units. Ask basic questions such as power costs for aeration, nitrogen removal, heating, and pumping. If specific questions are not answered satisfactorily, consult an engineer.

This is not to imply that all culture systems aside from ponds are inappropriate and that such systems cannot be used to rear catfish commercially. To the contrary, under certain conditions intensive fish culture systems can be operated economically. For example, in special circumstances heated water is readily available and pumping costs are low, such as from warm-water artesian wells, geothermal springs, or power plant condensers. Also, in urban areas where space is limited and markets exist for higher-priced fish, intensive culture systems may be a viable alternative for rearing catfish. Generally, these types of production facilities are limited to supplying catfish to a relatively small and localized market.

Culture systems used for rearing fish, with respect to water movement, may be static or flowing. The most common static system uses ponds as the culture unit. Cages for rearing catfish can be used in either static or flowing water. Flowing-water systems used for commercial culture of fish generally employ raceways for the culture unit, but tanks or silos can be used. In static systems, after the initial filling, water is only added to replace losses due to evaporation and seepage. Water in flowing-water systems continually enters and leaves the culture unit.

Fish culture systems may be further classed as open, closed, or semiclosed. In open systems the water leaving the culture unit is not reused. In closed systems the water is passed through a treatment process to improve water quality and then recirculated to the culture unit. In semiclosed systems part of the water is recirculated and part is replaced with new water.

Fish culture conducted in static ponds is considered to be extensive, although catfish are often cultured at relatively high densities (5,000 to 10,000 pounds/acre). Intensive fish culture is typically conducted in flowing-water systems and may reach densities equivalent to 1 million pounds/acre, based on available water surface area. But, when based on the volume of water used during the growing season, differences in production (per gallon used) between static and flowing-water systems are not nearly so great.

The culture of catfish in cages, raceways, and closed, water-recirculating systems will be discussed in the remainder of this chapter. These are the most commonly used alternative culture systems. The material presented should be considered as introductory, because, with the exception of cage culture, little is known of the feasibility of developing large-scale catfish culture based on these systems. Most raceway and closed systems used for culturing catfish have been small-scale systems, primarily used for research. Also, there are numerous effective system designs that may differ from those presented herein and systems are dependent on site-specific criteria.

CAGE CULTURE

Many species of fish and some aquatic invertebrates have been reared in cages throughout the world. Cage culture of fish is known to have existed in various countries before the beginning of this century. Presently cage culture represents a small fraction of the commercial catfish production in the United States, but about 40 percent of the total production of rainbow trout in Scotland and almost all yellowtail reared in Japan and Atlantic salmon in Western Europe are in cages (Beveridge 1987). Salmon are also reared in net pens (large cages made from netting) in the northern coastal regions of the United States and Canada. Although cage culture of catfish has been used in several states in the United States, most of the production (500,000 pounds annually) is in western Arkansas (Collins 1988a). Cage culture of catfish has not expanded greatly for a number of reasons, but primarily because large-scale culture of catfish is simply more economical in earthen ponds.

Cage culture of catfish may appear to be much easier than open pond culture, but the risks are perhaps greater in cage culture because the fish are confined to a small area and thus are more crowded than pond-reared fish. In general, the same husbandry principles apply to either type of culture. The two principal advantages of cage culture are that (1) aquatic environments that are not typically used for aquaculture (e.g., lakes, reservoirs, mining pits, streams, rivers, and ponds) can be used and (2) harvest is simplified, particularly in deep bodies of water; that is, the fish can easily be concentrated and dipped out or if the cages are small the entire cage can be removed from the water. Major disadvantages are that (1) feed cost is generally higher in cage culture because the levels of certain nutrients in the feed are usually increased to ensure that nutritional deficiencies will not occur since the fish do not have access to natural food; (2) low levels of dissolved oxygen may be a problem to manage since the fish are not free to move to areas of higher oxygen; however, this problem can be managed in large cages by using small aerators directly in the cage; (3) diseases can spread rapidly because the fish are closely confined; (4) vandalism or poaching is facilitated.

Cages

Several types of cages, including surface cages resting on the bottom, submerged cages, and floating cages, have been used for fish culture. Floating cages are best for culture of catfish (Fig. 13.1). Styrofoam is a

408 Alternative Culture Systems

FIGURE 13.1.

A floating cage. (Photo: Charles Collins)

good flotation device, though other materials (polyvinyl chloride [PVC], plastic or metal drums, or plastic bottles) can be used. The flotation material should be adequate to keep the cage top about 6 inches above the water surface.

Cages used for the culture of catfish have been of various sizes and shapes and have been constructed from many different types of materials. The shape of the cage does not appear to impact production, and cage size depends on production and marketing strategies. Popular sizes used for catfish culture include cylindrical cages 4 feet in diameter by 4 feet deep, square cages 4 feet wide by 4 feet long by 4 feet deep, and rectangular cages 3 feet wide by 4 feet long by 3 feet deep or 8 feet wide by 4 feet long by 4 feet deep. Collins (1988) recommends that cages used for production of catfish be at least 3 feet wide by 4 feet long by 3 feet deep.

Cage frames can be constructed from wood (redwood or cypress is best because of its resistance to decay), metal, Fiberglas, or PVC. If the frames are made from material that is subject to rust, they should be coated with a water-resistant substance. The top should be hinged and should contain an opening for feeding.

Cages should be constructed of vinyl-coated welded wire with a mesh size of at least 1/2 inch for good water circulation through the cage. A larger mesh may be used if large fingerlings are to be stocked. Other materials that are commonly used for cages include galvanized wire and plastic or nylon netting.

Feeding rings, which are enclosures built into the cage or cage cover, are needed to prevent feed from floating out of the cage. A feeding ring that encloses the entire perimeter of the cage is generally preferred, but smaller rings can be used. The ring should be constructed from 1/8-inch mesh rust-resistant material, and it should extend 10 to 12 inches below the water surface and 4 to 6 inches above the water surface.

Cage Placement

Water circulation is a major consideration in selection of a site for cage culture. A continuous water exchange is necessary to provide dissolved oxygen and to remove waste. In small static ponds, the cages should be placed in an open area away from any obstructions that could prevent good water circulation generated by wind and waves. In large reservoirs a site that shelters the cages from the main force of wind and waves yet allows adequate circulation is best. Cages should be placed far enough apart to provide for good circulation between cages (Fig. 13.2). There are few actual data to indicate an optimal spacing of cages, but spacing less than 3 feet is not recommended (Schmittou 1969). Collins (1988a) indicates that no particular spacing has been proved to be more effective than another but suggests that about 25 feet between cages is sufficient. If commercial fish culture is conducted in cages, aerators to supply oxygen should be considered.

Water depth is another factor in selecting a site for placement of cages. Collins (1988a) suggests that a minimum of 6 feet of water under the cage bottom is desirable so that waste products will not contaminate the site. Although 6 feet of water below the cages may be desirable, 3 feet or less of water below the cage bottoms will suffice.

Cages suspended in a body of water can be anchored in a number of ways. Bottom anchors to suspend an individual cage or a series of cages, a nylon rope or a steel cable placed across the water and attached to docks, or a cemented pipe on the bank works well. Cages can be tied to the line for easy removal. Regardless of the type of anchor system used, be sure that it is strong enough for the type of environment to which the cages will be exposed.

Stocking Density

To produce a 1.25- to 1.5-pound fish in about 180 days, a 7- to 8-inch fingerling catfish should be stocked at a rate of 8 to 10 fish/cubic foot

410 Alternative Culture Systems

FIGURE 13.2.

Cages suspended in a large body of water. (Photo: Charles Collins)

of cage. Stocking less than 6 fish/cubic foot may result in fighting among the fish (Collins 1988b), which in turn will result in higher mortality rate and decreased production. However, recall that the total number of fish that can be stocked into cages suspended in a body of water is dependent on the carrying capacity of the body of water. In static ponds where no supplemental aeration is used, no more than 1,500–2,000 fish/acre of available water should be stocked. If supplemental aeration is adequate or flowing water is used, higher stocking densities (4,000 to 5,000 fish/acre of water) may be used.

Fish Husbandry and Feeding

Generally the same husbandry techniques used in ponds apply to fish rearing in cages. These techniques have been discussed in detail in other chapters. Suffice it to say that rearing fish in cages is perhaps more difficult than in ponds because the fish are crowded into a small space and are unable to move to other areas if water quality deteriorates. Thus the effects of stress, such as outbreaks of infectious diseases, may be more severe in cage-reared fish.

Feeding catfish reared in cages differs little from feeding pond-reared catfish; thus the guidelines given in Chapter 10 apply. However, since the fish are confined in cages they cannot forage for natural foods and it is critical that all nutrients be provided in the diet. The same is generally true for catfish reared in ponds at high density, but pond-reared fish may meet at least part of their micronutrient requirements (vitamins, minerals, essential fatty acids) with natural foods. Even in pond culture the contribution of natural food is not accounted for when formulating feeds, so a complete feed is given. The primary difference in feeds used to rear catfish in cages and those used for catfish culture in ponds is the higher level of protein (36 versus 32 percent) in feeds for cage-reared fish. Some catfish producers who use cages may also prefer a feed that contains an excess of vitamins.

A feeding schedule similar to that given in Table 10.11 may be used for feeding catfish reared in cages. To simplify matters, channel catfish should be fed 3 percent of body weight daily until they reach a size of about 0.3 pound. Then they should be fed from 2.5 to 2.0 percent of their body weight daily for the rest of the growing season. A rule of thumb is to feed all that the fish will eat within 10 to 15 minutes when the water temperature is above 70°F. However, fish in cages are sometimes hesitant to feed immediately after feed is offered; thus feeding what the fish will eat in a specified time is often difficult. Regardless of the method used to arrive at the feeding rate, do not overfeed.

Disease Treatments

It is essential in cage culture to observe fish behavior, particularly during feeding, because change in behavior can be a sign of a disease outbreak. If the fish are diseased, dead or dying fish will begin to appear. Once a disease problem is suspected, get a live sample of fish and water and have it checked at a fish disease diagnostic laboratory. A list of these labs is given in the Appendices. Once the diagnosis is made, the fish will likely require treatment. Diseases that commonly affect catfish and methods for their treatment are discussed in Chapter 11.

Depending on what type of disease is diagnosed, it may be possible to feed a medicated feed or to add a chemical to the water to eliminate bacteria or parasites. Internal bacterial infections are generally treated by feeding a medicated feed. Parasites or external bacterial infestations are usually treated by adding a chemical to the water. If the pond is relatively small and fish in all cages are to be treated, then the chemical can be added to the entire pond. When fish in only some of the cages need to be treated, it may be possible to isolate the infected cages for treatment by using a plastic bag. If this is done, the infected cage or cages are covered with a plastic bag or other suitable material and the chemical is added to the water in the enclosure. Isolating a

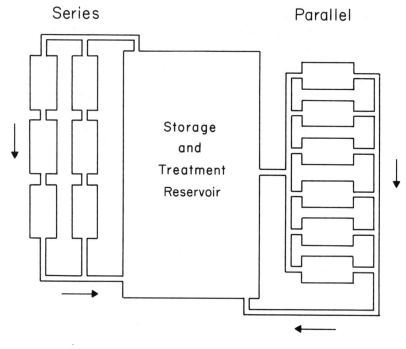

FIGURE 13.3.

Schematic of raceways arranged in series and in parallel. These raceways are part of a closed system using a storage and treatment reservoir.

cage for treatment can cause problems with dissolved oxygen; thus supplemental aeration should be available. Follow all precautions given in Chapter 11 when treating diseases.

RACEWAY CULTURE

A raceway is generally considered to be a linear channel with a continuous flow of water entering at one end and exiting at the opposite end. However, round culture chambers are often referred to as circular raceways. Raceways can be arranged in series or in parallel (Fig. 13.3). Water can be used once and then discharged (single-pass system) or race-

FIGURE 13.4.

Harvesting raceways.

ways can be part of a closed system utilizing a storage and water purification reservoir.

The primary requirement for culture of fish in raceways is an abundant supply of suitable water, particularly for a single-pass system. Energy costs for pumping water may prohibit profitable production unless artesian flow or surface water with suitable head is available. The major advantage of using raceways to culture fish is that production per unit of space is much greater for fish reared in raceways. For example, in Idaho 10,000 pounds of catfish is routinely raised in 770 cubic feet of space in raceways (Avault 1989). Assuming 4,000 pounds/acre average production of catfish in ponds, a 2.5-acre pond would be needed to produce that amount of fish. Other advantages include ease of fish grading, harvest, feeding, and disease treatment. Harvest and grading can easily be accomplished by using grading bars and crowding the fish to one end of the raceway (Fig. 13.4). The main disadvantages of rearing catfish in raceways are the volume of water needed and the associated cost of pumping if sites are not chosen carefully.

Raceways

Raceways may be constructed from various materials, such as earth, stone, metal, or plastic, but most raceways used to rear fish commercially are concrete. They may be of various sizes, but they should not be too long because the fish often tend to congregate near the water inlet and effectively use only a part of the raceway. Also, long raceways are difficult to flush, and waste from the fish and uneaten feed may accumulate. The optimal size of a raceway for catfish culture has not been determined through university research. What is known about relatively large-scale culture of catfish in raceways is based primarily on practical knowledge gained by producers who use them. A brief description of two commercial raceway systems is presented. One is a single-pass raceway system in which water is passed through a series of raceways and then discharged from the raceway system. The second is a closed system in which water is passed through a series of raceways before being discharged into "oxidation" ponds and then eventually recirculated through the raceways.

Single-pass raceway systems. The following description of Fish Breeders of Idaho, Incorporated, was taken from Ray (1981) and Avault (1989). Catfish, trout, and tilapia are reared in raceways supplied with geothermal water that is mixed with cooler spring water to maintain the proper temperature. Water flow through each raceway is 1,500 to 2,000 gallons/minute. The raceways are constructed from concrete and are divided into four sections that are arranged in a series with a 2-foot drop between each section. Each section is 24 feet long, 10 feet wide, and 3.5 feet deep. The amount of space used by the fish in each section is 770 cubic feet. The raceways are in pairs with a common center wall. Water enters the raceways and passes through four raceways (sixteen sections) before exiting. Catfish are reared in the upper raceways and tilapia in the lower raceways.

Stocking densities for large fingerling catfish (0.25 pound/fish) are from 2 to 4 pounds/cubic foot of space. Lower densities are suggested for smaller fish. Approximately 10,000 to 14,000 pounds of catfish is harvested from each raceway section. On the basis of water flow, production is 40,000–50,000 pounds/cubic foot/second (CFS).

Water quality parameters that are considered to be critical are dissolved oxygen, ammonia, and settlable solids. The drop between each raceway section replaces about 50 percent of the oxygen consumed by fish in the preceding section. Dissolved oxygen concentrations vary from about 7 ppm in water entering the first raceway section to about 3 ppm in water leaving the fourth section of raceway. Dissolved oxygen is replenished as the water enters a ditch and flows to a second set of raceways. The water is passed through five sets of raceways before finally being discharged.

About 0.2 ppm of total ammonia-nitrogen is produced per CFS and the total ammonia-nitrogen concentration in the discharge at the end of four sections of raceway is about 0.8 ppm. Nitrification of ammonia to nitrate by bacteria begins as the water flows through an earthen ditch to the next set of raceways.

Settable solids are a major concern since the United States Environmental Protection Agency stipulates that solids cannot be greater than 3.3 milliliters/liter in discharge water. Solids (sludge) are collected behind a screen and drained into settling basins, which are aerated. The sludge is pumped into raceways where tilapia are reared.

The success of Fish Breeders of Idaho, Incorporated, is due in part to access to gravity-flow geothermal water, which provides for a year-round growing season, and to use of relatively small raceway sections and good management techniques.

Closed (recirculating) raceway system. The raceway system described later is atypical in that it is part of a closed system in which water enters at one end of the raceway system and is discharged at the opposite end into a reservoir to remove settable solids and to oxidize nitrogenous waste. The water is then recirculated through the raceway system. Channel catfish are reared in the raceways and various types of filter-feeding or detritus-feeding fish are reared in the reservoirs (or "oxidation ponds").

The two raceway systems currently used for production of catfish in southern Arkansas are similar. The larger of the two systems (Fig. 13.5) consists of 108 raceway sections that are 16 feet wide, 36 feet long, and 4 feet deep. The sections are arranged in three descending tiers, creating three oxygen-producing waterfalls. Oxygen can also be added by using a diffused liquid oxygen aeration system. Each three-tier raceway is 108 feet long. Water flow to the raceways is by gravity flow from a header pond. Water exchange rate is about 6 to 7 minutes. The water from the raceways is discharged into a pond and then circulated through a series of ponds to settle solids, oxidize nitrogenous waste, and reoxygenate the water. Water is eventually pumped back into the header pond by three pumps capable of pumping a total of about 30,000 gallons of water/minute. Approximately 140 acres of ponds is used for water cleanup and storage.

Catfish weighing 0.25 pound or more are stocked at a rate of 10 fish/cubic foot or about 15,000 fish per raceway section. Large fish are stocked so that two crops of fish can be reared during a 200-day growing season. The fish are fed floating catfish feed using demand feeders that are loaded by an auger system (Fig. 13.6). Theoretically, using demand feeders

FIGURE 13.5.

Raceways used for catfish culture in Arkansas. (Photo: Charles Collins)

(which allow the fish to feed ad libitum) reduces the large peaks in oxygen consumption, carbon dioxide, and ammonia production generally observed when large quantities of feed are offered at one time. The feed used may differ from a typical pond feed in that it is usually higher in protein (35 percent or so); it may also be overfortified with vitamins.

The oxidation ponds are used to rear fish other than catfish. Various species of filter- or detritus-feeding fish have been used, including paddlefish, bighead carp, and silver carp. The species of choice depends on existing markets. Generally, the markets for these types of fish in the United States are rather small and limited in number.

Total production in a system such as described previously is about the same as production in heavily stocked ponds, because the system is essentially a pond system (at least in respect to environmental constraints) with a mechanism (the raceway sections) to concentrate fish for ease of harvest and size grading and to facilitate disease treatment. The same processes and limitations that occur in a typical pond system apply to the recirculating raceway system. That is, only so much oxidation of nitrogenous waste can occur; thus feeding (and production) is limited as in pond culture.

FIGURE 13.6.

Demand feeder used for feeding catfish reared in raceways.

CLOSED, WATER-RECIRCULATING SYSTEMS

Recirculating systems (also called water-reuse systems) for the culture of fish have been primarily used by researchers or in hatcheries for species that require stringent environmental conditions for hatching and growth during early stages of development. Although these systems can be built outdoors, most are indoor systems where temperature is easier to control. These systems are particularly useful on a small scale where a highly controlled environment is needed. Water-recirculating systems are receiving attention from potential aquaculturists for large-scale production of fish because they conserve water, provide for flexibility in site selection of aquacultural enterprises, and extend the growing season to year-round if temperature control is incorporated in their design. These are appealing advantages, but there is a lack of available practical information concerning the use of these systems for rearing of fish on a commercial scale. Presently it is improbable that catfish can be reared profitably in such a system, except perhaps on a limited basis for a specialized market that commands a high price for fish.

There are many designs for water-recirculating systems, but the major concerns are the same regardless of design. That is, methods to eliminate potentially toxic nitrogenous metabolites, to remove particulate matter, to provide sufficient oxygen, and to control temperature should be incorporated into any design. The basic components (culture chamber, primary settling chamber, biofilter, and final clarifier) of a typical water-recirculating system are represented schematically in Figure 13.7. These components may be separate, as depicted in Figure 13.7, or combined into one or two units. One pump is needed to move water in a typical water-recirculating system. A source of makeup water is needed to replace that lost during removal of solids or through evaporation.

Culture Chambers

Culture chambers used in water-recirculating systems to rear aquatic organisms can be of various shapes and constructed from a multitude of materials. The most common culture chambers are either rectangular or round and are generally fabricated from Fiberglas or plastic materials. Round tanks do not have areas of dead water, such as corners found in rectangular tanks, where wastes can accumulate and foul the water. Rectangular tanks better utilize available floor space. Each tank should have its own water supply, but a common drain can be used for a series of tanks. The size of a culture chamber may range from a tank capable of holding a few gallons of water to one that can hold several thousand gallons of water. The size is dependent on the intent of the culturist.

Primary Settling Chamber

Removal of particulate matter reduces the load on the biofilter and improves efficiency of the system. Particulate matter can be removed in several ways ranging from settling basins to various types of filters. Pressurized sand filters may be used, but they require frequent backwashing for optimal performance. One of the simplest and least troublesome methods is to use a primary settling chamber before the biofilter. In some designs this chamber is not included or is combined

Closed, Water-Recirculating Systems **419**

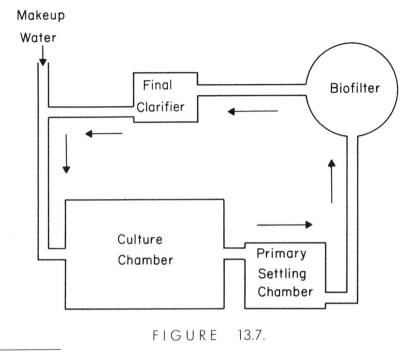

FIGURE 13.7.

Schematic of basic components of a water-recirculating system.

with the biofilter, and in some systems it is used in conjunction with a sand filter or other type of mechanical filter. The amount of sedimentation needed is dependent on the anticipated particulate load.

Water should enter and exit the primary settling chamber near the top to prevent resuspending wastes that have already settled. The chamber should be large enough so the water inflow and outflow are as far apart as possible to provide adequate water retention time to allow particulate to settle. A drain should be placed in the bottom of the chamber to allow removal of solids by draining. Sediments, which are composed primarily of feces, waste feed, and bacteria, may be discarded or collected and used as organic fertilizer or fed to certain species of fish (e.g., *Tilapia*).

Biofilters

Ammonia is the primary fish excretory product that degrades the quality of water in water-recirculating systems. Although it may be removed by air stripping or ion exchange, biofiltration is the most popu-

420 Alternative Culture Systems

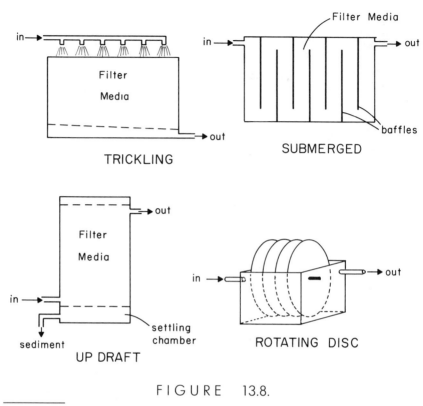

FIGURE 13.8.

Schematic diagrams of four types of nitrification biofilters.

lar method of removing it from water-recirculating systems because it is efficient, relatively inexpensive, and easy to operate (Lucchetti and Gray 1988). Biofilters work by providing extensive surface area and proper environmental conditions for colonization by bacteria that oxidize ammonia to nitrate. Nitrification is a two-step process with ammonia being first oxidized to nitrite by *Nitrosomonas* spp. and then to nitrate by *Nitrobacter* spp. These bacteria are widely distributed in soil and water. Bacterial nitrification is discussed in Chapter 9.

Several factors should be considered when designing a biofilter. These include providing maximum surface area for bacterial colonization, maintaining good oxygenation, maintaining uniform water flow through the filter, providing sufficient void space to prevent clogging, and sizing properly (McGee 1989). Variations in biofilter design are unlimited, but four basic types have been used, at least experimentally (Fig. 13.8). They may be classed as tricking, submerged, updraft, and rotating disc filters (Stickney 1979).

Closed, Water-Recirculating Systems   **421**

Water enters a trickling biofilter from the top, passes by gravity through the filter medium, and exits at the bottom. Water flow rate is such that the filter medium does not become completely submerged. Internally the filter is always wetted. Some trickling biofilters have rotating arms that distribute the incoming water evenly over the filter medium. Others use stationary water distribution systems.

Submerged biofilters are simply basins containing a filter medium that is completely covered with water at all times. They may contain baffles to route the water throughout the filter. Water flows into the filter near the top and exits at the opposite end near the top.

Updraft biofilters are designed so that water enters at the bottom, moves upward through the filter medium, and exits near the top. A settling chamber that will allow particulate matter in influent water to settle out and be removed by a sediment drain in the bottom of the filter can easily be incorporated into this design.

The rotating disc biofilter consists of a biofilter medium (circular disc) that is placed on a shaft and set in a tank with half of each disc submerged in the water and half exposed to the atmosphere. The shaft is slowly rotated (usually by an electric motor) a few revolutions per minute. The bacterial colonies established on the circular discs are alternately exposed to the water for nutrients and to the atmosphere for oxygen; this arrangement is an advantage of the rotating disc biofilter over other designs. A primary disadvantage is that a motor is needed to rotate the shaft containing the discs; thus mechanical failure can be a problem.

Biofilter size. It is difficult to define the optimal size for a particular biofilter precisely because size is dependent on expected ammonia production and on efficiency of ammonia removal. An estimate of ammonia production, which is proportional to the amount of feed put into the system, can be calculated (Colt and Armstrong 1981; Piper et al. 1982). However, accurately predicting the volume of medium needed to remove ammonia efficiently is difficult because of interaction between water quality and nitrifying bacteria. McGee (1989) suggests that a 3:1 ratio of filter volume to culture tank volume is usually sufficient.

Biofilter media. Biofilter media can be almost any substances that are nontoxic and provide a large surface area. Sand and gravel have been commonly used. Although these materials provide plenty of surface area per unit volume, they are dense and thus require rigid support. More importantly they contain less void space than lighter materials; thus they clog easily. Frequent backflushing may regenerate the filter, but backflushing can be problematic in a closed system. Channeling (water flow through the filter but along a restricted pathway) can occur in sand and gravel filters. A medium with less surface area but more void space is more efficient than sand and gravel.

Several types of media are available for biofiltration, and these can be purchased from various manufacturers. Also, common material such as polyvinyl chloride pipe cut into small sections has been used as a biofilter medium. Teflon and Styrofoam have also been used. The disc in rotating disc filters may be made from plastic, Fiberglas, or other suitable materials.

Seeding biofilters. A serious problem with using biofilters is the unpredictability of establishing and maintaining colonies of nitrifying bacteria. These bacteria are widespread and normally colonize the biofilter naturally, but this process may take several weeks. To ensure and hasten colonization, medium from an established biofilter may be added to a new filter or the filter can be inoculated with a commercially available bacterial suspension. Even with inoculation, the bacteria may take 3 or 4 weeks to colonize the filter in sufficient numbers to be effective. Generally the biofilter is allowed to condition for a few weeks before stocking fish into the culture chambers. During the conditioning period ammonia may be added periodically to provide substrate for the bacteria and to monitor the effectiveness of the biofilter.

Secondary Settling Chamber

A secondary settling chamber may be used in some water-recirculating systems as a final clarification process. The design of the secondary settling chamber is the same as that of the primary settling chamber. This chamber is not essential to the operation of water-recirculating systems, but it further reduces the load of particulate matter entering the culture chambers. Also, foam (which may be produced from dissolved organic materials) can be removed in this chamber or in other chambers if it is a problem.

Aeration

Good oxygenation is essential for proper functioning of the biofilter as well as providing oxygen for the fish. Dissolved oxygen concentration should be maintained above 60 percent saturation (McGee 1989). Oxygen should be provided in the culture chambers and in the system before or in the biofilter. Colt and Tchobanoglous (1981) and Speece (1981) provide insight into aeration methods for aquaculture systems. U-tube aeration and pure oxygen injection appear to be the most effi-

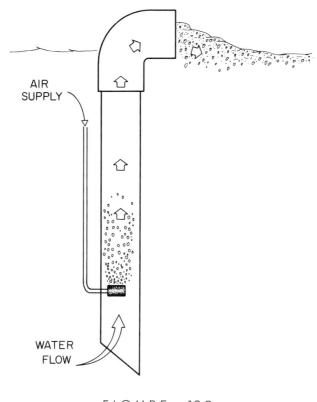

FIGURE 13.9.

Air lift pump.

cient methods (Colt and Tchobanoglous 1981). Trickling and rotating disc biofilters are self-aerating by design. Air lift pumps (Fig. 13.9), which are sometimes used to move water in water-recirculating systems, are another method of aeration. Air compressors and blowers may be used to provide aeration. Blowers are generally preferred because they provide large volumes of air at low pressure. If an air compressor is used to provide air, pressure regulators will be required to reduce pressure.

Maintaining pH

Accumulation of carbon dioxide and organic acids in water-reuse systems can depress the pH if the system is not properly buffered. Also, nitrification is an acid-producing process, and pH in closed systems tends to decline with time. Low pH can stress the fish or even cause death as well as adversely affect the microorganisms colonizing the biofilter. The pH for freshwater aquaculture systems should be maintained in the range of 7.0 to 8.5, depending on the total alkalinity of the water (see the section pH). Calcium carbonate is typically used as a buffer. Whole or crushed oyster shell and limestone are inexpensive forms of calcium carbonate that are used in water-recirculating systems. The amount of buffering agent to use is not precisely defined, but it does not appear to be critical. Stickney (1979) suggests that several pounds of buffering agent per cubic foot of filter capacity is adequate. Care should be taken when increasing pH because of increased ammonia toxicity.

Disease Control

Although most water-recirculating systems used in aquaculture do not incorporate sterilization techniques in the design, some sort of sterilization method is desirable since disease outbreaks are common in these systems. Dupree (1981) suggests that ultraviolet radiation, ozone, and chlorine (under certain conditions) have the most promise in controlling pathogens in water-recirculating systems. The most common sterilization method is to incorporate an ultraviolet lamp in the incoming water source. The proper size of ultraviolet filter and rate of water flow through the filter depend on the size and species of pathogen and on water clarity (Dupree 1981). Reducing turbidity by use of sand filters, and so on, increases the effectiveness of ultraviolet sterilization.

Ozone is effective in eliminating pathogens affecting fish; however, it is toxic to fish or to organisms on the biofilter. Lethal levels and sublethal effects are not known for most fish. Ozonization is discussed by Dupree (1981) and Lucchetti and Gray (1988).

Chlorine can be used for sterilization, but even low concentrations of residual chlorine or its by-products are toxic to fish. Chlorine is useful for sterilization of components of water-recirculating systems before setup or when the system is to be dismantled. However, it should not be used in an active system because of the danger to fish and organisms inhabiting the biofilter.

Temperature Control

One advantage of water-recirculating systems is that water temperature can be controlled if needed. The most efficient and economical method for temperature control varies from system to system. Some systems employ heat exchangers, and others use heating or cooling coils suspended into a water reservoir or into a part of the recirculating system. Several different types of temperature control systems have been described (Scott 1972; Syrett and Dawson 1972; Chavin 1973; Wurtsbaugh and Davis 1976; Robinson, Tash, and Holanor 1978; Lemke and Dawson 1979; Jenson 1980). Temperature control is costly; thus heat recovery should be used.

Stocking Density and Management

The stocking density for catfish in closed, water-recirculating culture systems is dependent on a number of factors, many of which are functions particular to the individual system. Thus there are no formulas to determine the rate at which catfish should be stocked into such systems precisely. Estimates of the optimal stocking density can be made from predicting ammonia production and relating this to biofilter capacity, but in reality stocking densities are generally determined by trial and error. Sample (1989) gives a brief review of production in indoor warm-water culture, systems in which he reports production of various species of fish up to 13 pounds/cubic foot. Tackett and Carter (1981) suggest stocking two to three fish/cubic foot. Broussard et al. (1973) reported production of catfish exceeding 7 pounds/cubic foot in a water-recirculating system. Although stocking rates generated for various recirculating systems can be used as guides for stocking other systems, so many variables influence stocking density that it is difficult to suggest a density for a particular system. Depending on the experience of the manager, it is usually best initially to stock at a safe density and increase the density as experience is gained with a particular system.

Regardless of design, the management of water-recirculating systems requires frequent monitoring of the performance of the system. A close check on ammonia and nitrite concentrations is required. Dissolved oxygen concentrations should be maintained above 60 percent saturation (McGee 1989). Alkalinity, hardness, and pH should be measured at regular intervals and adjusted as necessary. It is critical that the fish not be overfed. Feeding rates are as previously described in other sections and in Chapter 10. All things considered, use of a water-recirculating system for the intensive culture of fish requires a higher level of management than most other culture systems and thus requires a higher level of skill of the manager. The management required is proportionate to the complexity of the system as well as the stocking density. It is ironic that the novice aquaculturist is often attracted to such systems.

REFERENCES

Avault, J. W., Jr. 1989. How will fish be grown in the future? *Aquaculture Magazine* 15(1):57–59.
Beveridge, M. C. M. 1987. *Cage Aquaculture.* Farnham, England: Fishing News Books Ltd.
Chavin, W. 1973. A reliable water temperature control apparatus for open freshwater systems. *Progressive Fish-Culturist* 35:202–204.
Collins, C. 1988a. Rearing channel catfish in cages—part I. *Aquaculture Magazine* 14(1):53–55.
Collins, C. 1988b. Rearing channel catfish in cages—part II. *Aquaculture Magazine* 14(2):56–58.
Colt, J. E., and D. A. Armstrong. 1981. Nitrogen toxicity to fish, crustaceans and mollusks. In *Bioengineering Symposium for Fish Culture (FCS Publ. 1),* ed. L. J. Allen and E. C. Kinney, pp. 39–42. Bethesda, Md.: American Fisheries Society.
Colt, J. E., and G. Tchobanoglous. 1981. Design of aeration systems for aquaculture. In *Bioengineering Symposium for Fish Culture (FCS Publ. 1),* ed. L. J. Allen and E. C. Kinney, pp. 138–148. Bethesda, Md.: American Fisheries Society.
Dupree, H. K. 1981. An overview of the various techniques to control infectious diseases in water supplies and in water reuse aquacultural systems. In *Bioengineering Symposium for Fish Culture (FCS Publ. 1),* ed. L. J. Allen and E. C. Kinney, pp. 83–89. Bethesda, Md.: American Fisheries Society.
Jenson, N. J. 1980. System of individually temperature-regulated saltwater aquaria. *Progressive Fish-Culturist* 42:166–168.
Lemke, A. E., and W. F. Dawson. 1979. Temperature-monitoring and safety-control device. *Progressive Fish-Culturist* 41:165–166.
Losordo, T. M., J. E. Easley, and P. W. Westerman. 1989. The preliminary results of a feasibility study of fish production in recirculating aquaculture systems. Paper read at International Meeting of the American Society of Agricultural Engineers, 12–15 December, 1989, at New Orleans, Louisiana.
Lucchetti, G. L., and G. A. Gray. 1988. Water reuse systems: A review of principal components. *Progressive Fish-Culturist* 50:1–6.
McGee, M. 1989. Recirculation, filtration can sometimes cut water costs in small aquacultural units. *Water Farming Journal* 4:6–7.
Piper, R. G., I. B. McElwain, L. E. Orme, J. P. McCraren, L. G. Fowler, and J. R. Leonard. 1982. *Fish Hatchery Management.* Washington D.C.: U.S. Fish and Wildlife Service.
Ray, L. 1981. Channel catfish production in geothermal water. In *Bio-Engineering Symposium for Fish Culture (FCS Publ. 1),* ed. L. J. Allen and E. C. Kinney, pp. 192–195. Bethesda, Md.: American Fisheries Society.
Robinson, F. W., J. C. Tash, and S. H. Holanov. 1978. Cooling discontinuous waters with a single refrigeration unit. *Progressive Fish-Culturist* 40:15.
Sample, D. 1989. Indoor warmwater fish culture systems. *Aquaculture Magazine* 15(3):65–66.

References

Schmittou, H. R. 1969. Cage culture of channel catfish. *Proceedings of the 1969 Fish Farming Conference.* Texas A&M University, College Station: Texas Agricultural Extension Service, Department of Wildlife Science, and the College of Agriculture.

Scott, K. R. 1972. Temperature control system for recirculation fish-holding facilities. *Journal of the Fisheries Research Board of Canada* 29:1082–1083.

Speece, R. E. 1981. Management of dissolved oxygen and nitrogen in fish hatchery waters. In *Bioengineering Symposium for Fish Culture (FCS Publ. 1)*, ed. L. J. Allen and E. C. Kinney, pp. 53–62. Bethesda, Md.: American Fisheries Society.

Stickney, R. R. 1979. *Principles of Warmwater Aquaculture.* New York: John Wiley & Sons.

Syrett, R. F., and W. F. Dawson. 1972. An inexpensive electronic relay for precise water-temperature control. *Progressive Fish-Culturist* 34:241–242.

Tackett, D. L., and R. R. Carter. 1981. Catfish reuse systems. *Internal Paper of the Fish Farming Experimental Laboratory,* Stuttgart, Ark.: Technical Information Services, Fish Farming Experimental Laboratory.

Wurtsbaugh, W. A., and G. E. Davis. 1976. Laboratory apparatus for providing diel temperature regimes for aquatic animals. *Progressive Fish-Culturist* 38:198–199.

Glossary

ACUTE: Severe and of short duration. Acute epizootics are characterized by large losses over a short period.

AD LIBITUM: Implies that feed is available at all times, but in respect to the feeding of aquaculture animals it refers to the presentation of feed until satiation is reached, after which feeding is stopped.

AEROBIC: Free oxygen is present in the environment or (as an adjective) requiring oxygen.

ALGAE: Members of an artificial grouping of plants that lack true stems, leaves, and roots.

ALKALINITY: Measure of the capability of a water to neutralize acids. Alkalinity is due to the presence of bases, primarily bicarbonate and carbonate, in the water.

ALLELE: An alternate form of a gene.

AMBIENT: The environment surrounding an organism.

AMINO ACID: Simplest organic structure that serves as a structural unit of protein.

ANAEROBIC: Free oxygen is absent from the environment or (as an adjective) not requiring oxygen.

ANEMIA: A deficiency of red blood cells, hemoglobin, or both.

ANTIBODY: A protein produced by the body in response to an antigen. Antibodies inactivate or destroy the antigen as part of the immune response.

ANTIGEN: A substance not normally present in the body that produces an immune response once it enters the body.

ANTIOXIDANT: A substance that inhibits the oxidation of other substances.

APPETITE: A desire for food or water.

AQUACULTURE: The rearing of aquatic organisms under controlled or semi-controlled conditions.

BIOFILTER: Part of a closed recirculating water system that detoxifies certain dissolved compounds, primarily nitrogenous compounds.

BLOOM: The plankton community of a body of water.

BLUE-GREEN ALGAE: A group of primitive microorganisms that possess characteristics of both bacteria and plants (algae).

BREED: A group of fish that have a common origin and share certain distinguishing features that separate them from other such groups.

CAGE CULTURE: Rearing of aquatic organisms in enclosures generally constructed of wire or netting around rigid frames, floated or suspended in large bodies of water.

CARBOHYDRATE: Energy-yielding organic compounds containing carbon, hydrogen, and oxygen in the ratio of 1:2:1.

CARNIVORE: An animal that feeds only on the tissues of other animals.

CELLULOSE: Polymer of glucose in linkage resistant to digestive enzymes. Cellulose is considered to be indigestible by catfish.

CHLOROPHYLL: The green pigment in plants that is important in photosynthesis.

CHROMOSOMES: The structures on which genes are located. There are two types of chromosomes: autosomes and sex chromosomes.

CHRONIC: Lasting a long time or recurring often. Chronic epizootics are characterized by moderate mortality rate over a long period.

CLOSED SYSTEM: A water system used for the culture of aquatic animals in which water is reused. Water is added only to replace losses due to evaporation, and so on.

COMMUNITY: The organisms inhabiting a natural physiographic area.

COMPLETE DIET: A diet that provides all nutrients and energy required by a particular species.

CULTURE CHAMBER: Any container used to rear aquatic animals.

CRUDE FIBER: The insoluble carbohydrates remaining in a feed after boiling in acid and alkali. Represents the portion of the feed considered to be indigestible.

CRUDE PROTEIN: The nitrogen content in a feed or animal or plant tissue, and so on, multiplied by a factor, which is generally 6.25.

DEMAND FEEDER: A feeder that dispenses feed when activated by aquatic animals.

Glossary **431**

DENITRIFICATION: The chemical reduction of nitrate to elemental nitrogen by microorganisms.

DETRITUS: Finely divided settlable material suspended in the water column.

DIET: A selection or mixture of feedstuffs provided on a continuous or prescribed schedule.

DIGESTIBILITY: The percentage of a feed nutrient that is broken down into a form suitable for absorption.

DIPLOID (2*N*): A cell or fish in which chromosomes occur in pairs.

DISPENSABLE AMINO ACID: An amino acid that can be synthesized in animal tissues and thus does not have to be provided in the diet.

DISSOLVED OXYGEN (DO): The amount of elemental oxygen present in a solution.

DIURNAL: Actions that are completed within 24 hours (daily) or recur every 24 hours.

DOMINANT ALLELE: An allele whose phenotype is expressed regardless of the genotype.

DRY MATTER: The portion of a feed or tissue remaining after water is removed.

EDEMA: An abnormal accumulation of fluid in the intercellular spaces of the body.

ENVIRONMENT: The total of all internal and external conditions that may affect an organism.

ENZYME: A protein that acts as an organic catalyst in initiating or speeding up specific chemical reactions.

EPIZOOTIC: An outbreak or epidemic of a disease or parasite in a population.

ESSENTIAL FATTY ACID: A fatty acid that cannot be synthesized in animal tissues and thus must be provided in the diet.

ESTROGENS: Ovarian steroid hormones responsible for promoting estrus and the development and maintenance of secondary sex characteristics.

EUTROPHIC: Describing waters that contain abundant nutrients, resulting in high levels of organic production.

EXOPHTHALMIA ("POPEYE"): A condition in which the eyeballs begin to protrude from the head as a result of the accumulation of fluid or gases at the back of the eye.

EXTENSIVE CULTURE: Low-intensity aquaculture characterized by relatively low fish stocking densities, low energy inputs, and low yields.

EXTRUSION: The process by which feeds are prepared by passing the ingredients through a die under high temperature and pressure, resulting in a floating pellet.

F_1: The first filial generation. This is the first generation that is produced in a breeding program.

F_2: The second filial generation. This is the second generation that is produced in a breeding program.

F_n: The nth filial generation.

FAMILY: A group of fish that have either one or two parents in common.

FATTY ACID: Organic acids composed of carbon, hydrogen, and oxygen that combine with glycerol to form fats.

FECUNDITY: The number of eggs produced annually by a female animal or per unit body weight of a female.

FEED: Food for animals.

FEED CONVERSION RATIO: In aquaculture, the amount of feed fed divided by weight gain.

FEEDSTUFF: Any substance suitable for animal feed.

FILAMENTOUS ALGAE: Species of algae in which individual cells are connected end to end in long filaments.

FILTER MEDIUM: A substrate in a mechanical or biological filter that traps suspended particles or provides a surface for the attachment of microorganisms.

FINGERLING: A fish larger than a fry but not of marketable size. Fish of about 1 to 10 inches in length.

FLOATING FEED: Feed prepared by extrusion that remains on the water surface for extended periods.

FOOD CHAIN: A sequence of organisms, each of which provides food for the next, from primary producers to ultimate consumers, or top carnivores.

FRY: Newly hatched fish that externally are replicates of the adult.

GAMETE: An egg or sperm.

GAS-BUBBLE TRAUMA: Results from the supersaturation of atmospheric gases (particularly nitrogen) in water. Fish with gas-bubble trauma may exhibit exophthalmia and air bubbles under the skin. Mortality rate can be high.

GENE: The basic unit of inheritance.

GENOTYPE: An animal's genetic makeup.

GRAM-NEGATIVE: One of two large groups of bacteria (the other being gram-positive) classified by their reaction to a differential staining process called the Gram stain. Gram-negative bacteria of various genera are responsible for virtually all bacterial diseases of catfish.

GYNOGENESIS: Development of an ovum after sperm penetration, but without fusion of the gametes.

Glossary **433**

HAPLOID (N): A cell or individual that has only one homologue of each chromosome pair.

HARDNESS: Concentration of divalent ions (primarily calcium and magnesium) present in water.

HERBIVORE: An animal that feeds exclusively on plant material.

HETEROSIS: Hybrid vigor; the superiority or inferiority of hybrids compared to the parents.

HETEROZYGOUS: Condition in which one of the pair of genes responsible for a particular trait is dominant and the other is recessive.

HIGHLY UNSATURATED FATTY ACID (HUFA): Fatty acids containing several double bonds.

HOMEOTHERMIC: Warm-blooded animals.

HOMOZYGOUS: Condition in which both genes of a pair responsible for a particular trait are either dominant or recessive.

HORMONE: A secretion produced in the body (primarily by endocrine glands) that is carried in the bloodstream to other parts and has a specific effect.

HYBRID: Produced by mating two different breeds, strains, or species.

HYBRID VIGOR: Synonym for heterosis.

INBREEDING: The mating of relatives.

INDISPENSABLE AMINO ACID: An amino acid that cannot be synthesized by the animal or cannot be synthesized in quantities sufficient to meet the animal's requirements. Thus it must be supplied in the diet.

INTENSIVE CULTURE: The rearing of aquaculture organisms in extremely high densities.

INTERSPECIFIC HYBRID: A hybrid produced by mating fish from two different species.

INTRASPECIFIC HYBRID: A hybrid produced by mating fish from two strains, breeds, or races within a species.

KILOCALORIE: The quantity of heat required to raise the temperature of 1 kilogram of water 1°C (from 15 to 16°C).

LATENT: Dormant or concealed; present, but not expressed.

LESION: Any visible alteration in the normal structure of organs, tissues, or cells.

LOCUS (plural: loci): The location of a particular gene on a chromosome. The terms *locus* and *gene* are often used interchangeably.

MONOCULTURE: Rearing of a single species.

NITRIFICATION: The microbiological oxidation of nitrogen from ammonia through nitrite to nitrate.

NITROGEN FIXATION: The process by which certain bacteria and blue-green algae are able to transform elemental nitrogen into ammonia.

NUTRIENT: A chemical that nourishes, such as a protein, carbohydrate, mineral, or vitamin.

OMNIVORE: An animal that consumes both plant and animal material.

OOGENESIS: The process by which a mature ovum is produced.

OSMOREGULATION: The process by which organisms are able to maintain an internal salt balance different from that which occurs in the external medium.

PARASITE: An organism that lives in or on another organism (the host) and depends on the host for food. Parasites harm the host when present in large numbers.

PATHOGEN: An organism that causes disease.

PATHOGENIC: Causing disease.

pH: Expression of the acid-base relationship; defined as the logarithm of the reciprocal of the hydrogen-ion activity.

PHENOTYPE: The chemical or physical expression of a gene.

PHOTOSYNTHESIS: Process by which chlorophyll-containing cells in green plants convert sunlight to chemical energy, synthesize glucose from carbon dioxide and water, and release oxygen.

PHYTOPLANKTON: The plant constituents of the plankton community.

PLANKTON: Aquatic organisms that are suspended in the water column and do not have the capability of controlling their position.

POIKILOTHERMIC: Cold-blooded animals.

POLYCULTURE: Rearing of two or more species in the same culture chamber.

POLYPLOID: A state in which a cell or an individual contains three or more sets of chromosomes.

POLYUNSATURATED FATTY ACID (PUFA): Fatty acids containing two or more double bonds.

POPULATION: Group of fish.

QUALITATIVE PHENOTYPE: A phenotype that is described (albinism is an example).

QUANTITATIVE PHENOTYPE: A phenotype that is measured (weight is an example).

RECIPROCAL HYBRID: The two hybrids that can be produced by mating fish from two different groups. That is, A female × B male and B female × A male.

RECESSIVE ALLELE: An allele whose phenotype is expressed only when an individual is homozygous at that locus.

SAC FRY: A fish with an external yolk sac.

Glossary **435**

SALINITY: The measure of the total amount of dissolved salts in a sample of water.

SAPONIFICATION NUMBER: Used to characterize the carbon chain size of a fat, specifically the number of milligrams of alkali required to saponify 1.0 gram of fat.

SAPROPHYTE: An organism deriving nutriment from decaying organic matter.

SATIETY: The condition of being fully satisfied with food.

SATURATED FAT: Contains fatty acids that are devoid of unsaturated bonds (double bonds).

SELECTION: A breeding program in which the breeder chooses which fish will be the next generation's brood stock on the basis of some predetermined criteria.

SIBS OR SIBLINGS: Brothers and sisters.

SOMATIC: Cells other than reproductive cells in the body of an organism.

SPERMATOGENESIS: The maturation of sperm cells.

STANDING CROP: The biomass present in a body of water at a given time.

STARCH: Polymer of glucose readily hydrolyzed by digestive enzymes; cooking improves the digestibility of starch to catfish.

STRAIN: A group of fish that come from a particular location or are produced by a particular breeding program.

SUSPENDED SOLIDS: Particles larger than 0.45 micrometer that are found in the water column.

TETRAPLOID (4N): A cell or individual that has four sets of chromosomes.

TRIGLYCERIDE: A molecule composed of glycerol and three fatty acids.

TRIPLOID (3N): A cell or individual that has three set of chromosomes.

TURBIDITY: A degree to which the penetration of light in water is limited by the presence in the water of suspended or dissolved substances.

VIRULENCE: The capability of an infectious agent to produce disease.

VITAMIN: An organic compound that is required in small amounts for normal growth and health.

ZOOPLANKTON: The animal component of the plankton community.

ZYGOTE: The cell formed by the uniting of male and female gametes.

Appendices

A.1.
English–Metric Conversions

To Convert Column 1 to Column 2, Multiply by	Column 1	Column 2	To Convert Column 2 to Column 1, Multiply by
		LENGTH	
2.54	Inch	Centimeter (cm)	0.3937
0.3048	Foot	Meter (m)	3.281
30.48	Foot	Centimeter	0.0328
0.914	Yard	Meter	1.094
1.609	Mile	Kilometer (km)	0.6214
		AREA	
0.4047	Acre	Hectare (ha)	2.471
		VOLUME	
29.57	Ounce (fluid)	Milliliter (mL)	0.0338
0.9463	Quart (U.S., 32 ounces)	Liter (L)	1.057
3.785	Gallon (U.S., 4 quarts)	Liter	0.2642
1233.6	Acre-foot	Cubic meter (m^3)	0.000811
		WEIGHT	
28.35	Ounce (avoirdupois)	Gram (g)	0.0353
453.6	Pound (avoirdupois)	Gram	0.002205
0.4536	Pound (avoirdupois)	Kilogram (kg)	2.205
		OTHER CONVERSIONS	
1.12	Pounds/acre	Kilogram/hectare	0.892
760	Atmosphere	Millimeters of mercury	0.00132

A.2.
Fahrenheit–Celsius Temperature Conversions

°F	°C	°F	°C	°F	°C	°F	°C
32	0.0	52	11.1	72	22.2	92	33.3
33	0.6	53	11.7	73	22.8	93	33.9
34	1.1	54	12.2	74	23.3	94	34.4
35	1.7	55	12.8	75	23.9	95	35.0
36	2.2	56	13.3	76	24.4	96	35.6
37	2.8	57	13.9	77	25.0	97	36.1
38	3.3	58	14.4	78	25.6	98	36.7
39	3.9	59	15.0	79	26.1	99	37.2
40	4.4	60	15.6	80	26.7	100	37.8
41	5.0	61	16.1	81	27.2	101	38.3
42	5.6	62	16.7	82	27.8	102	38.9
43	6.1	63	17.2	83	28.3	103	39.4
44	6.7	64	17.8	84	28.9	104	40.0
45	7.2	65	18.3	85	29.4	105	40.6
46	7.8	66	18.9	86	30.0	106	41.1
47	8.3	67	19.4	87	30.6	107	41.7
48	8.9	68	20.0	88	31.1	108	42.2
49	9.4	69	20.6	89	31.7	109	42.8
50	10.0	70	21.1	90	32.2	110	43.3
51	10.6	71	21.7	91	32.8		

A.3.

Length–Weight Relationship for Channel Catfish

The following table describing the relationship between total length (in inches) and weight (in pounds) was developed from a regression equation fitted to data collected from nearly 9,000 channel catfish from commerical ponds in Mississippi. Fish were not fed during the 24-hour period before measuring length and weight. The regression equation for the relationship is ln (weight) = $-4.2044833 - 1.384549$ (length) $+ 1.601366$ (length)$^2 - 0.183297$ (length)^3.

Length (Inches)	Weight (Pounds)	Length (Inches)	Weight (Pounds)
5	0.05	17	1.76
6	0.07	18	2.11
7	0.11	19	2.50
8	0.16	20	2.94
9	0.23	21	3.43
10	0.32	22	3.98
11	0.43	23	4.61
12	0.57	24	5.31
13	0.74	25	6.13
14	0.94	26	7.07
15	1.18	27	8.18
16	1.44		

A.4.

Common and Scientific Names of Species

Common Name	Scientific Name
Atlantic salmon	*Salmo salar*
Bighead carp	*Aristichthys nobilis*
Bigmouth buffalo	*Ictiobus cyprinellus*
Blue catfish	*Ictalurus furcatus*
Blue tilapia	*Tilapia aurea*
Black buffalo	*Ictiobus niger*
Brown bullhead	*Ictalurus nebulosus*
Channel catfish	*Ictalurus punctatus*
Chinook salmon	*Oncorhynchus tshawytscha*
Common carp	*Cyprinus carpio*
Fathead minnow	*Pimephales promelas*
Flathead catfish	*Pylodictus olivaris*
Grass carp	*Ctenopharyngodon idella*
Gizzard shad	*Alosa cepedianum*
Green sunfish	*Lepomis cyanellus*
Japanese eel	*Anguilla japonica*
Largemouth bass	*Micropterus salmoides*
Mosquito fish	*Gambusia affinis*
Mozambique tilapia	*Tilapia mossambica*
Nile tilapia	*Tilapia nilotica*
Paddlefish	*Polydon spathula*
Rainbow trout	*Oncorhynchus mykiss*
Red drum	*Sciaenops ocellatus*
Silver carp	*Hypothalmicthys molotrix*
Smallmouth bass	*Micropterus dolomieui*
Walking catfish	*Clarius batrachus*
White catfish	*Ictalurus catus*
Yellowtail	*Seriola quinqueradiata*

A.5.

Fish Disease Diagnostic Laboratories

The following is a partial list of federal and state diagnostic laboratories in the major catfish-producing areas of the United States. Other diagnostic services, including private consultants, are available, and are often listed in the various trade journals serving the industry. Always contact the laboratory before submitting fish for diagnosis.

Alabama
  Alabama Fish Farming Center
  Highway 69 North
  Greensboro, Alabama 36744
  (205) 624-4016

  Southeastern Cooperative Fish Disease Laboratory
  Department of Fisheries and Allied Aquaculture
  Auburn University, Alabama 36849
  (205) 844-4786

Arkansas
  U.S. Fish and Wildlife Service
  Fish Farming Experimental Station
  P. O. Box 860
  Stuttgart, Arkansas 72160
  (501) 673-8761

  Fish Disease Laboratory
  434 S. Cokley
  Lake Village, Arkansas 71653
  (501) 265-5883

  Fish Disease Laboratory
  P. O. Box Drawer D
  Lonoke, Arkansas 72086
  (501) 676-3124

California
  Fish Disease Laboratory
  Department of Medicine
  School of Veterinary Medicine
  University of California
  Davis, California 95616
  (916) 762-3411

California Department of Fish and Game
Fish Disease Laboratory
2111 Nimbus Road
Rancho Cordova, California 95670
(916) 355-0811

Florida
College of Veterinary Medicine and
  Department of Fisheries and Aquaculture
Institute of Food and Agricultural Sciences (IFAS)
University of Florida
7922 NW 71st Street
Gainesville, Florida 32606
(904) 392-9617

Northwest Florida Aquaculture Farm
Institute of Food and Agricultural Sciences (IFAS)
P. O. Box 434
Blountstown, Florida 32424
(904) 674-8353

Kissimmee Diagnostic Laboratory
Florida Department of Agriculture and Commerce Services
2700 N. Bermuda Avenue
Kissimmee, Florida 32741
(407) 846-3185

Live Oak Diagnostic Laboratory
Florida Department of Agriculture and Consumer Services
P. O. Drawer O
Live Oak, Florida 32060
(904) 362-1216

Georgia
Athens Diagnostic Laboratory
College of Veterinary Medicine
University of Georgia
Athens, Georgia 30602
(404) 542-3473

Veterinary Diagnostic and Investigational Laboratory
P. O. Box 1389
Tifton, Georgia 31793
(912) 386-3340

Fish Disease Diagnostic Laboratories **443**

Extension Fishery Specialist
Food Science Building
University of Georgia Experiment Station
Griffen, Georgia 30223
(404) 228-7328

Extension Fishery Specialist
Barrow Hall
University of Georgia
Athens, Georgia 30602
(404) 542-1924

Idaho
Department of Fish and Wildlife Resources
University of Idaho
Moscow, Idaho 83843
(208) 885-6336

Idaho Department of Fish and Game
Fish Disease Laboratory
Hagerman State Hatchery
Hagerman, Idaho 83332
(208) 837-6672

Illinois
Fisheries Research Laboratory
Southern Illinois University
Carbondale, Illinois 62901
(618) 536-7761

Louisiana
Aquatic Animal Diagnostic Laboratory
School of Veterinary Medicine
Louisiana State University
Baton Rouge, Louisiana 70803
(504) 346-3312

Maryland
Fish Disease Laboratory
Department of Microbiology
University of Maryland
College Park, Maryland 20742
(301) 454-5411

Mississippi
Mississippi Cooperative Extension Service
Fish Disease Diagnostic Laboratory
Box 142
Stoneville, Mississippi 38776
(601) 686-9311

Mississippi Cooperative Extension Service
Fish Disease Diagnostic Laboratory
Box 631
Belzoni, Mississippi 39038
(601) 247-2917

Fish Diagnostic Laboratory
College of Veterinary Medicine
Drawer V
Mississippi State, Mississippi 39762
(601) 325-3432

Missouri
Missouri Department of Conservation
666 W. Primrose
Springfield, Missouri 65807
(417) 883-6677

North Carolina
Department of CASS
School of Veterinary Medicine
4700 Hillsborough St.
Raleigh, North Carolina 27606
(919) 829-4200

South Carolina
Extension Aquaculture Specialist
Department of Aquaculture, Fisheries, and Wildlife
G08 Lehotsky Hall
Clemson University, South Carolina 29634-0362
(803) 656-2810

Aquaculture Demonstration Center
Estill, South Carolina 29918
(803) 625-4525

Texas
> Extension Fish Disease Specialist
> Department of Wildlife and Fisheries
> Nagle Hall
> Texas A&M University
> College Station, Texas 77843
> (409) 845-7471

West Virginia
> National Fish Health Laboratory
> U.S. Fish and Wildlife Service
> P. O. Box 700
> Kearneysville, West Virginia 25430
> (304) 725-8461

Index

Achyla, 345
Acid death point, 50
Acousticolateralis system, 16
Actinomycetes, 261
Additive genetic variance, 35–38, 117–118
Aeration
 emergency, 240–241
 feeding rate relationship, 231–232
 nightly, 241–242
 practices, 238–242
 raceways, 414–415
 recirculating systems, 422–423
 transport tanks, 397–398
 turbidity relationship, 283
Aerators
 oxygen transfer, 220, 232–233
 performance, 232–233
 placement in ponds, 242
 size requirements, 242
 transport tanks, 397–398
 types, 233–238, 397–398, 422–423
Aeromonas, 324–345
Agricultural limestone, 245
Albinism, 32–34, 114–115
Algal toxins, 285–287
Algicides, 266–267, 281–282
Alkaline death point, 50

Alkalinity
 carbon dioxide, 50, 247, 249
 copper toxicity, 59, 245, 370
 definition, 48
 desired range, hatcheries, 156
 desired range, ponds, 49, 245, 249
 metal toxicity, 59
 pH, 50, 245, 249
Alum, 284–285
Ambiphrya, 352–353
Amino acids. See also Protein
 available, 293
 essential index, 82–83
 digestibility, 103–104
 dispensable, 77–78
 indispensable, 75
 interactions, 78
 requirements, 75–59, 300
 supplemental, 78–79, 294
Ammonia
 concentrations in ponds, 253–255
 effect on fish, 51–53
 excretion, 51–52
 hatcheries, 154, 156
 losses and transformations, 253–254
 management in ponds, 255–257
 measurement, 255
 nitrification, 252, 257

Index

Anatomy, 9–12
Androgens, 21–22, 132
Anemia, 48, 53, 359–360
Antibiotics, 363–365
Antibodies, 18
Antimycin A, 175
Aquaculture, 1, 2
Aquatic weeds. *See* Weeds
Ascorbic acid, 89–90, 298, 301, 302, 339

Bacterial diseases, 333–345, 362, 362–365
Bacterial egg rot, 362
Bar graders, 179–180
Bicarbonate. *See* Alkalinity
Bighead carp, 213, 282, 416
Bigmouth buffalo, 208–209
Biofilters, 419–422
Biotin, 87
Birds, 206–207
Blood
 anemia, 48, 53, 359–360
 circulation, 13–15
 functions, 14–15, 17–18
Blood meal, 296, 303
Blue catfish, 3, 126–130, 131
Blue-green algae
 dissolved oxygen, 221, 225, 269
 off-flavor, 261, 264, 269
 toxins, 285–287
Blue tilapia, 210–212
Breeding programs
 albinism, 114–115
 comparison, 130–131
 family selection, 117, 121–122
 guidelines, 113–114
 interspecific hybridization, 126–130
 intraspecific hybridization, 122–126
 mass selection, 117–118, 119–121, 126
 strain selection, 116–117
 unplanned, 111–113
Brood fish
 care of, 138–140
 feeding, 139, 313–314
 requirements, 140–142
 selection, 136–138
 sex ratios, 140
 stocking rates, 138–139
Brown-blood disease, 53–55, 257–260
Buffalo, 4, 208–209
Bullhead catfishes, 3

Cabomba, 272
Cage culture, 407–412

Calcium, 49, 155–156, 399. *See also* Hardness
Calcium chloride, 154, 250, 399
Calcium, dietary, 95
Calcium hydroxide, 244, 247–248
Calcium sulfate, 250, 284, 399
Carbohydrates, dietary, 66–68, 104–105, 296–297
Carbon dioxide
 concentrations in ponds, 246–247
 diurnal changes, 245–247
 effects on fish, 49, 50
 fish respiration, 13–14
 hatcheries, 154, 155
 measurement, 247–248
 pH, 50, 248–250
 removal, 154, 155, 247–248
Carp, common, 209, 275, 283
Carp pituitary extract, 150
Catfish offal meal, 296
Cat's clay, 250
Cattails, 272, 279, 280
Cellular immunity, 18
Ceratophyllum, 272
Channel catfish
 anatomy, 9–12
 commercially important traits, 3
 distribution, 19
 general life history, 19–20
 history of culture, 3–6
 hybrids, 122–130
 spawning behavior, 23–25
 strains, 116–117
Channel catfish reovirus, 333
Channel catfish virus disease, 117, 328–333
Chara, 269–270, 279
Chelated copper herbicides, 279
Chloride
 dietary, 96
 nitrite toxicoses, 54–55, 258–260
Chlorine, 161, 175
Choline, 88–89, 301
Chromosomes, 28, 31–32
Circulatory system, 13–15
Clinical signs, diseases, 321–324
Closed systems. *See* Recirculating systems
Cobalt, dietary, 96, 301
Columnaris, 339–342
Complement, 18
Coontail, 272, 279, 280
Copper
 chelated, 279

dietary, 98
toxicity, 58–59
Copper sulfate
  algicide, 266–267, 277–279
  disease therapeutant, 370
Cormorants, 206–207
Corn, 297
Corral seining, 395
Costia, 354–356
Costs
  bird depredation, 206
  infectious diseases, 317
  off-flavor, 264–265
  production, 195–199
Cottonseed meal, 294–295, 299, 302, 304
Cropping systems, ponds, 181–192
Crossbreeding, 122–131

2,4-D, 280
Diets. See Feeds
Denitrification, 253
Dieldrin, 169–170
Diffused aeration, 233–234
Diffusion, oxygen, 218–221
Digestibility, nutrient
  amino acids, 103, 104
  carbohydrates, 104, 105
  determination, 100–101
  energy, 105
  lipids, 104
  minerals, 105–106
  protein, 101–102
Digestible energy, 64–65
Digestion, 10, 98–100
Diquat, 279–280
Diseases, infectious
  Ambiphrya (Scyphidia), 352–353
  channel catfish reovirus, 333
  channel catfish virus disease, 328–333
  clinical signs, 321–324
  columnaris, 339–342
  costs of, 317
  diagnosis, 324–328
  eggs, 361–362
  enteric septicemia of catfish, 334–339
  fungal, 345–347
  gill flukes, 356
  Ichthyobodo (Costia) 354–355
  Ichthyophthirius, 348–351
  Ligictaluridus (Cleidodiscus), 356
  motile aeromonad septicemia, 342–345
  proliferative gill disease, 358–359
  role of the environment, 319–321
  treatments, 362–378
  Trichodina, 351–352
  Trichophrya, 354
  winter kill, 357–358
Diseases, treatments
  antibiotics, 363–365
  cage culture, 411–412
  rate calculations, 373–378
  recirculating systems, 424
  therapeutic chemicals, 365–371
Dissolved oxygen. See also Aeration, Aerators, Oxygen
  budget in ponds, 218–223, 226–227
  diffusion, 218–221
  diurnal changes, 223–226
  fry nursery ponds, 178
  hatcheries, 154, 155
  herbicide treatments, 281
  lethal levels, 46–48
  measurement, 227–231
  nighttime losses, 226–227
  photosynthesis, 221
  plankton respiration, 221–223
  required by fish, 46–47
  solubility in water, 219
  spawning, 47, 144
  transport tanks, 399
  turnovers, 224
Dominance genetic variance, 35–37, 122
Dress-out percentage, 3, 117, 118, 119, 122

Economics, 195–199
Edwardsiella ictaluri, 3, 89, 185, 334–339
Eggs
  age of, 146–147
  development, 25
  diseases, 153, 160–161, 361–362
  numbers produced, 24–25
  options for handling, 144–147
  parental care, 25
  transporting, 147, 401–402
Egrets, 206–207
Embryo development, 31
Emergency aeration, 240–241
Endothall, 280
Endrin, 169–170
Energy, dietary
  digestibility, 105
  expression, 62–65
  feedstuffs, 296–297
  partitioning, 63–65
  requirements, 65, 300

Enteric septicemia of catfish, 334–339
Epistatic genetic variance, 35–36
Essential amino acid index, 82–83
Essential fatty acids, 73–74
Estrogens, 24, 132–133
Evaporation, 202–203
Extruded feeds, 306–307

Family selection, 117, 121–122
Fanwort, 272, 280
Fatty acids
  essential, 73–74
  nomenclature, 68–71
Feeding
  allowances, 308–312
  ammonia production, 253
  brood fish, 139–140, 313–314
  cage culture, 410–411
  fingerlings, 176–177, 312–313
  food fish, 313
  fry, 161, 176–177, 312
  water quality, 178, 186, 188, 217–218, 231–232, 253, 257, 265–266, 308–312
  winter, 314
Feeds
  binders, 305
  extruded, 306–307
  fingerling, 304
  food fish, 304
  formulation, 297–305
  fry, 303
  meals and crumbles, 307
  medicated, 308, 363–365
  particle sizes, 308, 309
  pelleted, 305–306
  physical characteristics, 303–305
  processing, 305–307
  recommended nutrient levels, 298, 300–301
Feedstuffs
  amino acid availability, 103, 293
  blood meal, 296, 303
  carbohydrate digestibility, 104
  cottonseed meal, 294–295, 302, 304
  corn, 297, 304
  energy digestibility, 105
  fats and oils, 297
  fish meal, 295, 302, 303, 304
  full fat soybeans, 296
  grains, 296–297, 302, 304
  lipid digestibility, 104
  meat and bone meal, 295–296, 303, 304

  peanut meal, 295
  phosphorus availability from, 106
  protein digestibility, 102
  protein efficiency ratios, 292
  rice bran, 297, 304
  soybean meal, 294, 303, 304
  wheat, 297, 304
Fertilization, eggs, 25
Fertilization, ponds, 176, 274–275
Filamentous algae, 269–270
Fingerlings. *See also* Fry
  feeding, 176–177, 312–313
  feeds, 304
  grading, 179–180, 386–387, 391
  harvesting, 178–180, 384–395
  length–weight relationship, 180
  pond preparation, 174–175
  size desired, 181–182
  stocking rate calculation, 192–194
Fish meal, 295, 302, 303, 304
Flathead catfish, 3
Flexibacter columnaris, 339–342
Flukes, gill, 356
Fluridone, 280
Folic acid, 87–88, 300
Food fish
  cropping systems, 182–192
  feeding, 313
  feeds, 304
  harvest size distribution, 183–185
  stocking density, 186–190
  yields from ponds, 190–192
Formalin, 370–371
Freshwater prawns, 213–215
Fry. *See also* Fingerlings
  feeding, 161, 176–177, 312
  feeds, 303
  inventory methods, 162
  rearing, 160–161
  stocking densities in ponds, 176–177
  transporting, 401–402
Fungal diseases, 345–347, 361–362

Gametogenesis, 29–31, 43–44
Gas bubble trauma, 47, 56–58, 153–155
Genes, 28
Genetics. *See also* Breeding programs
  albinism, 32–34, 114–115
  breeding programs, 111–131
  chromosome replication, 28–30
  engineering, 133
  heritability, 36–37, 115, 118
  inbreeding, 37–38, 112, 114, 123, 126, 140

qualitative traits, 32–33, 114–115
quantitative traits, 33–38, 115–131
polyploidism, 28, 29, 31, 131–132
sex determination, 31–32
sex reversal, 31, 132–133
variance components, 34–38
Geosmin, 261, 264, 266, 267
Geothermal water, 166, 406, 414–415
Glyphosate, 280
Gonosomatic index, 20, 22
Gossypol, 294, 303
Grading, 179–180, 386–387, 391
Grass carp, 212, 275–277
Great blue herons, 206–207
Groundwater
    hatchery supplies, 152
    pond supplies, 167–168
Growth rate
    stocking density, 177, 186–192
    temperature, 41, 42–43, 44
Gynogenesis, 31
Gypsum, 250, 284, 399

Hamburger gill disease, 358–359
Hardness. See also Calcium
    definition, 48
    desired range, hatcheries, 155–156
    desired range, ponds, 245
    metal toxicity, 59
Harvesting
    custom, 384
    efficiency, 393–394
    fingerlings, 178–180
    general considerations, 382–383
    levee ponds, 384–394
    seines, 384–387
    watershed ponds, 394–395
Hatcheries
    design, 157–160
    practices, 160–161
    water quality, 153–157
    water supplies, 152–153
Herbicides, 277–281
Heritability, 36–37, 117–118
Hormones
    induced spawning, 150–151
    reproductive, 21–24
Human chorionic gonadotropin, 150–151
Humoral immunity, 18
Hybridization, 122–131
Hybrids
    channel × blue catfish, 126–130, 131
    interspecific, 126–130

intraspecific, 123–126
    Marion × Kansas strain, 123–126, 130–131
Hydrated lime, 244, 247
Hydrodictyon, 270
Hydrogen sulfide, 55, 157, 285
Hydrology, 199–205

Ichthyobodo, 354–356
Ichthyophthirius, 348–351
Immune system, 16–18, 42–43, 321
Inbreeding, 37–38, 112, 114, 123, 126, 140
Infectious diseases. See Diseases, infectious
Inflammation, 17
Inositol, 90
Insects, fry ponds, 175
Interspecific hybridization, 126–130
Intraspecific hybridization, 123–126
Investment requirements, 196
Iodine, dietary, 96–97
Iodine number of lipids, 70, 72
Iron
    dietary, 97–98
    hatchery water supplies, 156–157
    pond water supplies, 167

Lateral line, 10, 16
Least-cost feeds, 298–303
Length–weight relationship, 180, 439
Ligictaluridus, 356
Lipids, dietary
    classification and structure, 68–72
    digestibility, 104
    fatty acids, 68–74
    functions, 72, 73–74
    requirements, 72–73, 300
Lime, agricultural, 245
Lime, hydrated, 244, 247
Liquid oxygen, 397–398, 415
Livo cars, 385–392
Luteinizing hormone-releasing hormone, 151

Macrobrachium rosenbergii, 213–215
Magnesium, dietary, 95–96, 301
Manganese, dietary, 97, 301
Mass selection, 117–118, 119–121, 126
Meat and bone meal, 295–296
Medicated feeds, 308, 363–365
Meiosis, 29, 30
Methemoglobinemia, 53–55, 257–260
2-Methylisoborneol, 261, 264, 266, 267

Minerals, dietary, 93–98
Mitosis, 29
Motile aeromonad septicemia, 342–345
Mozambique tilapia, 210–212
Mucus, 17
Multiple-batch culture, 181–192
Myriophyllum, 272

Najas, 270–272, 276, 279, 280
Net protein utilization, 82
Niacin (nicotinic acid), 86–87, 300
Nightly aeration, 241–242
Nile tilapia, 210–212
Nitrate, 251, 252
Nitrification, 252, 257–258, 419–420
Nitrite
  concentrations in ponds, 257–258
  effects on fish, 48, 53–55
  management, 258–260
  measurement, 259–260
Nitrogen
  cycle in ponds, 251–253
  fixation, 251
  gas, 251
No-blood disease, 359–360
Nursery ponds
  aeration, 178
  fertilization, 176
  insect problems, 174–175
  trash fish problems, 174–175
  weed problems, 273
Nutrition. See also Feeding, Feeds, Feedstuffs
  carbohydrates, 66–68
  digestion, 98–106
  energy, 62–65
  lipids, 68–74
  minerals, 93–98
  proteins and amino acids, 74–83
  vitamins, 83–93

Off-flavor
  costs to producers, 264–265
  incidence, 262–264
  management, 265–268, 281–282
  types, 261–264
Offal, 296
Olfaction, 11
Oogenesis, 23, 29–31
Osmoregulation, 15, 40–41, 399
Ovaries, 22
Oxygen. See also Aeration, Dissolved oxygen
  consumption rates, fish, 45–46
  liquid, 397–398, 415
  meters, 227–30
  solubility in water, 219
  transfer rates, 232–233
Oxytetracycline, 363, 364–365

Paddlefish, 209–210, 416
Paddlewheel aerators, 236–238
Pantothenic acid, 86, 300
Parasites, 347–356
Parrotfeather, 272
Peanut meal, 295
Pelleted feeds, 305–306
Pesticides, 169–170, 287–288
pH
  alkalinity, 49, 50, 245, 249–250
  ammonia equilibrium, 51–53, 251, 254–256
  carbon dioxide, 50, 245–250
  definition, 50
  desired range, 50
  measurements, 250
  ponds, 248–250
  recirculating systems, 423–424
Phosphorus, dietary, 95, 105, 106
Photosynthesis, 221, 245–246
Phytoplankton
  ammonia assimilation, 253–254
  carbon dioxide, 245–247
  die-offs, 225–226, 254, 255–256
  dissolved oxygen, 221–227
  management, 281–282
  pH, 249
  toxins, 285
  weed problems, 269
Pithophora, 270, 277, 279
Plankton. See Phytoplankton, Zooplankton
Polyculture
  bighead carp, 213
  buffalo, 208–209
  common carp, 209
  freshwater prawns, 213–215
  grass carp, 212
  paddlefish, 209–210
  silver carp, 212–213
  tilapias, 210–212
Polygonum, 272, 276
Polypoid, 28, 131–132
Ponds
  levee type, 167–172
  water budgets, 199–205
  watershed type, 172–174
Potassium, dietary, 96

Potassium permanganate
  disease treatments, 367–370
  dissolved oxygen, 244
  rotenone detoxification, 175
  toxicity to fish, 369–370
Prawns, freshwater, 213–215
Precipitation, 199–205
Proliferative gill disease, 287, 358–359
Propeller-aspirator-pump aerators, 234–235
Protein, dietary
  digestibility, 101, 102
  feedstuffs, 294–296
  nutritional value, 81–82
  requirements, 79–81
  structure and function, 74–75
Protein efficiency ratios, 81, 292
Protozoan parasites, 347–356
Pump-sprayer aerator, 235–236
Pyridoxine, 86, 300

Qualitative traits, 32–34, 114–115
Quantitative traits, 33–38, 115–131

Raceway culture, 414–417
Recirculating systems
  aeration, 422–423
  biofilters, 419–422
  culture chambers, 418
  disease control, 424
  pH control, 423–424
  settling chambers, 418–419, 422
  stocking density, 425–426
Record keeping, 195
Reproduction, 20–24, 42–43. See also Spawning
Respiration, channel catfish, 13–15, 45–46
Respiration, plankton, 221–223, 245–246
Riboflavin, 85–86, 300
Rice bran, 297, 304
Romet®, 363, 365
Rotenone, 174–175

Salinity, 40–41, 155
Salix, 272
Salt
  disease treatment, 371
  nitrite protection, 54–55, 258–266
  seepage reduction, 204
  transportation aid, 399
Saponification number, 70, 72
Saprolegnia, 345–347, 361
Scyphidia, 352–353

Secondary sexual characteristics, 137–138
Seepage, 169, 202–204
Seines, 384–387
Selection
  correlated responses, 118–119
  family, 117, 121–122
  mass, 117–118, 119–121, 126
  unintentional, 111–113
Selenium, dietary, 92, 97
Sensory function, 16
Sex
  chromosomes, 31–32
  determination, 21, 22–23, 31, 137
  reversal, 31–32, 132–133
Silver carp, 212–213, 282, 416
Simazine, 279
Single-batch culture, 181–192
Smartweed, 272, 276, 280
Sodium bicarbonate, 248
Sodium chloride. See Salt
Sodium, dietary, 96
Sodium polyphosphates, 204
Soils, pond site, 169–170, 204
Solricin 135®, 266, 279
Soybean meal, 294, 303, 304
Soybeans, full fat, 296
Spawning. See also Reproduction
  aquarium methods, 150–151
  behavior, 20, 23–24
  containers, 142–144
  induced, 150–151
  pen methods, 148–149
  pond methods, 142–148
  temperature, 23, 43–44
Spermatogenesis, 20–21, 29–31
Spirogyra, 270, 276
Standard aeration efficiency, 232–233
Standard oxygen transfer rate, 232–233
Stocking densities
  cages, 409–410
  food fish ponds, 186–190
  fry nursery ponds, 176, 177
  raceways, 414–416
  recirculating systems, 425–426
Strains, channel catfish
  Auburn, 116, 117
  Kansas, 116, 117, 118, 123–126
  Marion, 116, 117, 118, 123–126
  Minnesota, 117
  performance comparison, 116–117
  Rio Grande, 116, 117, 118
  Tifton, 115
Stress response, 17, 39, 321

Sulfadimethoxine, 363, 365
Sulfur, dietary, 96
Surface water
    hatchery supply, 152–153
    pond supply, 168
Suspended solids, 55–56

Temperature
    acclimatization, 42, 43
    critical, 41–42
    feeding rates, 310, 314
    growth, 41, 42–43
    harvesting and grading, 390, 393–394
    hatcheries, 153
    immune function, 42–43
    recirculating systems, 424–425
    reproduction, 43–44
    transporting, 398–400
Tempering, 42, 268, 399, 400, 402
Terramycin®, 363, 364–365
Testes, 20, 21
Tetraploid, 28, 31, 131–132
Therapeutants, 365–372
Thiamin, 85, 300
Tilapia
    algae control, 282
    brood fish forage, 139–140, 212
    off-flavor abatement, 212, 267
    polyculture with catfish, 210–212
    weed control, 275
Total gas pressure, 56, 153
Toxaphene, 169–170
Toxic algae, 285–287
Transporting fish
    aeration, 397–398, 401–402
    eggs and fry, 401–402
    equipment, 396–398
    fish loading rates, 400–401
    water quality, 398–400
Treatment rate calculations, 372–378
Trichodina, 351–352
Trichophrya, 354
Triploid, 28, 29, 131–132
Turbidity
    hatcheries, 56
    management, 282–285
    tolerance by fish, 55–56
Turnovers, 224
Typha, 272

Vaccines
    channel catfish virus disease, 333
    enteric septicemia of catfish, 339

Vertical pump aerators, 234
Viral diseases, 328–333
Vitamins
    A, 90–91, 301, 302
    B_{12}, 88, 301
    biotin, 87
    C, 89–90, 298, 301, 302, 309
    choline, 88–89
    D, 91–92, 301
    deficiency signs, 84
    E, 92–93, 301, 302
    fat soluble, 90–93
    folacin, 87–88, 300, 302
    inositol, 90
    K, 93, 301
    niacin (nicotinic acid), 86–87, 300
    pantothenic acid, 86, 300
    pyridoxine, 86, 300, 302
    requirements, 84, 300–301
    riboflavin, 85–86, 300
    thiamin, 85, 300, 302
    water soluble, 83–90

Water budgets, ponds, 199–205
Water exchange, 205
Water quality
    feeding rates, 178, 186, 188, 217–218, 231–232, 253, 257, 265–266, 308–312
    hatcheries, 153–157
    infectious diseases, 321
    nursery ponds, 177–178
    recirculating systems, 418–426
    requirements, 39–40
    transporting fish, 398–400
Water supplies
    hatcheries, 152–157
    ponds, 167–169, 172–174
Weeds
    biological control, 275–277
    chemical control, 277–281
    occurrence, 272–273
    prevention, 273–275
    types, 269–273
Wheat, 297, 304
White catfish, 3
Willows, 272, 280
Winter feeding, 314
Winter kill, 346, 357–358

Zinc, 58–59
Zinc, dietary, 97, 301
Zooplankton, 213, 282